T0299892

Stochastic Structural Optimization

Stochastic Structural Optimization presents a comprehensive picture of robust design optimization of structures, focused on nonparametric stochastic-based methodologies. Good practical structural design accounts for uncertainty, for which reliability-based design offers a standard approach, usually incorporating assumptions on probability functions which are often unknown. By comparison, a worst-case approach with bounded support used as a robust design offers simplicity and a lower level of sensitivity. Linking structural optimization with these two approaches by a unified framework of non-parametric stochastic methodologies provides a rigorous theoretical background and high level of practicality. This text shows how to use this theoretical framework in civil and mechanical engineering practice to design a safe structure which accounts for uncertainty.

- Connects theory with practice in the robust design optimization of structures.
- Advanced enough to support sound practical designs.

This book provides comprehensive coverage for engineers and graduate students in civil and mechanical engineering.

Makoto Yamakawa is a Professor at Tokyo University of Science, and a member of the Advisory Board of the 2020 Asian Congress of Structural and Multidisciplinary Optimization.

Makoto Ohsaki is a Professor at Kyoto University, Japan, treasurer of the International Association for Shell & Spatial Structures and former President of the Asian Society for Structural and Multidisciplinary Optimization.

Stochastic Structural Optimization

Makoto Yamakawa
Tokyo University of Science
Makoto Ohsaki
Kyoto University

CRC Press is an imprint of the
Taylor & Francis Group, an **informa** business

Cover image: Makoto Yamakawa and Makoto Ohsaki

First edition published 2024
by CRC Press
6000 Broken Sound Parkway NW, Suite 300, Boca Raton, FL 33487-2742

and by CRC Press
4 Park Square, Milton Park, Abingdon, Oxon, OX14 4RN

CRC Press is an imprint of Taylor & Francis Group, LLC

Library of Congress Cataloging-in-Publication Data

Names: Yamakawa, Makoto, author. | Ōsaki, Makoto, 1960- author.
Title: Stochastic structural optimization / Makoto Yamakawa and Makoto Ohsaki.
Description: First edition. | Boca Raton, FL : CRC Press, 2024. | Includes bibliographical references and index.
Identifiers: LCCN 2023002898 | ISBN 9780367720391 (hbk) | ISBN 9780367720407 (pbk) | ISBN 9781003153160 (ebk)
Subjects: LCSH: Structural optimization. | Stochastic processes.
Classification: LCC TA658.8 .Y36 2024 | DDC 624.1/7713--dc23/eng/20230124
LC record available at https://lccn.loc.gov/2023002898

ISBN: 978-0-367-72039-1 (hbk)
ISBN: 978-0-367-72040-7 (pbk)
ISBN: 978-1-003-15316-0 (ebk)

DOI: 10.1201/9781003153160

Typeset in CMR10
by KnowledgeWorks Global Ltd.

Publisher's note: This book has been prepared from camera-ready copy provided by the authors.

Contents

Preface

Structural optimization is an established field of research that has attracted researchers and engineers especially since the development of high-performance computing technology in 1990s. Recently, large-scale optimization approaches are available for practical design. However, in the real-world design process, uncertainty unavoidably exists in various aspects including material property, design load, geometry of parts, and so on. Therefore, an optimal structure under certain deterministic constraints may be far from optimum when realistic uncertainty is considered.

Among the design methods considering uncertainty, reliability-based design (RBD) is one of the most standard approaches. RBD usually incorporates assumptions on probability functions for the parameters, which are often unknown. Due to its simplicity and small sensitivity of the solution to parameter variation, a worst-case approach with bounded support, which is called *robust design*, is also widely used. Recent studies have linked structural optimization with robust design and RBD by a unified framework of non-parametric stochastic methodologies, which provide rigorous theoretical background and high level of applicability to practical problems.

Although those theoretical results are ready to be transferred to practical use, the methodologies have so far not been well organized in the form of a comprehensive textbook for engineers. The authors have been working together on the topics of design under uncertainty for more than twelve years. This new monograph is a successor and a more advanced work to the introductory monograph of structural optimization written by the second author twelve years ago, and can contribute to the development of the community of researchers and engineers in this field. With the remarkable progress of design methodologies in recent years, many researchers, engineers and graduate students in the fields of civil engineering and mechanical engineering are getting increasingly interested in the subject related to design under uncertainty. This monograph will help them obtain the comprehensive knowledge they need to design a safe structure which takes account of uncertainty. The monograph is organized as follows:

In Chapter 1, basic concepts of stochastic and probabilistic approaches are explained, and some related illustrative examples are introduced. Real-world design problems essentially have uncertainty and hence uncertainty should be appropriately considered in the process of designing structures. The typical structural optimization problem is formally formulated to minimize an objective function representing the structural cost under constraints on the structural performance measures, or maximize structural performance under constraint on the structural cost. Various sources and measures of uncertainty in structural optimization problems are categorized, and types of solution methods including deterministic, probabilistic and possibilistic methods are briefly explained.

In Chapter 2, theoretical fundamentals and materials of stochastic optimization methods are summarized. Probability theory plays a crucial role in the methods and hence theoretical concepts of stochasticity and randomness are introduced. This chapter may be skipped or referred later by those whose interests focus on practical application and engineering problems. However, it is quite difficult to attain practical efficiency without theoretical framework. Indeed, due to the curse of dimensionality, some heuristic methods should be used for a practical optimization problem with moderately large number of variables.

In Chapter 3, two types of random search-based optimization methods are explained for problems with integer variables and parameters, namely, global random search and random multi-start local search. Worst-case approach is a well-known robust design method, where the structures are designed considering the worst possible scenario. The process of finding the worst value for unknown-but-bounded uncertainty is referred to as worst-case analysis, which can be regarded as a deterministic optimization process. Some results are presented for the worst-case analysis and design problems using the random search methods.

In Chapter 4, robust optimization problem is regarded as a two-level optimization problem, and is linked to random search with order statistics in the lower-level problem, that is, the worst-case analysis. A stopping rule of random search is described on the basis of order statistics theory. Numerical examples of design problem with dynamic analysis under uncertainty are considered. The results imply that the order statistics can be used for practical design problems considering uncertainty of parameters including those of seismic motions.

In Chapter 5, simultaneous geometry and topology optimization of truss and frame structures is extensively studied. One of the drawbacks for geometry and topology optimization is convergence to an unstable structure with possible existence of very long/short and/or overlapping members. Accordingly, the optimal solution may not have enough robustness. To overcome this difficulty, quantile-based approach is introduced. Some results in topology optimization imply that the order statistics and quantile-based approach improves the robustness of the structures.

In Chapter 6, the design problem is formulated as a multi-objective optimization problem to find several candidate solutions as Pareto optimal solutions. Trade-off relations of the objective functions and other performance measures are incorporated in the approach. Some results are presented for minimizing the mean and standard deviation of the uncertain function. Two objective functions representing exploration and exploitation may be maximized in the process of Bayesian optimization. It is demonstrated that robustness level can be represented by the order of the solution, and Pareto optimal solutions correspond to various levels of robustness are generated.

In Chapter 7, reliability-based design optimization is considered and efficiently solved by Bayesian optimization, also well-known as the *efficient global optimization*, which is regarded as a sequential process consisting of Bayesian regression model called Gaussian process regression and maximization of an acquisition function. The method is classified into surrogate-assisted method and successfully applied to the multi-objective reliability-based design optimization problems.

Some problem formulations, methodologies and numerical examples in this monograph are compilations of the authors' work during the period 2010–2022. The authors are grateful for collaborations on the studies that appear as valuable contents in this monograph to Dr. Wei Shen at Nanyang Technological University; Mr. Bach Do at the Department of Architecture and Architectural Engineering, Kyoto University; Ms. Kana Watanabe at Kozo Keikaku Engineering Inc. The authors would also like to thank again Mr. Bach Do at Kyoto University for checking the details of the manuscript. The first author also would like to acknowledge supervision by Prof. Fred van Keulen at Delft University of Technology, and advices by Prof. Matthijs Langelaar and Prof. Hans Goosen at Delft University of Technology during his sabbatical leave in the Netherlands.

The assistance of CRC Press and Taylor & Francis in bringing the manuscript to its final form is heartily acknowledged.

January 2023 Makoto Yamakawa and Makoto Ohsaki

1 Basic Concepts and Examples

Structural optimization is an application of mathematical optimization to structural design. Recent studies have linked structural optimization with robust design by a unified framework of non-parametric stochastic-based methodologies, which provides a rigorous theoretical background and a high level of practicality. The purpose of this monograph is to provide a comprehensive perspective of the development in recent years in the field of robust design optimization of structures. In particular, we will focus on stochastic-based methodologies. In this chapter, basic concepts of stochastic and probabilistic approaches are explained, and some related illustrative examples are introduced.

1.1 OVERVIEW OF STOCHASTIC STRUCTURAL OPTIMIZATION

Designers and/or engineers are required to make various decisions in the design process. They often have to consider cost, serviceability, constructability, durability, and aesthetic property of the product in addition to mechanical property. The requirements may conflict with each other, and all of them cannot be satisfied simultaneously. Inevitably, they need to consider trade-off relations between two or more conflicting objectives. How can they make decisions and design the products as good as possible in view of those conflicting measures? Mathematical optimization may help them in such situation. Applying mathematical optimization to structural design is referred to as *structural optimization* [Ohsaki, 2010]. The typical structural optimization problem is formally formulated to minimize an objective function representing the structural cost under constraints on the structural performance measures, or maximize structural performance under constraint on the structural cost. Structural optimization is definitely a useful tool for structural designers. At the same time, it has a limitation in application and some drawbacks for designing real-world structures. As a first step in starting structural optimization, designers have to formulate the design problem by mathematical formulations. Solvability of the problem depends on the types of the variables and complexity of the formulations. However, it is not an easy task to express design conditions as mathematical equalities or inequalities in advance. Furthermore, real-world design problems essentially have uncertainty. For example, for design problems in civil and architectural engineering, we cannot exactly predict the occurrence and properties of earthquakes in current technology. This fact implies that we should treat earthquakes as uncertain phenomena. Unfortunately, designs obtained by standard deterministic structural

DOI: 10.1201/9781003153160-1

optimization methods are often sensitive to variation of the given design conditions. This means that structures may be vulnerable to unexpected loads or accidental events. Thus, uncertainty should be appropriately considered in the process of designing structures.

Several strategies have been proposed for incorporating uncertainty in the design process. Robust design is known as an effective method for designing structures that are insensitive to environment and other uncontrollable factors [Knoll and Vogel, 2009; Elishakoff and Ohsaki, 2010]. In the 1950s, Taguchi developed the foundation of robust design for producing high-quality products [Taguchi and Phadke, 1989; Taguchi, 1957, 1958]. However, as stated above, a design solution obtained by deterministic structural optimization often lacks *robustness*. Therefore, linking robust design with structural optimization is a significant challenge, and extensive research has been carried out for design methods such as robust optimization [Beyer and Sendhoff, 2007; Zang et al., 2005]. However, there exist many definitions of robustness, and if the worst response is considered, computational cost for obtaining the exact worst (extreme) value is usually very large even when simple interval variables are used for representing uncertainty [Ohsaki and Katsura, 2012; Dimarogonas, 1995; Qiu, 2005]. Reliability-based design (RBD) is a standard approach to incorporate various types of uncertainty in the process of structural design [Noh et al., 2011]. Although approximation methods such as dimension reduction method [Xu and Rahman, 2004] are available, computational cost for finding the failure point or the most probable point is still large even for small-scale structures, if limit state functions corresponding to multiple failure modes are considered. Thus, in the approach based on reliability index with the failure probability, the stochastic variables are often implicitly assumed to be normally distributed or evaluated through various approximations [Der Kiureghian and Liu, 1986; Nowak and Collins, 2012]. Recent studies have linked structural optimization with robust design as well as RBD by a unified framework of non-parametric stochastic-based methodologies, which provide a rigorous theoretical background and high level of practicality for design methods considering uncertainty. The purpose of this monograph is to provide a comprehensive perspective of the development in recent years in the field of robust design optimization of structures mainly concerning stochastic-based methodologies.

1.2 STRUCTURAL OPTIMIZATION

1.2.1 BASICS OF STRUCTURAL OPTIMIZATION

Let us consider a simple structural optimization problem of an elastic structure subjected to static loads, where the design variables have real values, and the objective and constraint functions are continuous and differentiable with respect to the variables. Let $\boldsymbol{a} = (a_1, \cdots, a_m)^\top \in \mathbb{R}^m$ denote the vector of m design variables, which may represent the cross-sectional areas, cross-sectional dimensions and other properties of the structure. For simplicity, we consider the case where a_i represents the cross-sectional area of the ith member. All vectors are assumed to be column vectors in this section.

The vector of state variables representing the nodal displacements is denoted by $\boldsymbol{u} = (u_1, \cdots, u_n)^\top \in \mathbb{R}^n$, where n is the number of degrees of freedom. In most of the design problems in various fields of engineering, the design requirements for structural responses such as stresses and displacements are given with inequality constraints, which are formulated as

$$g_j(\boldsymbol{a}, \boldsymbol{u}(\boldsymbol{a})) \leq 0 \quad (j = 1, \cdots, l), \tag{1.1}$$

where l is the number of constraints. The constraint function $g_j(\boldsymbol{a}, \boldsymbol{u}(\boldsymbol{a}))$ depends on the design variables implicitly through the displacement (state variable) vector $\boldsymbol{u}(\boldsymbol{a})$ and also directly on the design variables. For example, the axial force t_i of the ith member of a truss is given using a constant vector $\boldsymbol{b}_i \in \mathbb{R}^n$ defining the stress-displacement relation as

$$t_i = a_i \boldsymbol{b}_i^\top \boldsymbol{u}(\boldsymbol{a}), \tag{1.2}$$

which depends explicitly on a_i and implicitly on $\boldsymbol{a}$ through $\boldsymbol{u}(\boldsymbol{a})$.

The upper and lower bounds, which are denoted by a_i^{U} and a_i^{L}, respectively, are given for the design variable a_i. The objective function to be minimized is denoted by $f(\boldsymbol{a}, \boldsymbol{u}(\boldsymbol{a}))$. Then the structural optimization problem is formulated as

$$\begin{aligned}
&\text{Minimize} && f(\boldsymbol{a}, \boldsymbol{u}(\boldsymbol{a})) && && \text{(1.3a)}\\
&\text{subject to} && g_j(\boldsymbol{a}, \boldsymbol{u}(\boldsymbol{a})) \leq 0 && (j = 1, \cdots, l), && \text{(1.3b)}\\
& && a_i^{\mathrm{L}} \leq a_i \leq a_i^{\mathrm{U}} && (i = 1, \cdots, m). && \text{(1.3c)}
\end{aligned}$$

The constraints (1.3c) are called side constraints, bound constraints or box constraints, which are treated separately from the general inequality constraints (1.3b) in most of the optimization algorithms. If there is no misunderstanding, for convenience, the functions with and without $\boldsymbol{u}(\boldsymbol{a})$ are denoted by the same symbol as

$$f(\boldsymbol{a}) = f(\boldsymbol{a}, \boldsymbol{u}(\boldsymbol{a})), \qquad g_j(\boldsymbol{a}) = g_j(\boldsymbol{a}, \boldsymbol{u}(\boldsymbol{a})). \tag{1.4}$$

In addition to this, we will often omit the variable $\boldsymbol{a}$, and write $f(\boldsymbol{a})$ and $f(\boldsymbol{a}, \boldsymbol{u}(\boldsymbol{a}))$ simply as f.

1.2.2 STRUCTURAL COMPLIANCE OPTIMIZATION

The *structural compliance* is equivalent to the external work by static loads, which is also equivalent to the twice of the strain energy if the linear elastic response is considered. For the sake of simplicity, we shall refer to the structural compliance as *compliance*. For the case without forced displacement, a smaller compliance leads to a stiffer structure against the specified loads. For a truss with m members and n degrees of freedom, let $\boldsymbol{K}(\boldsymbol{a}) \in \mathbb{R}^{n \times n}$ denote the $n \times n$ stiffness matrix, which is a function of the design variable vector $\boldsymbol{a}$. The stiffness (equilibrium) equation for computing $\boldsymbol{u}(\boldsymbol{a})$ against the constant load vector $\boldsymbol{p} \in \mathbb{R}^n$, which is assumed to be independent of the design variables, is written as

$$\boldsymbol{K}(\boldsymbol{a})\boldsymbol{u}(\boldsymbol{a}) = \boldsymbol{p}. \tag{1.5}$$

Using (1.5), we can write the compliance $w(\boldsymbol{a})$ as

$$\begin{aligned} w(\boldsymbol{a}) &= \boldsymbol{p}^\top \boldsymbol{u}(\boldsymbol{a}) \\ &= \boldsymbol{u}(\boldsymbol{a})^\top \boldsymbol{K}(\boldsymbol{a})\boldsymbol{u}(\boldsymbol{a}) \\ &= -2\Big(\frac{1}{2}\boldsymbol{u}(\boldsymbol{a})^\top \boldsymbol{K}(\boldsymbol{a})\boldsymbol{u}(\boldsymbol{a}) - \boldsymbol{p}^\top \boldsymbol{u}(\boldsymbol{a})\Big). \end{aligned} \tag{1.6}$$

It is seen that the compliance is also equivalent to the total potential energy multiplied by -2. The total structural volume $v(\boldsymbol{a})$ is defined as

$$v(\boldsymbol{a}) = \sum_{i=1}^{m} a_i \ell_i = \boldsymbol{\ell}^\top \boldsymbol{a}, \tag{1.7}$$

where ℓ_i is the length of the ith member and $\boldsymbol{\ell} = (\ell_1, \cdots, \ell_m)^\top \in \mathbb{R}^m$. Let w^{U} denote the specified upper bound for $w(\boldsymbol{a})$. Then the optimization problem for minimizing the total structural volume under compliance constraint is formulated as

$$\begin{aligned} \text{Minimize} \quad & v(\boldsymbol{a}) = \boldsymbol{\ell}^\top \boldsymbol{a} && \text{(1.8a)} \\ \text{subject to} \quad & w(\boldsymbol{a}) = \boldsymbol{p}^\top \boldsymbol{u}(\boldsymbol{a}) \leq w^{\mathrm{U}}, && \text{(1.8b)} \\ & a_i \geq a_i^{\mathrm{L}} \quad (i = 1, \cdots, m), && \text{(1.8c)} \end{aligned}$$

where upper bound is not given for a_i for simplicity. We can intuitively recognize that minimization of compliance under volume constraint is equivalent, after an appropriate scaling of the objective function, to minimization of the volume under compliance constraint, because the compliance is a decreasing function and the total structural volume is an increasing function of the cross-sectional areas. Therefore, utilizing these properties, many approaches have been presented for equivalent reformulation of the problem. Thus, let us consider the following problem of minimizing the compliance under constraint on the total structural volume:

$$\begin{aligned} \text{Minimize} \quad & w(\boldsymbol{a}) = \boldsymbol{p}^\top \boldsymbol{u}(\boldsymbol{a}) && \text{(1.9a)} \\ \text{subject to} \quad & v(\boldsymbol{a}) = \boldsymbol{\ell}^\top \boldsymbol{a} \leq v^{\mathrm{U}}, && \text{(1.9b)} \\ & a_i \geq a_i^{\mathrm{L}} \quad (i = 1, \cdots, m), && \text{(1.9c)} \end{aligned}$$

where v^{U} is the specified upper bound for the total structural volume.

The $n \times n$ stiffness matrix of a truss can be written as a linear function of $\boldsymbol{a}$ using constant matrices $\boldsymbol{K}_i \in \mathbb{R}^{n \times n}$ $(i = 1, \cdots, m)$ as

$$\boldsymbol{K}(\boldsymbol{a}) = \sum_{i=1}^{m} a_i \boldsymbol{K}_i, \quad \boldsymbol{K}_i = \frac{\partial \boldsymbol{K}}{\partial a_i}. \tag{1.10}$$

If $a_i^{\mathrm{L}} = 0$ for all members, the problem corresponds to the topology optimization problem that will be extensively studied in Chap. 5. We can also use the approach of *simultaneous analysis and design* for plane frames, as follows, with the cross-sectional areas and the nodal displacements as variables [Bendsøe and Sigmund, 2003] by explicitly incorporating the stiffness equation (1.5) that is reformulated using (1.10) into the constraints:

$$\text{Minimize} \quad w(\boldsymbol{a}, \boldsymbol{u}) = \boldsymbol{p}^\top \boldsymbol{u} \tag{1.11a}$$

$$\text{subject to} \quad \sum_{i=1}^{m} a_i \boldsymbol{K}_i \boldsymbol{u} = \boldsymbol{p}, \tag{1.11b}$$

$$v(\boldsymbol{a}) = \boldsymbol{\ell}^\top \boldsymbol{a} \leq v^{\mathrm{U}}, \tag{1.11c}$$

$$a_i \geq 0 \quad (i = 1, \cdots, m), \tag{1.11d}$$

where the variables are both of $\boldsymbol{a}$ and $\boldsymbol{u}$.

1.2.3 ILLUSTRATIVE EXAMPLE: COMPLIANCE MINIMIZATION OF A TWO-BAR STRUCTURE

We will start to introduce the basic concepts of structural optimization through a simple illustrative example of the two-bar structure, as shown in Fig. 1.1, which is fixed at node 3 and subjected to horizontal forces p_1 and p_2 at nodes 1 and 2, respectively. The bars have the same length ℓ and Young's modulus E. The design variables are the cross-sectional areas a_1 and a_2. For this simple problem, the displacements u_1 and u_2 of the free nodes 1 and 2 are taken as the state variables and expressed as explicit functions of the design

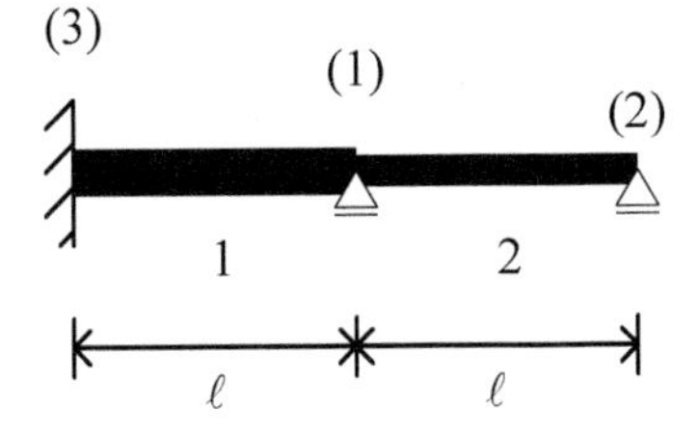

(a) Node and member numbers

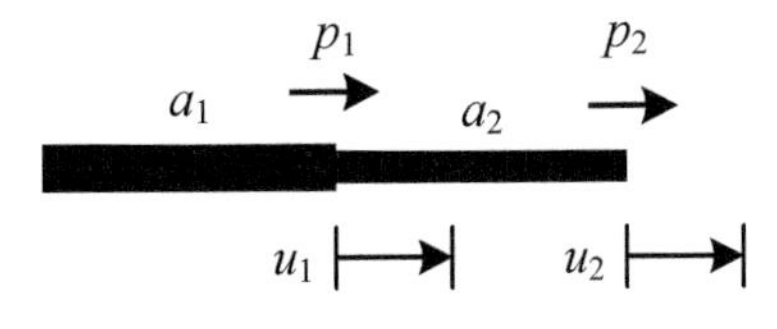

(b) Forces and displacements

Figure 1.1 A two-bar structure.

variables. From the equilibrium equations, we have

$$u_1 = u_1(a_1) = \frac{\ell}{E}\left(\frac{p_1 + p_2}{a_1}\right),$$

$$u_2 = u_2(a_1, a_2) = \frac{\ell}{E}\left(\frac{p_1 + p_2}{a_1} + \frac{p_2}{a_2}\right).$$

The corresponding compliance is written as

$$\begin{aligned} w(a_1, a_2) &= p_1 u_1(a_1) + p_2 u_2(a_1, a_2) \\ &= \frac{\ell}{E}\left[\frac{(p_1 + p_2)^2}{a_1} + \frac{p_2^2}{a_2}\right]. \end{aligned} \tag{1.12}$$

We consider a constraint on the volume of the structure, which is given by

$$v(a_1, a_2) = a_1 \ell + a_2 \ell \leq v^{\mathrm{U}}. \tag{1.13}$$

For physical reasons, the design variables must be non-negative, i.e.,

$$a_1 \geq 0, \qquad a_2 \geq 0. \tag{1.14}$$

Thus, the problem is formulated as

$$\begin{aligned} &\text{Minimize} && w(a_1, a_2) && (1.15\text{a}) \\ &\text{subject to} && v(a_1, a_2) \leq v^{\mathrm{U}}, && (1.15\text{b}) \\ &&& a_1 \geq 0,\ a_2 \geq 0. && (1.15\text{c}) \end{aligned}$$

For $p_1 > 0$ and $p_2 > 0$, one finds the solution of (1.15) as

$$a_1^* = \frac{v^{\mathrm{U}}}{\ell}\left(\frac{p_1 + p_2}{p_1 + 2p_2}\right), \qquad a_2^* = \frac{v^{\mathrm{U}}}{\ell}\left(\frac{p_2}{p_1 + 2p_2}\right), \tag{1.16}$$

and the corresponding optimal compliance as

$$w^* = w(a_1^*, a_2^*) = \frac{(p_1 + 2p_2)^2 \ell^2}{E v^{\mathrm{U}}}. \tag{1.17}$$

For $p_1 = p_2 = p > 0$, the optimal solution and the corresponding optimal compliance are respectively given by

$$a_1^* = \frac{2}{3} a_0, \quad a_2^* = \frac{1}{3} a_0, \quad w^* = 9 w_0, \tag{1.18}$$

where

$$a_0 = \frac{v^{\mathrm{U}}}{\ell}, \qquad w_0 = \frac{p^2 \ell^2}{E v^{\mathrm{U}}}. \tag{1.19}$$

The solution (1.18) and the corresponding functions are shown graphically in Fig. 1.2. As intuitively and expectedly, the volume constraint (1.15b) is satisfied with equality at the optimal solution, i.e., $v(a_1^*, a_2^*) = v^{\mathrm{U}}$. As can be seen from Fig. 1.2, the contour line of the objective function $w(a_1, a_2) = w^*$ is tangent to the volume constraint $v(a_1, a_2) = v^{\mathrm{U}}$ at the optimal solution (a_1^*, a_2^*). It is also confirmed that strains of the two members at their optimal values have the same value $3p/(Ea_0)$, which is well known as the optimality condition of the compliance optimization problem of a truss structure.

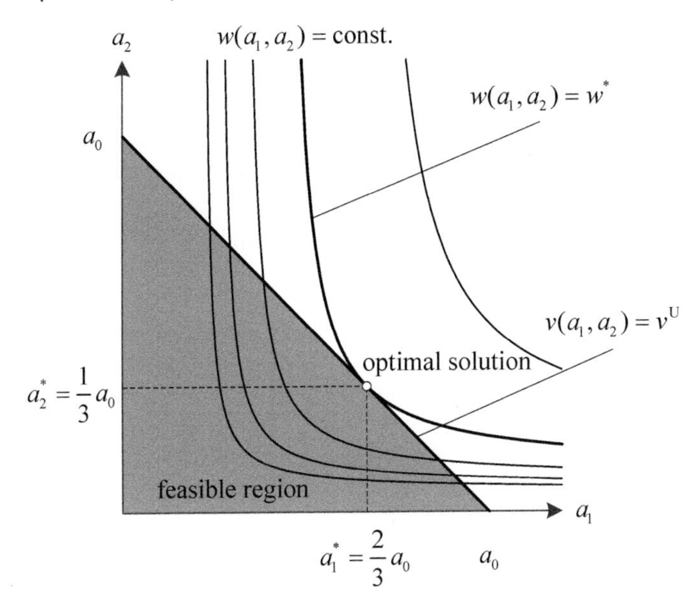

Figure 1.2 Geometric illustration of the two-bar structure problem for $p_1 = p_2 = p$.

1.3 STOCHASTICITY AND ROBUSTNESS

1.3.1 CATEGORIES OF UNCERTAINTY

In the process of designing a structure, it is desirable to determine its geometry and stiffness considering uncertainty in design loads, material parameters, etc. [Ben-Tal et al., 2009; Elishakoff and Ohsaki, 2010]. Therefore, in the field of structural optimization, extensive research has been carried out for development of design methods that are referred to as *robust optimization* [Zang et al., 2005; Beyer and Sendhoff, 2007; Gabrel et al., 2014]. In the process of robust optimization, a function associated with *robustness* is maximized; however, there exist many definitions of robustness and robustness measures. According to Taguchi, who is one of the pioneers of robust design,

> *"robustness is the state where the technology, product, or process performance is minimally sensitive to factors causing variability (either in the manufacturing or user's environment) and aging at the lowest unit manufacturing cost."* [Taguchi et al., 2000]

It should be noted that a deterministic optimization problem using the nominal values, i.e., a structural optimization without uncertainty, can be solved by using any of the probabilistic, stochastic and randomized approaches. These approaches are often incorporated into metaheuristics, e.g., evolutionary algorithm and simulated annealing [Blum and Roli, 2003], and are known to be effective to obtain a solution with good performance by escaping from a local optimum and eventually approaching a better solution. However, this topic is not a main concern of this book. For example, the reader is referred to Rao [2019]. In the following, we briefly classify the definitions and related topics on structural optimization problem considering uncertainty. This type of problem and its solution are referred to as *robust optimization problem* and *robust solution*, respectively.

Consider a response function of some parameters of a structure, e.g., the cross-sectional properties, material parameters, geometry of the structure, external force, environmental parameters and so forth. The controllable parameters are referred to as design variables that are combined into a vector $\boldsymbol{a} \in \mathbb{R}^m$ of m components. On the other hand, the vector consisting of input parameters, which correspond to environmental and/or uncontrollable properties, are denoted by $\boldsymbol{p} \in \mathbb{R}^r$ that has r components. Thus, the response function may be denoted by

$$f = f(\boldsymbol{a}; \boldsymbol{p}) = f\left(\boldsymbol{a}, \boldsymbol{u}(\boldsymbol{a}); \boldsymbol{p}\right), \tag{1.20}$$

where $\boldsymbol{u}(\boldsymbol{a})$ is the state vector depending on the design variables as mentioned in Sec. 1.2.1. A semicolon is used to divide the design variables and parameters in the arguments of a function. In the real world design, the structure has to face different kinds of uncertainty. The uncertainty in the design variable vector $\boldsymbol{a}$ and/or the parameter vector $\boldsymbol{p}$ are modeled using the uncertain parameter vector denoted by $\boldsymbol{\theta} \in \Omega$, where Ω is the pre-specified uncertainty set. For simplicity, we may assume that the uncertain parameter set is bounded. Various types of uncertainty are propagated into the response function f. We classify uncertainty in design optimization problem, as follows, into four types, see e.g., Beyer and Sendhoff [2007]:

(A) Uncertainty in the uncontrollable parameters. This corresponds to variability in the environmental and operating conditions. In many cases, we cannot know the exact external forces to be applied to the structure, and have to estimate their nominal values from engineering judgment. Uncertainty can be supposed to enter the structure via parameters $\boldsymbol{p}$ in (1.20) as

$$f^{(\mathrm{A})}(\boldsymbol{a}; \boldsymbol{p}) = f(\boldsymbol{a}; \boldsymbol{p} + \boldsymbol{\theta}) = f\left(\boldsymbol{a}, \boldsymbol{u}(\boldsymbol{a}); \boldsymbol{p} + \boldsymbol{\theta}\right). \tag{1.21}$$

(B) Uncertainty in the design variables. Uncertainty may exist in production tolerance and manufacturing accuracy. A structure is manufactured based on the design variables to a certain degree of accuracy. This type of uncertainty can be considered to enter the structure via the design variable vector $\boldsymbol{a}$ in (1.20) as

$$f^{(\mathrm{B})}(\boldsymbol{a}; \boldsymbol{p}) = f(\boldsymbol{a} + \boldsymbol{\theta}; \boldsymbol{p}) = f\left(\boldsymbol{a} + \boldsymbol{\theta}, \boldsymbol{u}(\boldsymbol{a} + \boldsymbol{\theta}); \boldsymbol{p}\right). \tag{1.22}$$

In general, $\boldsymbol{\theta}$ may depend on $\boldsymbol{a}$. For example, $\boldsymbol{\theta} = \varepsilon \boldsymbol{a}$ models relative manufacturing tolerances.

(C) Uncertainty in the structural response. This uncertainty is due to imprecision in the evaluation of the structural responses and performance measures. This kind of uncertainty includes measurement errors and all kinds of approximation errors in the modeling and simulation of real physical properties. The actually observed response denoted by $\mathcal{F}$ can be considered to be a functional of f in terms of the probability distribution of errors as

$$f^{(\mathrm{C})} = \mathcal{F}\left[f(\boldsymbol{a};\boldsymbol{p});\boldsymbol{\theta}\right]. \tag{1.23}$$

(D) Feasibility uncertainty. Uncertainty in the constraints is different from types (A)–(C). This kind of uncertainty has effects on the design space $\mathcal{A}$, which may be denoted by

$$\boldsymbol{a} \in \mathcal{A}(\boldsymbol{\theta}). \tag{1.24}$$

Especially when the uncertainty enters into inequality constraints as formulated in (1.3b), the uncertainty can be considered along the same lines of types (A)–(C) using a penalty function. For example, the feasible set corresponding to (1.3b) with uncertain parameter $\boldsymbol{\theta}$ may be expressed by

$$\mathcal{A}(\boldsymbol{\theta}) = \left\{\boldsymbol{a} \in \mathbb{R}^m \;\middle|\; g_j\left(\boldsymbol{a}+\boldsymbol{\theta}, \boldsymbol{u}(\boldsymbol{a}+\boldsymbol{\theta});\boldsymbol{p}\right) \le 0 \quad (j=1,\cdots,l)\right\}. \tag{1.25}$$

In this case, problem (1.3) with feasibility uncertainty may be given by

$$\underset{\boldsymbol{a}}{\text{Minimize}} \quad f(\boldsymbol{a};\boldsymbol{p}) \tag{1.26a}$$

$$\text{subject to} \quad \boldsymbol{a} \in \mathcal{A}(\boldsymbol{\theta}), \tag{1.26b}$$

$$a_i^{\mathrm{L}} \le a_i \le a_i^{\mathrm{U}} \qquad (i=1,\cdots,m). \tag{1.26c}$$

By using a function representing the constraint violation

$$g_{\max}(\boldsymbol{a},\boldsymbol{u}(\boldsymbol{a});\boldsymbol{p}) = \sum_{j=1}^{l} \max\left\{g_j\left(\boldsymbol{a}+\boldsymbol{\theta},\boldsymbol{u}(\boldsymbol{a}+\boldsymbol{\theta});\boldsymbol{p}\right),0\right\},$$

we can reformulate problem (1.26), e.g., to the following problem:

$$\text{Minimize} \quad f^{(\mathrm{D})}(\boldsymbol{a};\boldsymbol{p}) = f(\boldsymbol{a},\boldsymbol{u}(\boldsymbol{a});\boldsymbol{p}) + \rho g_{\max}(\boldsymbol{a},\boldsymbol{u}(\boldsymbol{a});\boldsymbol{p}) \tag{1.27a}$$

$$\text{subject to} \quad a_i^{\mathrm{L}} \le a_i \le a_i^{\mathrm{U}} \quad (i=1,\cdots,m), \tag{1.27b}$$

with a sufficiently large penalty parameter $\rho > 0$. The problem can be formulated in a similar manner as types (A)–(C) by replacing $f(\boldsymbol{a},\boldsymbol{u}(\boldsymbol{a});\boldsymbol{p})$ with $f^{(\mathrm{D})}(\boldsymbol{a};\boldsymbol{p}) = f(\boldsymbol{a},\boldsymbol{u}(\boldsymbol{a});\boldsymbol{p}) + \rho g_{\max}(\boldsymbol{a},\boldsymbol{u}(\boldsymbol{a});\boldsymbol{p})$.

Furthermore, there are following three different methodologies to mathematically quantify the uncertainty, which are deterministic, probabilistic and possibilistic methods:

(I) Deterministic method. Domain of uncertain parameters where they can vary is deterministically defined. This approach is also referred to as *unknown-but-bounded* model [Qiu and Elishakoff, 1998; Ben-Haim, 1994].

(II) Probabilistic method. Probability measures are defined and the likelihood for occurrence of a certain event is considered. This type of methodology is standard in the field of RBD [Nowak and Collins, 2012].

(III) Possibilistic method. Fuzzy measures are defined for describing the possibility or membership grade for a certain event to be plausible or believable [Klir and Folger, 1988].

In mechanical engineering literatures, design optimization problems incorporating probabilistic approaches are sometimes distinguished from *robust optimization* and specifically referred to as *reliability-based design optimization* (RBDO). However, there is no general consensus that RBDO should be distinguished from robust optimization methodologies [Beyer and Sendhoff, 2007; Gabrel et al., 2014]. Indeed, Park et al. wrote:

> *"The distinctions are made based on the methods of the early stages. These days, the methods are fused, therefore, it may not be possible to distinguish them in some applications."* [Park et al., 2006]

In a similar fashion, Elishakoff and Ohsaki wrote:

> *"The probabilistic and anti-optimization approaches lead to the same result! Not only is there no antagonism between them, but they both tell us the same thing!"* [Elishakoff and Ohsaki, 2010]

Note that *anti-optimization* in Elishakoff and Ohsaki [2010] is one of the deterministic methods. In the remainder of this section, on the basis of above classification, we will introduce some robustness measures.

1.3.2 DETERMINISTIC MEASURES: WORST-CASE APPROACH

Deterministic approach is regarded as the most popular approach in the field of robust optimization. It considers unknown-but-bounded uncertainty, which may be written as

$$\boldsymbol{\theta} \in \Omega(\boldsymbol{a}, \boldsymbol{p}, \epsilon),$$

where $\boldsymbol{\theta}$ is an uncertain parameter vector and $\Omega(\boldsymbol{a}, \boldsymbol{p}, \epsilon)$ is an uncertainty set at the design $\boldsymbol{a}$ and the parameter $\boldsymbol{p}$, the size of which depends on the parameter ϵ. This set of uncertain parameters is referred to as *scenario*. For given objective function $f(\boldsymbol{a}; \boldsymbol{p})$ to be minimized as defined in (1.3a), the corresponding *robust counterpart function* for type (B) uncertainty may be defined as

$$f^{\mathrm{D}}(\boldsymbol{a}; \boldsymbol{p}, \epsilon) = \sup_{\boldsymbol{\theta} \in \Omega(\boldsymbol{a}, \boldsymbol{p}, \epsilon)} f(\boldsymbol{a} + \boldsymbol{\theta}; \boldsymbol{p}), \tag{1.28}$$

where f^{D} is the objective function to be minimized in robust optimization problem and the following equation should be satisfied:

$$\lim_{\epsilon \to 0} f^{\mathrm{D}}(\boldsymbol{a}; \boldsymbol{p}, \epsilon) = f(\boldsymbol{a}; \boldsymbol{p}).$$

Similarly, the function for type (A) uncertainty may be defined as

$$f^{\mathrm{D}}(\boldsymbol{a}; \boldsymbol{p}, \epsilon) = \sup_{\boldsymbol{\theta} \in \Omega(\boldsymbol{a}, \boldsymbol{p}, \epsilon)} f(\boldsymbol{a}; \boldsymbol{p} + \boldsymbol{\theta}).$$

Thus, the robust optimization using a deterministic measure is interpreted as finding the worst value of the function for a given scenario, i.e., a given set of uncertain parameters. This type of formulations is referred to by various names, e.g., worst-case approach, robust counterpart approach, unknown-but-bounded data model or anti-optimization approach, and the method of optimization by minimizing f^{D} is called, e.g., bilevel optimization or minimax optimization. The model has significant methodological advantage; however, it has been also pointed out that the worst-case-oriented approach might be impractical because of its expensive computational cost, high conservativity, etc. The worst value is obtained by solving a maximization problem (more accurately, finding supremum) in the general case. If the worst structural response is considered, computational cost for obtaining the exact worst (extreme) value is very large.

1.3.3 EXPECTANCY MEASURES

Robustness measures can be formulated probabilistically regarding $\boldsymbol{\theta}$, and accordingly f in (1.21)–(1.23), as random variables, which are linked with randomness and probability distribution. We avoid explaining further details of random variables here for an intuitive understanding. Typical mathematical definitions are given on the basis of probability theory in Sec. 2.2.

We use a capital letter for a random variable, like $\boldsymbol{\Theta}$ and F for $\boldsymbol{\theta}$ and f, respectively, to avoid confusion with the deterministic variables; see also Sec. 1.4 for some terminologies of probability theory. The probability distribution function (pdf) of the random variable is denoted by $p(\cdot)$. The expectancy measures are defined as the conditional expectation of a utility function $\mathcal{U}(\cdot)$:

$$f^{\mathrm{E}}(\boldsymbol{a}) = \mathbb{E}\left[\mathcal{U}(F)|\boldsymbol{a}\right].$$

When the utility function is given by $\mathcal{U}(F) = F^k$, one obtains momentum measures. For example, the special case $k = 1$ for type (B) uncertainty gives the expectation measure as

$$f_1^{\mathrm{E}}(\boldsymbol{a}) = \mathbb{E}\left[F|\boldsymbol{a}\right] = \int_{\Omega} f(\boldsymbol{a} + \boldsymbol{\Theta}) p(\boldsymbol{\Theta}) d\boldsymbol{\Theta}.$$

For the utility function

$$\mathcal{U}(F) = \left(F - \mathbb{E}[F|\boldsymbol{a}]\right)^2,$$

one has the variance measure

$$\begin{aligned} f_2^{\mathrm{E}}(\boldsymbol{a}) &= \mathbb{E}\left[(F - \mathbb{E}[F|\boldsymbol{a}])^2 \,|\boldsymbol{a}\right] \\ &= \mathbb{E}\left[F^2|\boldsymbol{a}\right] - \left(\mathbb{E}\left[F|\boldsymbol{a}\right]\right)^2 \\ &= \int_\Omega f(\boldsymbol{a}+\boldsymbol{\Theta})^2 p(\boldsymbol{\Theta}) d\boldsymbol{\Theta} - \left(f_1^{\mathrm{E}}(\boldsymbol{a})\right)^2 \\ &= \mathrm{Var}\left[F|\boldsymbol{a}\right]. \end{aligned}$$

In the standard formulation of robust design, f_1^{E} and f_2^{E} are minimized simultaneously. One may aggregate them into a single function, e.g., using a weighted sum of the objective functions as

$$f_3^{\mathrm{E}}(\boldsymbol{a}) = \alpha f_1^{\mathrm{E}}(\boldsymbol{a}) + \beta\sqrt{f_2^{\mathrm{E}}(\boldsymbol{a})} = \alpha\mathbb{E}\left[F|\boldsymbol{a}\right] + \beta\sqrt{\mathrm{Var}\left[F|\boldsymbol{a}\right]},$$

where α and β are the given positive real numbers. Alternatively, various multi-objective optimization approaches are also applicable. Details of these approaches will be described in Chap. 6.

1.3.4 PROBABILISTIC THRESHOLD MEASURES

Instead of considering the expected values of utility functions, one can utilize threshold measures of the distribution. Let us consider a random variable $F = F(\boldsymbol{\Theta})$, which may correspond to, e.g.,

$$f(\boldsymbol{a}+\boldsymbol{\Theta};\boldsymbol{p}), \quad f(\boldsymbol{a};\boldsymbol{p}+\boldsymbol{\Theta}), \quad \mathcal{F}[f(\boldsymbol{a};\boldsymbol{p});\boldsymbol{\Theta}],$$

as described in Sec. 1.3.1. By using the distribution of F, we define a threshold dependent probability measure for a given threshold q as

$$f_1^{\mathrm{P}} = f_1^{\mathrm{P}}(\boldsymbol{a};q) = \Pr\{F \le q\}.$$

This measure corresponds to the relative frequency of satisfying the inequality and should be as large as possible because F represents the response function and hopefully less than q. When it is difficult to determine an appropriate value of the threshold q, a proportion of the population corresponding to the specified distribution can be alternatively specified. For a given proportion denoted by γ, a threshold measure is defined as

$$f_2^{\mathrm{P}} = f_2^{\mathrm{P}}(\boldsymbol{a};\gamma) = \inf\left\{q : \Pr\{F \le q\} \ge \gamma\right\}. \tag{1.29}$$

which is generally called *γth quantile,* and should be as small as possible because q is the response value that is to be minimized. The cumulative distribution function (cdf) of the random variable F is denoted by

$$\mathcal{F}(q) = \Pr\{F \le q\}.$$

When $\mathcal{F}(\cdot)$ is continuous, these two measures are simply expressed as

$$f_1^{\mathrm{P}} = \mathcal{F}(q), \qquad f_2^{\mathrm{P}} = \mathcal{F}^{-1}(\gamma),$$

where $\mathcal{F}^{-1}(\cdot)$ is the inverse function of $\mathcal{F}(\cdot)$.

1.3.5 ILLUSTRATIVE EXAMPLE: WORST-CASE APPROACH

Consider again the two-bar structure as shown in Fig. 1.1. Let θ_1 and θ_2 denote the uncertain parameters for horizontal forces p_1 and p_2, respectively, which are added to the nominal values, i.e., $\tilde{p}_1 = p_1 + \theta_1$ and $\tilde{p}_2 = p_2 + \theta_2$. Thus, the uncertainty is propagated to the displacements as

$$\tilde{u}_1 = \tilde{u}_1(a_1; \theta_1, \theta_2) = \frac{\ell}{E}\left(\frac{p_1 + \theta_1 + p_2 + \theta_2}{a_1}\right),$$

$$\tilde{u}_2 = \tilde{u}_2(a_1, a_2; \theta_1, \theta_2) = \frac{\ell}{E}\left(\frac{p_1 + \theta_1 + p_2 + \theta_2}{a_1} + \frac{p_2 + \theta_2}{a_2}\right).$$

The compliance considering uncertainty can be expressed as

$$\begin{aligned} \tilde{w} &= \tilde{w}(a_1, a_2; \theta_1, \theta_2) \\ &= \frac{\ell}{E}\left[\frac{(p_1 + \theta_1 + p_2 + \theta_2)^2}{a_1} + \frac{(p_2 + \theta_2)^2}{a_2}\right], \end{aligned} \tag{1.30}$$

which is to be minimized. Among several concepts for modeling uncertainty as explained in Sec. 1.3.1, we will use a deterministic measure, i.e., worst-case approach in Sec. 1.3.2. Suppose the uncertain parameters are bounded by the intervals:

$$\Omega = \{(\theta_1, \theta_2) \mid -\alpha p \leq \theta_1 \leq \alpha p, -\alpha p \leq \theta_2 \leq \alpha p\},$$

where p is a positive constant of the force level and $\alpha \in [0, 1]$ is the specified parameter corresponding to the level of uncertainty. For given a_1 and a_2, the compliance is also bounded as

$$\tilde{w}^{\mathrm{L}}(a_1, a_2) \leq \tilde{w}(a_1, a_2; \theta_1, \theta_2) \leq \tilde{w}^{\mathrm{U}}(a_1, a_2) \text{ for all } (\theta_1, \theta_2) \in \Omega,$$

where $\tilde{w}^{\mathrm{L}}(a_1, a_2)$ and $\tilde{w}^{\mathrm{U}}(a_1, a_2)$ denote respectively the lower and upper bounds of the compliance, which are defined as

$$\tilde{w}^{\mathrm{L}}(a_1, a_2) = \min_{(\theta_1, \theta_2) \in \Omega} \tilde{w}(a_1, a_2; \theta_1, \theta_2),$$

$$\tilde{w}^{\mathrm{U}}(a_1, a_2) = \max_{(\theta_1, \theta_2) \in \Omega} \tilde{w}(a_1, a_2; \theta_1, \theta_2).$$

Especially, the upper bound $\tilde{w}^{\mathrm{U}}(a_1, a_2)$ is regarded as the worst response function for possible scenarios. By minimizing the worst response function, we can reduce the undesirable effects of uncertainty. Thus, the problem (1.15) under load uncertainty may be formulated as a standard nonlinear programming problem:

$$\begin{aligned} &\text{Minimize} && \tilde{w}^{\mathrm{U}}(a_1, a_2) && (1.31\mathrm{a}) \\ &\text{subject to} && v(a_1, a_2) \leq v^{\mathrm{U}}, && (1.31\mathrm{b}) \\ & && a_1 \geq 0,\ a_2 \geq 0. && (1.31\mathrm{c}) \end{aligned}$$

This type of problem is classified into *robust optimization problem* and the solution is referred to as *robust solution* as mentioned in Sec. 1.3.1. For $p_1 > 0$ and $p_2 > 0$, by assigning the worst-case point $(\theta_1, \theta_2) = (\alpha p, \alpha p)$, one finds the solution of (1.31) as

$$\tilde{a}_1^* = \frac{v^{\mathrm{U}}}{\ell}\left(\frac{p_1 + p_2 + 2\alpha p}{p_1 + 2p_2 + 3\alpha p}\right), \tag{1.32a}$$

$$\tilde{a}_2^* = \frac{v^{\mathrm{U}}}{\ell}\left(\frac{p_2 + \alpha p}{p_1 + 2p_2 + 3\alpha p}\right), \tag{1.32b}$$

and the corresponding worst response function as

$$\tilde{w}^{\mathrm{U}*} = \tilde{w}^{\mathrm{U}}\left(\tilde{a}_1^*, \tilde{a}_2^*\right) = \frac{(p_1 + 2p_2 + 3\alpha p)^2 \ell^2}{E v^{\mathrm{U}}}. \tag{1.33}$$

For $p_1 = p_2 = p > 0$, from (1.32), we have

$$\tilde{a}_1^* = \frac{2}{3} a_0, \qquad \tilde{a}_2^* = \frac{1}{3} a_0, \tag{1.34}$$

where a_0 is defined in (1.19). From comparison of (1.18) and (1.34), one sees that the robust solution coincides with the nominal optimal solution in this case. However, this is not always true. For $p_1 = 2p$ and $p_2 = 0$, the nominal optimal solution is given by

$$a_1^* = a_0, \qquad a_2^* = 0, \tag{1.35}$$

and the robust solution is given by

$$\tilde{a}_1^* = \frac{2 + 2\alpha}{2 + 3\alpha} a_0, \qquad \tilde{a}_2^* = \frac{\alpha}{2 + 3\alpha} a_0. \tag{1.36}$$

From (1.35) and (1.36), it is seen that the robust solution is different from the nominal optimal solution when $\alpha \neq 0$. The robust solution (1.36) is plotted with respect to α in Fig. 1.3. It is seen that missing a cross-section is prevented by assigning a non-zero value for uncertainty parameter α.

1.4 PROBABILITY-BASED STRUCTURAL OPTIMIZATION

There are many methods for considering uncertainty of parameters in the design process. The worst-case approach with bounded support is most widely used for robust design, because its formulation is very simple and can be solved using a standard method of nonlinear programming. This approach is usually classified into type I of deterministic method in Sec. 1.3.1 but at the same time yields the same results obtained by type II of probabilistic methodologies under appropriate assumptions [Elishakoff and Ohsaki, 2010].

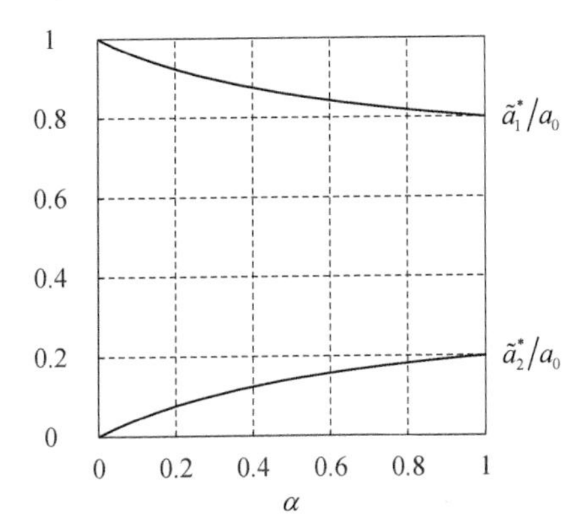

Figure 1.3 Robust solution of the two-bar structure for $p_1 = 2p$ and $p_2 = 0$ depending on the level of uncertainty α.

In some special cases, we may find the exact or moderately accurate worst values of structural responses; however, in many practical problems, finding such worst value is impossible or requires large computational cost even when simple interval variables are used for representing uncertainty; hence, we must use approximate worst values [Ben-Tal et al., 2009; Ohsaki and Katsura, 2012]. Note that the exact worst value may correspond to an extremely rare event that has probability 0 from the viewpoint of probability theory. For example, it is impossible and/or unreasonable to ensure safety of a building structure against any natural disasters with extremely small probability of occurrence. For some cases, it may be important to minimize the median and/or quantile values of the response. Hence, we can justify relaxing from the worst value to an approximate worse value.

RBD is also a standard approach that can incorporate various types of uncertainty including those in material, geometry and external loads. The structural responses are regarded as random variables, and the corresponding safety measure of the structure is described by a reliability index, which should meet a target value associated with a threshold failure probability of the component or the system within a reference period. The RBD usually involves assumptions on the type of pdf. In this monograph, RBD is regarded as one of the probabilistic methodologies. The worst response has theoretically very small probability of occurrence, which is sensitive to the parameter variation due to the behavior of the tail of the distribution. Thus, it is not an easy task to handle the worst response also by the RBD methodologies. For these reasons, we will focus on the nonparametric methods, and whenever possible, we avoid making specific assumptions as much as possible on stochasticity, e.g., normal distribution, time-wise stationarity, space-wise homogeneity, ergodicity of the process and the Markov property.

1.4.1 CENTRAL MOMENTS

Let X be a real-valued random variable. The cdf is defined by

$$F(x) = \Pr\{X \leq x\}.$$

$F(x)$ is an increasing function, and $0 \leq F(x) \leq 1$ for all x. If $F(x)$ is a strictly increasing function, the equation $F(x) = \gamma$ $(0 < \gamma < 1)$ has a unique solution denoted by $Q(\gamma) = F^{-1}(\gamma)$, which is called the (population) γth quantile; γ is referred to as the proportion of population (or content), and the function $Q(\gamma)$ is referred to as the quantile function. For example, $Q(1/2)$ is the median of the distribution. Even if F is not strictly increasing, we can define the γth quantile or quantile function as

$$Q(\gamma) = \inf\{y|F(y) \geq \gamma\},$$

and obviously $F(Q(\gamma)) = \gamma$ holds. If $F(x)$ is differentiable, its derivative $f(x) = dF(x)/dx$ is the pdf. The expectation of the random variable X is defined as

$$\mathbb{E}[X] = \int_{-\infty}^{\infty} x dF(x) = \int_{-\infty}^{\infty} x f(x) dx,$$

if this integral exists. We may also write, via the transformation $x = Q(u)$ and $du = dF(x)$,

$$\mathbb{E}[X] = \int_{0}^{1} Q(u) du.$$

A function $g(X)$ is itself a random variable and has expectation

$$\mathbb{E}[g(X)] = \int_{-\infty}^{\infty} g(x) dF(x) = \int_{-\infty}^{\infty} g(x) f(x) dx = \int_{0}^{1} g(Q(u)) du. \tag{1.37}$$

The shape of a probability distribution can be generally described by the moments of the distribution. The first moment is mean, which is usually denoted by

$$\mu = \mathbb{E}[X].$$

The mean is the center of location of the distribution. The dispersion of the distribution about its center is measured by the second moment, which is referred to as the variance obtained by

$$\mu_2 = \mathrm{Var}(X) = \mathbb{E}[(X-\mu)^2],$$

or the standard deviation

$$\sigma = \mu_2^{1/2} = \left\{\mathbb{E}[(X-\mu)^2]\right\}^{1/2}.$$

The higher moments are given by

$$\mu_r = \mathbb{E}[(X-\mu)^r] \qquad (r = 2, 3, \cdots).$$

These moments are especially referred to as *central moments* when they are distinguished from other moments.

Analogous quantities can be computed from a set of data samples $x_1, \cdots, x_n$. The sample mean

$$\bar{x} = \frac{1}{n}\sum_{i=1}^{n} x_i$$

is the natural estimator of μ. The higher sample moments

$$m_r = \frac{1}{n}\sum_{i=1}^{n}(x_i - \bar{x})^r$$

are reasonable estimators of μ_r. Some unbiased estimators of the central moments are derived from the sample moments. For example, σ^2, μ_3 and the fourth cumulant $\kappa_4 = \mu_4 - 3\mu_2^2$ are respectively estimated without bias by

$$\tilde{\sigma}^2 = \frac{1}{n-1}\sum_{i=1}^{n}(x_i - \bar{x})^2 = \frac{n}{n-1}m_2,$$

$$\tilde{\mu}_3 = \frac{n^2}{(n-1)(n-2)}m_3,$$

$$\tilde{\kappa}_4 = \frac{n^2}{(n-2)(n-3)}\left[\left(\frac{n+1}{n-1}\right)m_4 - 3m_2^2\right].$$

1.4.2 ORDER STATISTICS AND TOLERANCE INTERVALS

Order statistics are developed for evaluating the statistical properties of the kth best/worst value in a set of specified number of random variables [David and Nagaraja, 2003]. Tolerance intervals and confidence intervals of quantiles are predicted by order statistics, the unique feature of which is its independence from the type of distribution of random parameter. It also ensures that the bound of quantile value can be obtained with smaller computational cost than the bound of the worst value. Such property is known as distribution-free tolerance interval. Therefore, the order statistics can be effectively utilized for obtaining approximate bounds of the worst responses of structures subjected to various types of uncertainty. The bound of quantile value of response for specified confidence can be quantitatively evaluated using the theory of order statistics. The bound of the quantile value decreases in accordance with decrease of the corresponding order statistics. Therefore, if the structure is designed such that the order statistics is less than the specified value, then the structure has a specified probability of its representative response to be less than the specified value.

Let $\boldsymbol{\Theta} = (\Theta_1, \cdots, \Theta_t) \in \Omega$ denote a t-dimensional random variable vector, which corresponds to uncertain parameters, e.g., yield stress, Young's modulus, cross-sectional area and external load, where $\Omega \subset \mathbb{R}^t$ is the prescribed continuous and bounded set. The representative response of a structure is denoted by $X = g(\boldsymbol{\Theta})$. Let $\boldsymbol{\Theta}_1, \cdots, \boldsymbol{\Theta}_n$ denote a set of n random variable vectors that are assumed to be independent and generated using the same probability distribution on Ω. Accordingly, the representative responses denoted by $X_1 = g(\boldsymbol{\Theta}_1), \cdots, X_n = g(\boldsymbol{\Theta}_n)$ are also assumed to be independent and identically distributed random variables (iidrvs), for which the cdf is assumed to be continuous and denoted by

$$F(x) = \Pr\{X \leq x\} = \Pr\{\boldsymbol{\Theta} \in \Omega | g(\boldsymbol{\Theta}) \leq x\}.$$

These random variables $X_1, \cdots, X_n$ are renumbered in decreasing order as $X_{1:n} \geq \cdots \geq X_{k:n} \geq \cdots \geq X_{n:n}$ $(1 \leq k \leq n)$, where the kth largest response among samples is referred to as the kth order statistics. Note that they are renumbered in increasing order in usual order statistics. We sort them in decreasing order because we are interested in large responses.

A proportion of a population estimated by order statistics $F(X_{k:n})$ is also regarded as a random variable. It is known that $F(X_{k:n})$ is equal in distribution to the kth order statistics for the n samples from a standard uniform distribution denoted by $U(0, 1)$. The pdf of $F(X_{k:n})$ is given by

$$f_{k:n}(t) = \frac{t^{n-k}(1-t)^{k-1}}{B(n-k+1, k)} \qquad (0 \leq t \leq 1),$$

where $B(\cdot, \cdot)$ denotes the beta function that is defined as

$$B(a, b) = \int_0^1 t^{a-1}(1-t)^{b-1} dt. \tag{1.38}$$

Readers can consult with, e.g., David and Nagaraja [2003] for details. Expectation and variance of $F(X_{k:n})$ are respectively given by

$$\mathbb{E}[F(X_{k:n})] = \frac{n-k+1}{n+1}, \qquad \mathrm{Var}[F(X_{k:n})] = \frac{k(n-k+1)}{(n+1)^2(n+2)}.$$

Furthermore, for a given real number γ $(0 < \gamma < 1)$, the following equation is derived:

$$\Pr\{F(X_{k:n}) \geq \gamma\} = \int_\gamma^1 f_{k:n}(t) dt = 1 - I_\gamma(n-k+1, k), \tag{1.39}$$

where $I_\gamma(\cdot, \cdot)$ is the incomplete beta function that is defined by

$$I_\gamma(a, b) = \frac{\int_0^\gamma t^{a-1}(1-t)^{b-1} dt}{\int_0^1 t^{a-1}(1-t)^{b-1} dt}. \tag{1.40}$$

If we select k and n that satisfy (1.39) for given α and γ $(0 < \alpha < 1, 0 < \gamma < 1)$ as

$$\Pr\{F(X_{k:n}) \geq \gamma\} \geq \alpha \Leftrightarrow I_\gamma(n-k+1, k) \leq 1 - \alpha, \tag{1.41}$$

Table 1.1
Minimum Sample Size Requirements $n(k)$ for $\alpha = \gamma = 0.9$

k	1	2	3	4	5	6	7	8	9	10
$n(k)$	22	38	52	65	78	91	104	116	128	140

Table 1.2
Proportion for Order Statistics of $\alpha = 0.9$, $n = 150$

k	1	2	3	4	5	6	7	8	9	10
γ	1.000	1.000	1.000	1.000	0.999	0.998	0.994	0.986	0.969	0.940
k	11	12	13	14	15	16	17	18	19	20
γ	0.894	0.829	0.746	0.647	0.540	0.432	0.331	0.242	0.169	0.113

we state that the probability of $g(\boldsymbol{\Theta})$ so that its proportion γ is less than $X_{k:n}$ is with α *confidence*. Note that the right hand side of (1.41) does not depend on the type of $F(\cdot)$, which means that the interval $(-\infty, X_{k:n})$ is distribution-free. Equation (1.41) is rewritten as

$$\Pr\{Q(\gamma) \leq X_{k:n}\} \geq \alpha \Leftrightarrow I_\gamma(n-k+1, k) \leq 1-\alpha. \tag{1.42}$$

As stated above, (1.42) holds independently of the distribution. Although the value of the γth quantile $Q(\gamma)$ is unknown without information of distribution, we can observe the sample value of $X_{k:n}$ corresponding to the upper bound of the γth quantile with confidence $100\alpha\%$. This formulation is known as one-sided nonparametric tolerance interval [Krishnamoorthy and Mathew, 2008]. Some results of minimum sample size $n(k)$ satisfying requirement (1.41) are summarized in Tables 1.1 and 1.2.

1.4.3 ILLUSTRATIVE EXAMPLE: PROBABILISTIC WORST-CASE APPROACH

Similarly to the compliance minimization problem (1.31) (see also Fig. 1.1), we are to formulate a design problem considering variations in design variables by probabilistic approach. Consider the iidrvs Θ_1 and Θ_2 that follow a uniform distribution between $-\Delta a$ and Δa as

$$\Theta_i \sim U(-\Delta a, \Delta a) \qquad (i = 1, 2), \tag{1.43}$$

where Δa is a specified value for the range of uncertainty of the cross-sectional areas a_1 and a_2. The uncertain parameters θ_1 and θ_2 denote the observed sample values of iidrvs Θ_1 and Θ_2, respectively. In this case, the pdf and the cdf of the random variables are respectively given by

$$f(\theta_i) = \frac{1}{2\Delta a}, \qquad F(\theta_i) = \Pr\{\Theta_i \leq \theta_i\} = \frac{\theta_i + \Delta a}{2\Delta a} \qquad (i = 1, 2),$$

for $-\Delta a \leq \theta_i \leq \Delta a$. Because the random variables Θ_1 and Θ_2 are associated with the cross-sectional areas a_1 and a_2, respectively, which are the design variables, the uncertainty may be treated for physical reasons as

$$A_i = \max\{a_i + \Theta_i, 0\} \qquad (i = 1, 2).$$

The uncertainty is propagated into the compliance as

$$W = W(a_1, a_2) = w(A_1, A_2) = \frac{\ell}{E}\left[\frac{(p_1 + p_2)^2}{A_1} + \frac{p_2^2}{A_2}\right]. \tag{1.44}$$

Note that A_1, A_2 and W are also random variables, and we can regard $W = W(a_1, a_2)$ as a stochastic function. One may straightforwardly formulate a compliance minimization problem of the two-bar structure under the cross-sectional area uncertainty as a stochastic optimization problem:

$$\text{Minimize} \quad W(a_1, a_2) \tag{1.45a}$$
$$\text{subject to} \quad v(a_1, a_2) \leq v^{\mathrm{U}}, \tag{1.45b}$$
$$a_1 \geq 0,\ a_2 \geq 0. \tag{1.45c}$$

In such a direct approach, we encounter a difficulty: How should we evaluate the stochastic function $W(a_1, a_2)$? Fortunately (or unfortunately), there are various measures to evaluate the stochasticity of the function. One of them is the quantile.

For $p_1 = p_2 = p > 0$, the range of the uncertain compliance $W = W(a_1, a_2)$ is given by

$$w_0\left(\frac{4a_0}{\tilde{a}_1^{\mathrm{U}}} + \frac{a_0}{\tilde{a}_2^{\mathrm{U}}}\right) \leq W \leq w_0\left(\frac{4a_0}{\tilde{a}_1^{\mathrm{L}}} + \frac{a_0}{\tilde{a}_2^{\mathrm{L}}}\right),$$

where a_0 and w_0 are defined in (1.19), and $\tilde{a}_i^{\mathrm{U}}$ and $\tilde{a}_i^{\mathrm{L}}$ are respectively given by

$$\tilde{a}_i^{\mathrm{U}} = a_i + \Delta a, \quad \tilde{a}_i^{\mathrm{L}} = \max\{a_i - \Delta a, 0\} \qquad (i = 1, 2).$$

For example, for $a_1 = 2a_0/3$, $a_2 = a_0/3$ and $\Delta a = a_0/10$, we simply have

$$\frac{2250}{299} w_0 \leq W \leq \frac{1350}{119} w_0. \tag{1.46}$$

When $\frac{1530}{161} w_0 \leq \bar{w} \leq \frac{1350}{119} w_0$, the closed form cdf of W is derived as

$$F(\bar{w}) = \Pr\{W \leq \bar{w}\} = 1 - \Pr\{W \geq \bar{w}\}$$
$$= 1 - \frac{1}{(2\Delta a)^2} \iint_{\bar{w} \leq w(a_1, a_2) \leq \frac{1350}{119} w_0} da_1 da_2$$
$$= 1 - \frac{100}{(\bar{w}/w_0)^2} \log \frac{3600}{(30 - 7w/w_0)(120 - 17w/w_0)} - \frac{119w/w_0 - 1350}{36w/w_0}, \tag{1.47}$$

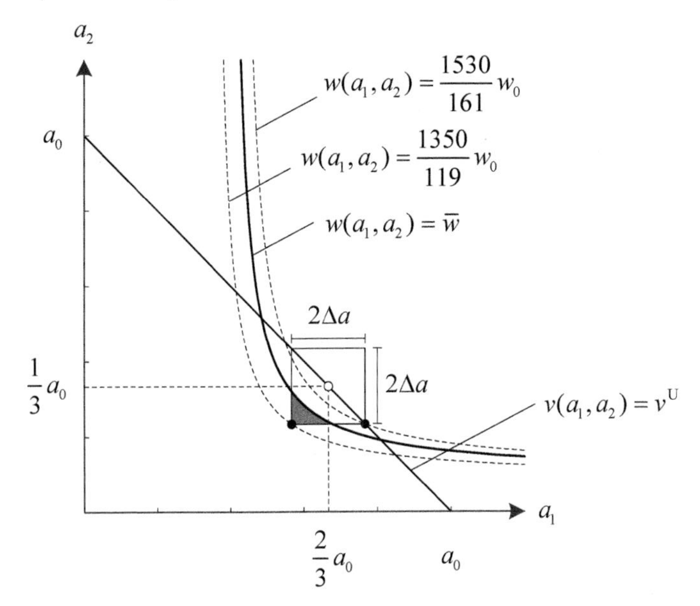

Figure 1.4 Geometric illustration of the stochastic two-bar structure problem for $p_1 = p_2 = p$, $a_1 = 2a_0/3$, $a_2 = a_0/3$ and $\Delta a = a_0/10$.

where the compliance $w(a_1, a_2)$ is obtained as

$$w(a_1, a_2) = \left(\frac{4a_0}{a_1} + \frac{a_0}{a_2} \right) w_0.$$

In this case, feasible regions of cross-sectional areas of the two-bar structure problem with probabilistic uncertainty and the corresponding contours of compliance for $\frac{1530}{161} w_0 \leq \bar{w} \leq \frac{1350}{119} w_0$ are shown graphically in Fig. 1.4. The gray shaded region in Fig. 1.4 represents the region where the compliance exceeds $\bar{w}$, and its area ratio to $(2\Delta a)^2$ corresponds to $\Pr\{W \geq \bar{w}\}$. The bottom left of the square region corresponds to the worst value in (1.46). The γth quantile of the uncertain compliance W is defined by

$$w_\gamma = \inf \{ w \mid \Pr\{W \leq w\} \geq \gamma \} \quad (0 \leq \gamma \leq 1),$$

where γ denotes a content ratio, which is the proportion of the population with the uniform distribution; see also Appendix A.2. We can find the γth quantile by solving $F(w_\gamma) = \gamma$ for w_γ with (1.47). The obtained γth quantile is shown in Fig. 1.5. For example, the 0.90th quantile is numerically found as

$$w_{0.90} \approx 10.19 w_0,$$

which is plotted as a black circle in Fig. 1.5. The upper bound of the γth quantile is obtained from (1.46), which corresponds to the value of $\gamma = 1.00$,

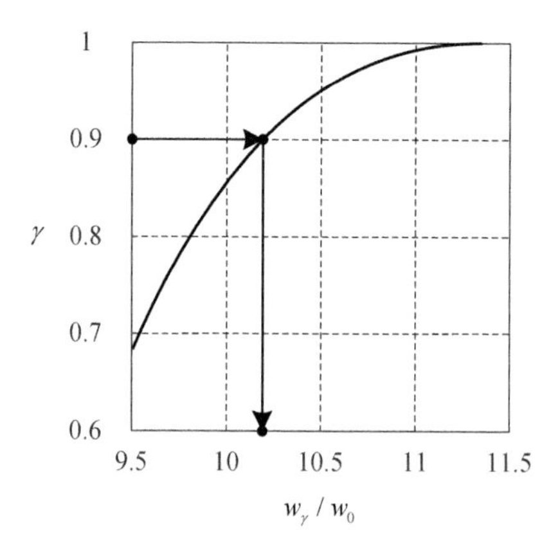

Figure 1.5 γth quantile of the two-bar structure problem for $p_1 = p_2 = p$, $a_1 = 2a_0/3$, $a_2 = a_0/3$ and $\Delta a = a_0/10$.

and hence we have

$$w_{1.00} = \frac{1350}{119} w_0 \approx 11.34 w_0.$$

In a similar fashion to Fig. 1.4, the corresponding contours of compliance are shown in Fig. 1.6. The gray shaded region in Fig. 1.6 represents the region where the compliance exceeds the threshold w_γ, whose area is denoted by s_γ, for $\gamma = 0.90$. In this case, the relationship between the proportion of the population γ and the area s_γ is interpreted as

$$\frac{s_\gamma}{(2\Delta a)^2} = 1 - \gamma.$$

Clearly, the γth quantile depends on the nominal values of a_1 and a_2. For example, for $p_1 = p_2 = p > 0$, $a_1 = 9a_0/10$, $a_2 = a_0/10$ and $\Delta a = a_0/10$, the γth quantile is derived as

$$\begin{aligned} F(\bar{w}) &= \Pr\{W \leq \bar{w}\} \\ &= 1 - \frac{5}{\bar{w}/w_0} + \frac{100}{(\bar{w}/w_0)^2} \log \frac{5\bar{w}/w_0 - 20}{4\bar{w}/w_0 - 20}, \end{aligned} \tag{1.48}$$

which is plotted in Fig. 1.7. The 0.90th quantile is plotted as a black circle in Fig. 1.7. In this case, some γth quantiles are obtained as follows:

$$w_{0.90} = 54.46 w_0, \quad w_{0.95} = 104.46, \quad w_{1.00} = \infty.$$

As stated before, $w_{1.00}$ is the worst value and diverges at this point. This means that the deterministic worst-case approach is inappropriate for this problem and the probabilistic approach has potentially broad applicability.

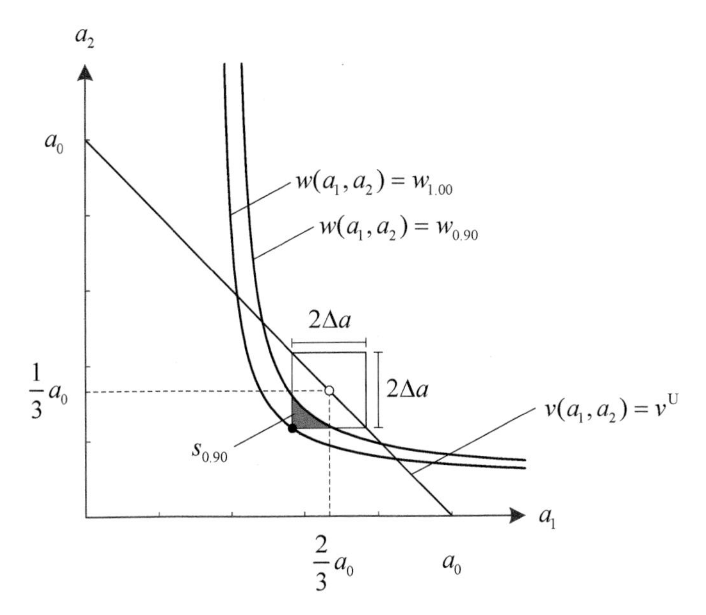

Figure 1.6 Geometric illustration of the γth quantile of the two-bar structure problem for $p_1 = p_2 = p$, $a_1 = 2a_0/3$, $a_2 = a_0/3$ and $\Delta a = a_0/10$.

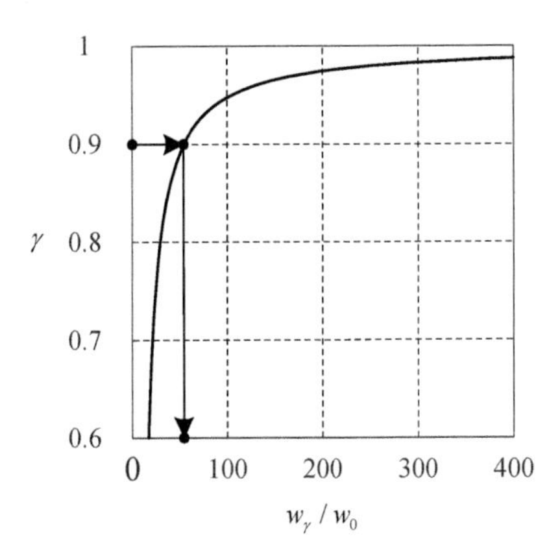

Figure 1.7 γth quantile of the two-bar structure problem for $p_1 = p_2 = p$, $a_1 = 9a_0/10$, $a_2 = a_0/10$ and $\Delta a = a_0/10$.

Thus, a stochastic compliance minimization problem of the two-bar structure under the cross-sectional area uncertainty may be formulated as

$$\begin{aligned} &\text{Minimize} && w_\gamma(a_1, a_2) && \text{(1.49a)}\\ &\text{subject to} && v(a_1, a_2) \leq v^{\mathrm{U}}, && \text{(1.49b)}\\ &&& a_1 \geq 0,\ a_2 \geq 0. && \text{(1.49c)} \end{aligned}$$

An approach to handle the γth quantile has potential advantages; however, its direct application may be intractable, and finding the exact γth quantile is generally difficult. Therefore, instead of the direct approach, *Monte Carlo simulation* (MCS) and/or approximation schemes are usually used. Order statistics can also be used for such purpose.

A set of n random variable vectors are denoted by $\{\boldsymbol{\Theta}_1, \cdots, \boldsymbol{\Theta}_n\}$ that are assumed to be independent and uniformly generated on $\Omega_2 = (-\Delta a, \Delta a) \times (-\Delta a, \Delta a) \subset \mathbb{R}^2$, i.e., $\boldsymbol{\Theta}_i \in \Omega_2\,(i = 1, \cdots, n)$. A set of the corresponding compliance values is denoted by $\{W_1, \cdots, W_n\}$. The kth order statistic denoted by $W_{k:n} = W_{k:n}(a_1, a_2)$ for a certain (a_1, a_2) is obtained as the kth largest value among the set $\{W_1, \cdots, W_n\}$. Thus, the following problem may be considered instead of problem (1.49):

$$\begin{aligned} &\text{Minimize} && w_{k:n}(a_1, a_2) && \text{(1.50a)}\\ &\text{subject to} && v(a_1, a_2) \leq v^{\mathrm{U}}, && \text{(1.50b)}\\ &&& a_1 \geq 0,\ a_2 \geq 0, && \text{(1.50c)} \end{aligned}$$

where $w_{k:n} = w_{k:n}(a_1, a_2)$ is an observation (sample value) of $W_{k:n}(a_1, a_2)$. For example, we select $k = 3$ and $n = 52$ for $\alpha = \gamma = 0.9$ from Table 1.1. This means $w_{0.90} \leq w_{3:52}$ holds with 90% confidence, which implies $w_{0.90}$ decreases with 90% confidence when $w_{3:52}$ is minimized.

For verification, we perform MCS for $\Delta a = a_0/10$, generating 10^4 different sample sets of $\{\boldsymbol{\Theta}_1, \cdots, \boldsymbol{\Theta}_{52}\}$. The observed sample sets are denoted by $\{\boldsymbol{\theta}_1^{(j)}, \cdots, \boldsymbol{\theta}_{52}^{(j)}\}_{j=1}^{10^4}$ and the third order statistics of the corresponding compliance values are denoted by $\{w_{3:52}^{(j)}\}_{j=1}^{10^4}$. The cumulative probability and the histogram of $F(w_{3:52}^{(j)})$ $(j = 1, \cdots, 10^4)$ for $a_1 = 2a_0/3$ and $a_2 = a_0/3$, which is referred to as *design A*, are shown in Fig. 1.8. In this case, the cdf is known as shown in (1.47) and the number of cases such that "the cumulative probability of the observed order statistics $F(w_{3:52}^{(j)})$ $(j = 1, \cdots, 10^4)$ is more than 0.9" is counted as 9093, i.e.,

$$\begin{aligned} \frac{\text{The number of } \{w_{3:52}^{(j)} \geq w_{0.90}; j = 1, \cdots, 10^4\}}{\text{Total number of the observations}} &= \frac{9093}{10^4}\\ &= 0.9093 \ (> 90\%). \end{aligned}$$

This implies the observed order statistics can predict the upper bound of the 0.90th quantile of the uncertain compliance with confidence of 90%. In the

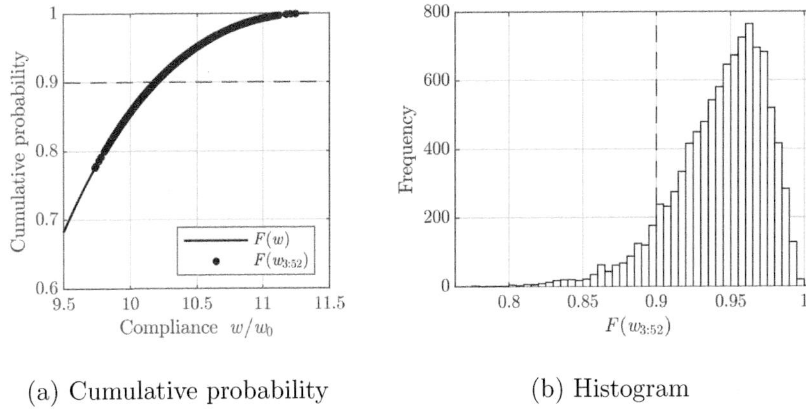

(a) Cumulative probability (b) Histogram

Figure 1.8 Cumulative probability and histogram of the observed order statistics of compliance for $a_1 = 2a_0/3$, $a_2 = a_0/3$ and $\Delta a = a_0/10$ (design A).

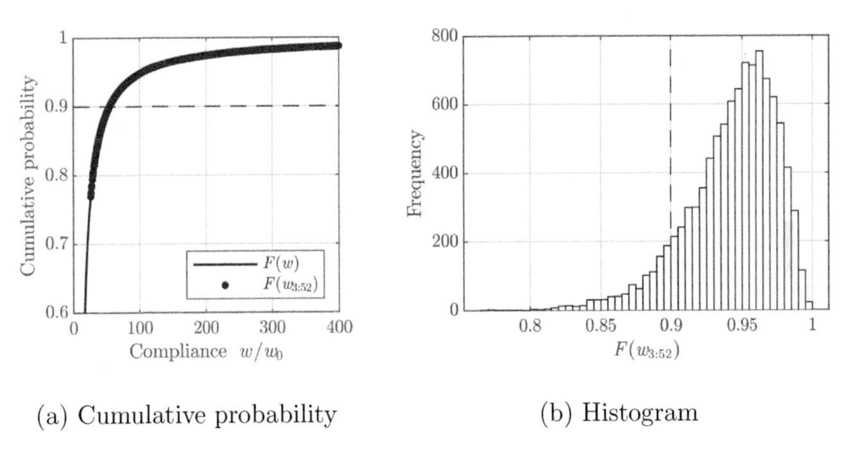

(a) Cumulative probability (b) Histogram

Figure 1.9 Cumulative probability and histogram of the observed order statistics of compliance for $a_1 = 9a_0/10$, $a_2 = a_0/10$ and $\Delta a = a_0/10$ (design B).

same way, those of cumulative probability and histogram for $a_1 = 9a_0/10$ and $a_2 = a_0/10$, which is referred to as *design B*, are shown in Fig. 1.9. In this case, we have the similar result as

$$\begin{aligned}\frac{\text{The number of } \{w_{3:52}^{(j)} \geq w_{0.90}; j = 1, \cdots, 10^4\}}{\text{Total number of the observations}} &= \frac{9020}{10^4} \\ &= 0.9020 \ (> 90\%).\end{aligned}$$

From comparison between Figs. 1.8 and 1.9, compliance of design B substantially increases in accordance with increase of the cumulative probability. The reason for this is that design B tends to be more unstable than design A due to parameter variation; namely, the compliance of design A is less sensitive to the parameter variation than design B. In this case, we will say that design A is more robust than design B. In this way, we can justify the use of order statistics with a certain confidence instead of finding the exact quantile.

1.4.4 L-MOMENTS

L-moments are alternatively used for describing the shapes of probability distributions. Historically, they arose as modifications of the *probability weighted moments* (PWM) [Greenwood et al., 1979]. The PWMs of a random variable X with cdf $F(\cdot)$ are defined as

$$M_{p,r,s} = \mathbb{E}\left[X^p\{F(X)\}^r\{1 - F(X)\}^s\right]. \tag{1.51}$$

Particularly useful special cases are the PWMs $\alpha_r = M_{1,0,r}$ and $\beta_r = M_{1,r,0}$. For a distribution that has a quantile function $Q(u)$, (1.37) and (1.51) give

$$\alpha_r = \mathbb{E}\left[X\{1 - F(X)\}^r\right] = \int_{-\infty}^{+\infty} x\{1 - F(x)\}^r dF(x) = \int_0^1 (1-u)^r Q(u)du,$$

$$\beta_r = \mathbb{E}\left[X\{F(X)\}^r\right] = \int_{-\infty}^{+\infty} x\{F(x)\}^r dF(x) = \int_0^1 u^r Q(u)du.$$

These equations may be contrasted with the definition of the ordinary moments, which may be written as

$$\mathbb{E}[x^r] = \int_0^1 [Q(u)]^r du.$$

Thus, the PWMs may be regarded as the integrals of $Q(u)$ weighted by the polynomials u^r or $(1-u)^r$. Furthermore, we define polynomials $P_r^*(u)$ $(r = 0, 1, \cdots)$ as follows:

(i) $P_r^*(u)$ is a polynomial of degree r in u.

(ii) $P_r^*(1) = 1$.

(iii) $\int_0^1 P_r^*(u)P_s^*(u)du = 0$ if $r \neq s$.

These conditions define the shifted Legendre polynomials, which have the explicit form

$$P_r^*(u) = \sum_{s=0}^{r} p_{r,s}^* u^s,$$

where

$$p_{r,s}^* = (-1)^{r-s}\begin{pmatrix} r \\ s \end{pmatrix}\begin{pmatrix} r+s \\ s \end{pmatrix} = \frac{(-1)^{r-s}(r+s)!}{(s!)^2(r-s)!}. \tag{1.52}$$

For a random variable X with quantile function $Q(u)$, we now define the L-moments λ_r of X as

$$\lambda_r = \int_0^1 P_{r-1}^*(u)Q(u)du. \tag{1.53}$$

In terms of the PWMs, the first four L-moments are given by

$$\lambda_1 = \alpha_0 = \beta_0, \tag{1.54a}$$
$$\lambda_2 = \alpha_0 - 2\alpha_1 = -\beta_0 + 2\beta_1, \tag{1.54b}$$
$$\lambda_3 = \alpha_0 - 6\alpha_1 + 6\alpha_2 = \beta_0 - 6\beta_1 + 6\beta_2, \tag{1.54c}$$
$$\lambda_4 = \alpha_0 - 12\alpha_1 + 30\alpha_2 - 20\alpha_3 = -\beta_0 + 12\beta_1 - 30\beta_2 + 20\beta_3, \tag{1.54d}$$

and in general

$$\lambda_{r+1} = (-1)^r \sum_{s=0}^{r} p_{r,s}^* \alpha_s = \sum_{s=0}^{r} p_{r,s}^* \beta_s.$$

Following Hosking and Wallis [1997], the kth smallest observation from a sample set of size n is denoted by $X_{k:n}$, so that the samples are ordered as $X_{1:n} \leq X_{2:n} \leq X_{n:n}$. Note that $X_{k:n}$ $(k = 1, \cdots, n)$ are renumbered in increasing order. "L" in L-moments emphasizes the construction of L-moments from linear combination of the order statistics. The first four L-moments are defined by

$$\begin{aligned}
\lambda_1 &= \mathbb{E}[X_{1:1}], \\
\lambda_2 &= \frac{1}{2}\mathbb{E}[X_{2:2} - X_{1:2}], \\
\lambda_3 &= \frac{1}{3}\mathbb{E}[X_{3:3} - 2X_{2:3} + X_{1:3}], \\
\lambda_4 &= \frac{1}{4}\mathbb{E}[X_{4:4} - 3X_{3:4} + 3X_{2:4} - X_{1:4}],
\end{aligned}$$

and in general

$$\lambda_r = \frac{1}{r}\sum_{j=0}^{r-1}(-1)^j \binom{r-1}{j} \mathbb{E}[X_{r-j:r}]. \tag{1.55}$$

Note that the two definitions (1.53) and (1.55) are consistent. The expectation of order statistics can be written as

$$\mathbb{E}[X_{r:n}] = \frac{n!}{(r-1)!(n-r)!}\int_0^1 Q(u)u^{r-1}(1-u)^{n-r}du. \tag{1.56}$$

Thus λ_r can be written as an integral of $Q(u)$ multiplied by a polynomial of u in (1.55), which can be shown to be equivalent to $P_{r-1}^*(u)$.

So far, L-moments have been defined for a given probability distribution; however, in practice they must often be estimated from a finite sample set.

Thus, we assume that estimation is based on a sample set of size n arranged in ascending order as $x_{1:n} \leq x_{2:n} \leq \cdots \leq x_{n:n}$. It is convenient to begin with an estimator of the PWM β_r. An unbiased estimator of β_r is denoted by b_r and obtained as

$$b_r = \frac{1}{n} \sum_{j=r+1}^{n} \begin{pmatrix} j-1 \\ r \end{pmatrix} x_{j:n} \Big/ \begin{pmatrix} n-1 \\ r \end{pmatrix}. \tag{1.57}$$

This may be alternatively written as

$$\begin{aligned}
b_0 &= \frac{1}{n} \sum_{j=1}^{n} x_{j:n}, \\
b_1 &= \frac{1}{n} \sum_{j=2}^{n} \begin{pmatrix} j-1 \\ 1 \end{pmatrix} x_{j:n} \Big/ \begin{pmatrix} n-1 \\ 1 \end{pmatrix} = \frac{1}{n} \sum_{j=2}^{n} \frac{j-1}{n-1} x_{j:n}, \\
b_2 &= \frac{1}{n} \sum_{j=3}^{n} \begin{pmatrix} j-1 \\ 2 \end{pmatrix} x_{j:n} \Big/ \begin{pmatrix} n-1 \\ 2 \end{pmatrix} = \frac{1}{n} \sum_{j=3}^{n} \frac{(j-1)(j-2)}{(n-1)(n-2)} x_{j:n},
\end{aligned}$$

and in general

$$\begin{aligned}
b_r &= \frac{1}{n} \sum_{j=r+1}^{n} \frac{(j-1)(j-2)\cdots(j-r)}{(n-1)(n-2)\cdots(n-r)} x_{j:n} \\
&= \frac{1}{n} \sum_{j=r+1}^{n} \frac{(j-1)!(n-r-1)!}{(n-1)!(j-r-1)!} x_{j:n}.
\end{aligned}$$

Analogously to (1.54), the first four sample L-moments are defined by

$$\begin{aligned}
l_1 &= b_0, \\
l_2 &= -b_0 + 2b_1, \\
l_3 &= b_0 - 6b_1 + 6b_2, \\
l_4 &= -b_0 + 12b_1 - 30b_2 + 20b_3,
\end{aligned}$$

and in general they are expressed as

$$l_{r+1} = \sum_{s=0}^{r} p^*_{r,s} b_s \qquad (r = 0, 1, \cdots, n-1), \tag{1.58}$$

where the coefficients $p^*_{r,s}$ are defined in (1.52). The sample L-moment l_r is an unbiased estimator of λ_r. From (1.52) and (1.58), l_r is a linear combination of the ordered sample values $x_{1:n}, \cdots, x_{n:n}$, and we can write

$$l_r = \frac{1}{n} \sum_{j=1}^{n} \omega^{(r)}_{j:n} x_{j:n},$$

where $\omega^{(r)}_{j:n}$ is the discrete Legendre polynomial [Neuman and Schonbach, 1974].

1.4.5 MAXIMUM ENTROPY PRINCIPLE

Jaynes [1957] presented the maximum entropy principle as a rational approach for finding a consistent probability distribution, among all possible distributions, which contains minimum spurious information. The principle states that the most unbiased estimate of a probability distribution maximizes the entropy subject to constraints formulated based on the available information, e.g., moments of a random variable. The distribution so obtained is regarded as the most unbiased, because its derivation involves a systematic maximization of uncertainty about the unknown information.

The consistency is a fundamental requirement in mathematical analysis, i.e., if a quantity can be found in more than one way, the results obtained by different methods must be the same. To ensure this, the method must satisfy some basic properties, referred to as consistency axioms. Shore and Johnson [1980] postulated four such axioms.

1. *Uniqueness:* The result should be unique.

2. *Invariance:* The choice of coordinate system should not matter.

3. *System independence:* It should not matter whether one accounts for independent information about independent systems separately in terms of different densities or together in terms of a joint density.

4. *Subset independence:* It should not matter whether one treats an independent subset of system states in terms of a separate conditional density or in terms of the full system density.

Given the moment constraints only, Shore and Johnson [1980] proved that the entropy maximization is a uniquely correct method of probabilistic inference that satisfies all the consistency axioms, which is referred to as maximum entropy method (MEM). The entropy of a continuous probability distribution with cdf $F(x)$ and pdf $f(x) = dF(x)/dx$ is computed as

$$H[f] = \int_{-\infty}^{\infty} [-\log f(x)] f(x) dx.$$

If a set of moments $\mathbb{E}[X^r]$ $(r = 1, \cdots, R)$ up to the Rth order is specified, the distribution that has the maximum entropy has probability density

$$f(x) \propto \exp\left(\sum_{r=1}^{R} \zeta_r x^r\right)$$

for suitable constants ζ_r $(r = 1, \cdots, R)$.

The main problem here is the derivation of the distribution that has the maximum entropy for specified values of its first R L-moments. The rth L-moment λ_r of a random variable X with quantile function Q is given in

(1.55) and (1.56). Suppose a performance function under uncertainty is a continuous random variable $Z = g(\boldsymbol{d}; \boldsymbol{\theta})$ with cdf $F(z)$ and pdf $f(z)$. Let $Q(u)$ and $q(u)$ denote the quantile function of Z and its corresponding derivative for $0 < u < 1$, respectively. Since $Q(u)$ is the inverse function of $F(z)$, i.e., $F(Q(u)) = u$, differentiation of both sides with respect to u leads to

$$F'(Q(u)) = \frac{1}{Q'(u)} \quad \Rightarrow \quad f(z) = \frac{1}{q(u)},$$

where $z = Q(u)$. Hence, the entropy denoted by H can be written in terms of q as follows:

$$\begin{aligned} H &= -\int_{-\infty}^{+\infty} f(z) \log f(z) dz \\ &= -\int_{-\infty}^{+\infty} \left[\log \frac{1}{q(u)} \right] f(z) dz \\ &= \int_0^1 \log q(u) du. \end{aligned} \tag{1.59}$$

Now consider the problem of finding the function q that maximizes the entropy H in (1.59) subject to a set of constraints of the form

$$\int_0^1 P_{r-1}^*(u) Q(u) du = c, \tag{1.60}$$

where c is a given parameter. Note that (1.60) represents constraints on L-moment as defined in (1.53). By integrating (1.60) by parts, we obtain

$$\int_0^1 P_{r-1}^*(u) Q(u) du = -\left[K_r(u) Q(u) \right]_0^1 + \int_0^1 K_r(u) q(u) du = c, \tag{1.61}$$

where $K_r(u) = \int_u^1 P_{r-1}^*(v) dv$. Furthermore, we rewrite (1.61) as

$$\int_0^1 K_r(u) q(u) du = h_r, \tag{1.62}$$

where $h_r = c + \left[K_r(u) Q(u) \right]_0^1 = c - K_r(0) Q(0)$. From the constraints in the form (1.62), we can determine $q(u)$ by solving the following problem:

$$\begin{aligned} &\underset{q}{\text{Maximize}} && \int_0^1 \log q(u) du \\ &\text{subject to} && \int_0^1 K_r(u) q(u) du = h_r \quad (r = 1, \cdots, S), \end{aligned} \tag{1.63}$$

where $K_1, \cdots, K_S$ are linearly independent polynomials, and S is the maximum order of the L-moments. The maximization is done over functions $q(u)$

that are strictly positive on $(0, 1)$. If there exist constants a_s $(s = 1, \cdots, S)$ that satisfy

$$\int_0^1 \frac{K_r(u)}{\sum_{s=1}^S a_s K_s(u)} du = h_r \qquad (r = 1, \cdots, S)$$

and

$$\sum_{s=1}^{S} a_s K_s(u) > 0 \quad (0 < u < 1), \tag{1.64}$$

then the problem has the solution for $q(u)$ as

$$q(u) = q(u; \boldsymbol{a}) = \frac{1}{\sum_{r=1}^S a_r K_r(u)}, \tag{1.65}$$

where $\boldsymbol{a} = (a_1, \cdots a_S)$ is a parameter vector. Among functions $q(u)$ that satisfy the constraints (1.63), we have the following inequality for the functional $H[q]$

$$H[q] \leq D(\boldsymbol{a}) - 1 \tag{1.66}$$

for any parameter vector $\boldsymbol{a}$ that satisfies (1.64), where

$$H[q] = \int_0^1 \log q(u) du, \tag{1.67}$$

$$D(\boldsymbol{a}) = -\int_0^1 \log \left(\sum_{r=1}^{S} a_r K_r(u) \right) du + \sum_{r=1}^{S} a_r h_r. \tag{1.68}$$

The function $D(\boldsymbol{a})$ is regarded as a potential function to be maximized, which is explained later. We write the constraint (1.53) in the form of (1.62). For $S = 4$, integration by parts gives

$$\lambda_1 - L = \int_0^1 P_0^*(u) Q(u) du = \int_0^1 (1 - u) q(u) du, \tag{1.69a}$$

$$\lambda_2 = \int_0^1 P_1^*(u) Q(u) du = \int_0^1 u(1 - u) q(u) du, \tag{1.69b}$$

$$\lambda_3 = \int_0^1 P_2^*(u) Q(u) du = \int_0^1 u(1 - u)(2u - 1) q(u) du, \tag{1.69c}$$

$$\lambda_4 = \int_0^1 P_3^*(u) Q(u) du = \int_0^1 u(1 - u)(5u^2 - 5u + 1) q(u) du. \tag{1.69d}$$

Thus, (1.69) can be reformulated by replacing L-moments λ_r with sample L-moments ℓ_r as

$$\int_0^1 K_r(u) q(u) du = h_r \quad (r = 1, \cdots, 4), \tag{1.70}$$

where

$$K_1(u) = 1 - u, \qquad h_1 = \ell_1 - Q(0), \tag{1.71a}$$
$$K_2(u) = u(1 - u), \qquad h_2 = \ell_2, \tag{1.71b}$$
$$K_3(u) = u(1 - u)(2u - 1), \qquad h_3 = \ell_3, \tag{1.71c}$$
$$K_4(u) = u(1 - u)(5u^2 - 5u + 1), \quad h_4 = \ell_4, \tag{1.71d}$$

and $Q(0)$ is the lower endpoint of the distribution if it is finite. From (1.66), by minimization of $D(\boldsymbol{a})$, we can obtain the upper bound of $H[q]$, or to be exact, $H[q] + 1$. Thus, $D(\boldsymbol{a})$ is regarded as the potential function and the quantile function (1.65) is computed with the optimized parameter vector $\boldsymbol{a}$ as

$$Q(u) = Q(0) + \int_0^u q_0(u; \boldsymbol{a})du = Q(0) + \int_0^u \frac{1}{\sum_{r=1}^{S} a_r K_r(u)} du, \tag{1.72}$$

which corresponds to the distribution that has the maximum entropy among all possible distributions for the given sample L-moments, which are defined in (1.58).

1.4.6 ILLUSTRATIVE EXAMPLE: L-MOMENTS AND MAXIMUM ENTROPY METHOD

Here again, let us consider the stochastic compliance (1.44). For $a_i \geq \Delta a$, $\Theta_i \sim U(-\Delta a, \Delta a)$ and $p_1 = p_2 = p > 0$, we simply have

$$W = W(a_1, a_2) = w_0 \left(\frac{4a_0}{a_1 + \Theta_1} + \frac{a_0}{a_2 + \Theta_2} \right). \tag{1.73}$$

The cdf of W is denoted by $F(w) = \Pr\{W \leq w\}$. We perform MCS for the sample size $n = 200$. A sample set of the random variable vectors $\{\boldsymbol{\Theta}_1, \cdots, \boldsymbol{\Theta}_{200}\}$ and the corresponding compliance $\{W_1, \cdots, W_{200}\}$ are generated and their observations are respectively denoted by $\{\boldsymbol{\theta}_1, \cdots, \boldsymbol{\theta}_{200}\}$ and $\{w_1, \cdots, w_{200}\}$ in the same way as in Sec. 1.4.3. We set the maximum order of the L-moments for $S = 4$ and minimize the potential function (1.68). In practice, the exact value of the lower bound $Q(0)$ in (1.71a) is unknown. Therefore, we approximate $Q(0)$ by order statistics $w_{1:200}$ that is the minimum observation (sample value) among $w_1, \cdots, w_{200}$; thus, instead of (1.71a), the following relation is used:

$$h_1 = \ell_1 - Q(0) \approx \ell_1 - w_{1:200}. \tag{1.74}$$

Obviously the values of the sample L-moments change for every observations. We compute the sample L-moments for different 200 sample sets and minimize the potential function (1.68) for each observation. The quantiles so obtained are shown in Figs. 1.10 and 1.11. The cumulative and exceedance probability

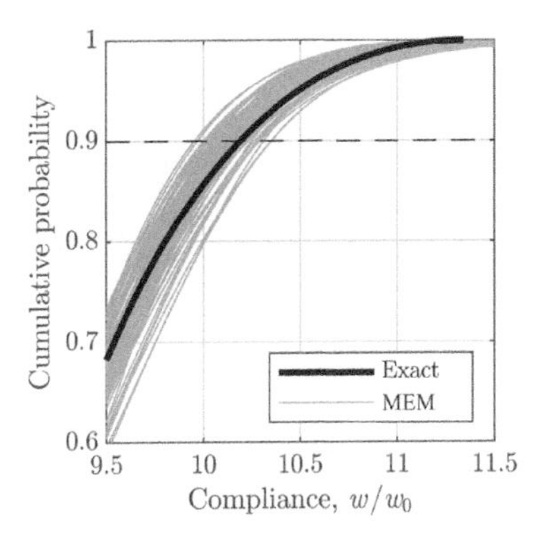

(a) Cumulative probability

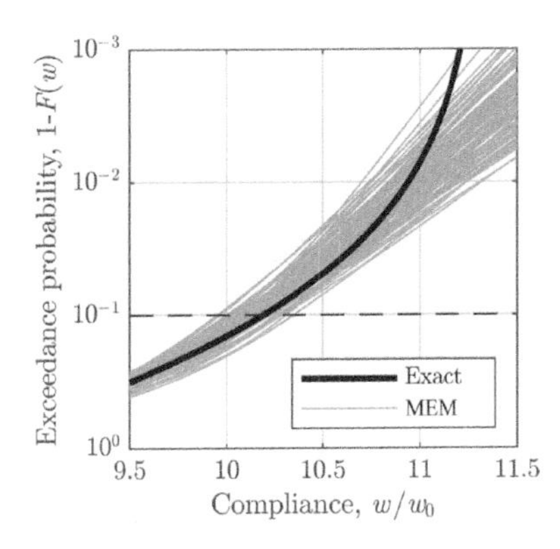

(b) Exceedance probability

Figure 1.10 Cumulative and exceedance probability of the compliance by the sample L-moments for $a_1 = 2a_0/3$, $a_2 = a_0/3$ and $\Delta a = a_0/10$ (design A).

of the compliance for $a_1 = 2a_0/3$, $a_2 = a_0/3$ and $\Delta a = a_0/10$ (design A) are shown in Fig. 1.10. Those for $a_1 = 9a_0/10$ and $a_2 = a_0/10$ (design B) are shown in Fig. 1.11. The gray lines in Figs. 1.10 and 1.11 represent the identified quantiles by MEM with the sample L-moments. In this example, the exact distributions are derived as (1.47) and (1.48), which are plotted with solid black lines in Figs. 1.10 and 1.11. It is observed from these figures that the identified distributions depend on the sample set, and relatively well approximate the exact distribution in the area where the cumulative probability is less than 0.9. However, we cannot have confidence level by this approach. This is a disadvantage particularly when we need higher reliability for the region where cumulative probability is more than 0.9.

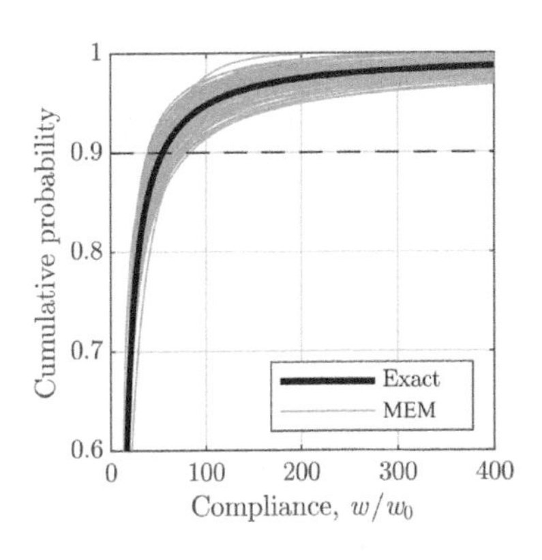

(a) Cumulative probability

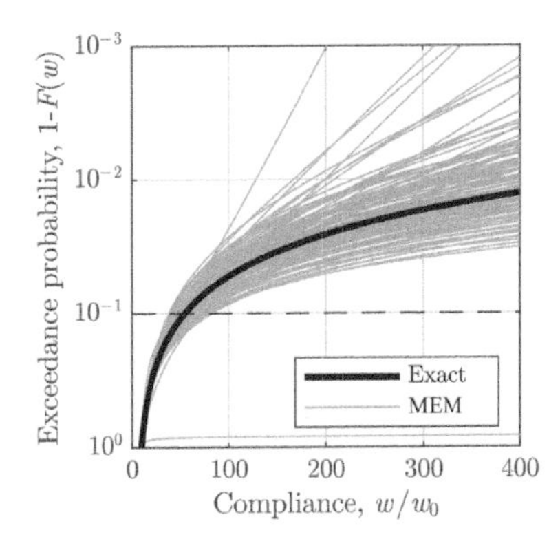

(b) Exceedance probability

Figure 1.11 Cumulative and exceedance probability of the compliance by the sample L-moments for $a_1 = 9a_0/10$, $a_2 = a_0/10$ and $\Delta a = a_0/10$ (design B).

1.5 SUMMARY

Stochastic and probabilistic approaches to structural optimization has been introduced mainly for application to worst case analysis and design problems in the field of structural engineering. Various sources and measures of uncertainty in structural optimization problems have been categorized, and types of solution methods including deterministic, probabilistic and possibilistic methods have been briefly explained. Basics of probability theory, order statistics, L-moments, etc., have been summarized for reference in the later chapters. Illustrative examples have been given using the compliance minimization problem of a two-bar structure.

Although we focus on structural analysis and design problems incorporating parameter uncertainty, it should be noted that probabilistic approaches can be effectively applied for solving deterministic optimization problems including worst case problems. Difficulty for obtaining the exactly worst value (extreme value) can be successfully alleviated by using order statistics and L-moments, which will be demonstrated in the following chapters.

2 Stochastic Optimization

In this chapter, we summarize the theoretical fundamentals and materials of stochastic optimization methods. In this monograph, stochastic algorithms for structural optimization considering uncertainty are focused on, which can be viewed as algorithms controlled by random variables and random processes. Thus, probability theory plays a crucial role, and hence some theoretical concepts of stochasticity and randomness are introduced. As a general introduction to the methods, only mathematical minimization problems are considered throughout this chapter. In later chapters, some maximization problems will be considered as worst-case analysis; however, minimization and maximization problems are essentially the same in theory. Indeed, maximization problems are transformed to minimization problems by changing the sign of the objective function. This chapter may be skipped or referred later by those whose interests focus on practical application and engineering problems. However, it is quite difficult to attain practical efficiency without theoretical framework.

2.1 INTRODUCTION

Many of classical optimization methods are classified into local optimization methods, which can often quickly find a local minimum (and a local solution). In terms of *globality of search*, the points that we evaluate the functions should be spread all over the feasible region. However, it is known that there is a fundamental trade-off between algorithm efficiency and algorithm robustness. There can never be a universally best search algorithm just as there is rarely a universally best solution to any general problem; see, e.g., Spall [2005]. The *no free lunch theorems* in Wolpert and Macready [1997] state *any elevated performance over one class of problems is offset by performance over another class.* An algorithm that is effective on one class of problems is likely to be ineffective on another class. Therefore, we should balance between *globality* and *locality* in the search process, and achieving the right balance depends on available information about functions and feasible region of the problem. Basic concepts of the theoretical framework [Zhigljavsky and Žilinskas, 2008, 2021; Spall, 2005; Papoulis and Pillai, 2002] will be introduced and thus there is no new findings in this chapter. Although readers interested in applications may skip this chapter, it is quite difficult to attain practical efficiency without theoretical framework for solving the problem.

DOI: 10.1201/9781003153160-2

General minimization problem

A general minimization problem is formulated as

$$\text{Minimize} \quad f(\boldsymbol{x}) \tag{2.1a}$$
$$\text{subject to} \quad \boldsymbol{x} \in \mathcal{X}, \tag{2.1b}$$

where $f(\cdot)$ is an objective function, $\boldsymbol{x} \in \mathbb{R}^m$ is a design variable vector, and $\mathcal{X}$ is the feasible domain. Let $f^* = \min_{\boldsymbol{x}\in\mathcal{X}} f(\boldsymbol{x})$ be the *minimum* of $f(\cdot)$ in $\mathcal{X}$. A solution of problem (2.1) is denoted by $\boldsymbol{x}^*$ in $\mathcal{X}$ such that $f(\boldsymbol{x}^*) = f^*$. We construct a sequence of points $\boldsymbol{x}_1, \boldsymbol{x}_2, \cdots$ in $\mathcal{X}$ such that the sequence of values $y_n = \min_{i=1,\cdots n} f(\boldsymbol{x}_i)$ approaches f^* as n increases, i.e.,

$$y_n \to f^* \quad \text{as} \quad n \to \infty.$$

The value y_n is called the *record value* or simply the *record*.

A point $\boldsymbol{z} \in \mathcal{X}$ is called a *local solution* or *local minimizer* of problem (2.1) if there exists a neighborhood $U_{\boldsymbol{z}}$ of $\boldsymbol{z}$ such that

$$f(\boldsymbol{z}) \leq f(\boldsymbol{x}) \text{ for all } \boldsymbol{x} \in U_{\boldsymbol{z}},$$

where the neighborhood of $\boldsymbol{z}$ is defined by $U_{\boldsymbol{z}} = \{\boldsymbol{x} \in \mathcal{X} : ||\boldsymbol{x} - \boldsymbol{z}|| < \varepsilon\}$ for a given $\varepsilon > 0$. The value $f(\boldsymbol{z})$ is called a *local minimum*. When the minimum is distinguished from local minimum, $\boldsymbol{x}^*$ and $f(\boldsymbol{x}^*) = f^*$ are referred to as a *global solution* (or a *global minimizer*) and the *global minimum*, respectively.

Local optimization algorithms, e.g., nonlinear programming and local search, usually search the neighborhood of a trial point sequentially to find a local solution. In structural optimization, response functions do not have convexity or special analytical form in many cases, and hence application of an algorithm leads to a local solution that most likely is not the global solution. The property of quickly finding a local solution is referred to as *locality of search*. Hence, local optimization algorithms are regarded as methods based on locality of search. On the other hand, it is known that without strong assumptions the following general result is valid [Zhigljavsky and Žilinskas, 2008]:

Theorem 2.1 *Let the feasible domain $\mathcal{X}$ be compact and the objective function be continuous in the neighborhood of a global solution. Then a global minimization algorithm converges in the sense $y_n \to f^*$ as $n \to \infty$ iff the algorithm generates a sequence of points $\boldsymbol{x}_i$ that is dense everywhere in $\mathcal{X}$.*

A set such that every open covering of it contains a finite subcover is called *compact*; see, e.g., Apostol [1974]. To increase the chance of finding the global minimum, the points $\boldsymbol{x}_i$ has to be spread all over the feasible domain $\mathcal{X}$. The property of spreading the trial points and reducing the unexplored area is referred to as *globality of search*, which is linked to efficiency of global optimization algorithms. Stochastic methods are available for acquiring such globality of search, which are sometimes referred to as stochastic global optimization methods.

2.2 BASIC THEORY OF STOCHASTICITY AND RANDOMNESS

A stochastic algorithm can be viewed as an algorithm that is partially controlled by a random process [Hromkovič, 2005]. Spall [2005] pointed out that stochasticity is considered in different aspects in stochastic optimization. We will focus on linking structural optimization with stochasticity and/or randomness, and properties in this case are categorized as follows:

Property A There is a random noise in the response function.

Property B There is a random choice made in the search direction as the algorithm iterates toward a solution.

Regarding Property A, noise is relevant to optimization under uncertainty, and it fundamentally changes the optimization process. The optimization algorithm will receive misleading information throughout the process. In Sec. 1.3, we categorized uncertainty into some mathematical models and methodologies. Such type of uncertainty can be linked with Property A.

On the other hand, it is sometimes beneficial to artificially inject randomness into the search process, which is classified into Property B. This injected randomness is usually generated by computer-based pseudorandom number generators and urges to search unexplored areas that may contain an unexpectedly good solution. Such a method is also referred to as *randomized algorithm* [Hromkovič, 2005], which can be regarded as a special case of stochastic algorithms.

This monograph deals with both of Properties A and B for stochastic structural optimization. In particular, probability-based methods are focused on. For this purpose, random process plays a crucial role, e.g., construction of algorithms, convergence and stopping rules, among those categorized in Sec. 1.3.

2.2.1 CONCEPTS OF RANDOM VARIABLE

The theory of probability describes and predicts statistics of mass phenomena. The probability of an event A is defined as some given probability measure on A denoted by $\Pr\{A\}$ or $P(A)$. A function and/or a number $X = X(\omega)$ is assigned to every outcome ω of an experiment. The experiment is specified by the domain Ω and the field of subsets of Ω called events. The function $X = X(\omega)$ defined in domain Ω, i.e., $\omega \in \Omega$, is called *random variable*. All random variables will be written in capital letters. The notation $\{X \leq x\}$ represents a subset of Ω of all outcomes ω such that $X(\omega) \leq x$. Note that $\{X \leq x\}$ is not a set of numbers but a set of experimental outcomes. The probability $\Pr\{X \leq x\}$ of the event $\{X \leq x\}$ is a number that depends on x, which is denoted by

$$F_X(x) = \Pr\{X \leq x\} = \Pr\{\omega \in \Omega : X(\omega) \leq x\} \tag{2.2}$$

and is called the cumulative distribution function (cdf) of the random variable X. If there is no possibility of ambiguity, we will sometimes omit the subscript X in (2.2) and simply write $F(x)$ instead of $F_X(x)$. If $x_1 < x_2$, the event $\{X \leq x_1\}$ is a subset of the event $\{X \leq x_2\}$, because, if $X(\omega) \leq x_1$ for some ω, then $X(\omega) < x_2$. Hence, $\Pr\{X \leq x_1\} \leq \Pr\{X \leq x_2\}$ holds, which implies that $F_X(x)$ is a nondecreasing function of x.

The derivative of the cdf $F_X(x)$ is called the probability distribution function (pdf) of the random variable X and is denoted by

$$\phi_X(x) = \frac{dF_X(x)}{dx}.$$

As the reverse relationship, we have

$$F_X(x) = \int_{-\infty}^{x} \phi_X(u)du.$$

From the monotonically nondecreasing nature of $F_X(x)$, it follows that

$$\phi_X(x) = \frac{dF_X(x)}{dx} = \lim_{\Delta x \to 0} \frac{F_X(x + \Delta x) - F_X(x)}{\Delta x} \geq 0.$$

If X is a continuous-type random variable, $\phi_X(x)$ will be a continuous function. If x is a discrete-type random variable, then its pdf has the following form

$$\phi_X(x) = \sum_i p_i \delta(x - x_i),$$

where x_i is a discontinuity point in $F_X(x)$ and $\delta(\cdot)$ denotes Dirac's delta function.

Suppose that X is a random variable and $g(x)$ is a function of the real variable x. The expression

$$Y = g(X)$$

is a new random variable and a number $Y(\omega) = g(X(\omega))$ is assigned to the random variable Y. The cdf $F_Y(y)$ of the random variable Y is defined by

$$\begin{aligned} F_Y(y) &= \Pr\{Y \leq y\} = \Pr\{g(X) \leq y\} \\ &= \Pr\{\omega \in \Omega : g(X(\omega)) \leq y\}. \end{aligned} \tag{2.3}$$

A *random variable vector* is defined by a vector

$$\boldsymbol{X} = (X_1, \cdots, X_m)$$

whose components X_i are random variables. The probability that $\boldsymbol{X}$ is in a region $\mathcal{D}$ of the n-dimensional space equals the integration of a function in $\mathcal{D}$:

$$\Pr\{\boldsymbol{X} \in \mathcal{D}\} = \int_{\mathcal{D}} \phi_{\boldsymbol{X}}(\boldsymbol{x})d\boldsymbol{x},$$

where $\boldsymbol{x} = (x_1, \cdots, x_m)$ and

$$\phi_{\mathbf{X}}(\boldsymbol{x}) = \phi_{\mathbf{X}}(x_1, \cdots, x_m) = \frac{\partial^n F_{\mathbf{X}}(x_1, \cdots, x_m)}{\partial x_1 \cdots \partial x_m}$$

is the joint pdf of the random variables X_i and

$$F_{\mathbf{X}}(\boldsymbol{x}) = F_{\mathbf{X}}(x_1, \cdots, x_m) = \Pr\{X_1 \leq x_1, \cdots X_m \leq x_m\}$$

is their joint cdf.

2.2.2 CONCEPTS OF RANDOM PROCESS

A random variable X is a function for assigning every outcome ω of an experiment to a number $X(\omega)$ as mentioned earlier. A *random process* (or *stochastic process*) X_t is a family of time functions depending on the parameter ω or a function of t and ω, i.e., $X_t = X(t, \omega)$. The domain of ω is the set of all experimental outcomes Ω and the domain of t is a set T of real numbers. The set T is called the index set of the random process. If T is the real interval, then X_t is referred to as a *continuous-time random process*. If T is a set of integers, then X_t is referred to as a *discrete-time random process*. In summary, X_t has the following multiple interpretations [Papoulis and Pillai, 2002]:

1. It is a family of functions $X(t, \omega)$, where t and ω are variables.
2. It is a function of time $X_t = X(t; \omega)$, where t is a variable and ω is fixed.
3. If t is fixed and ω is variable, then X_t is a random variable equal to the state of the given process at time t.
4. If t and ω are fixed, then X_t is a number.

A discrete-time random process is equal to a sequence of random variables and is also denoted by

$$X_1, X_2, \cdots \tag{2.4}$$

for $t = 1, 2, \cdots$. In this monograph, (2.4) is mainly considered and simply referred to as a random process.

2.3 STOCHASTIC METHODS

2.3.1 RANDOM SEARCH METHODS

In this section, we summarize random search (RS) methods. RS or randomized algorithm [Buot, 2006] has been studied extensively for knowledge discovery [Domingo et al., 2002], estimation of average and worst computational costs of an algorithm, and finding an approximate optimal solution of a combinatorial problem. Many RS algorithms are simple and often insensitive to the

Table 2.1
Principles of RS Algorithms

P1: random sampling of points at which $f(\cdot)$ is evaluated,
P2: covering of the space (exploration),
P3: combination with local optimization techniques (exploitation),
P4: use of different heuristics including cluster-analysis techniques to avoid clumping of points around particular local minimizers,
P5: use of statistical inference,
P6: more frequent selection of new trial points in the vicinity of good previous points, and
P7: reduction of randomness in the selection rules for the trial points.

Table 2.2
Popular Classes of RS Algorithms with Simple Updating Rules for the Distributions P_j

PRS (Pure Random Search). Random points $\boldsymbol{x}_1, \boldsymbol{x}_2, \cdots$ are independent and have the same distribution: $\boldsymbol{x}_j \sim P$ (so that $P_j = P$ for all j).
MGS (Markovian Global Search). The distribution P_j depends on $\boldsymbol{x}_{j-1}$ and the objective function value at this point but does not depend on the values of $f(\cdot)$ computed earlier.
PAS (Pure Adaptive Search). P_j is uniform on the set $\mathcal{X}_j = \{\boldsymbol{x} \in \mathcal{X} : f(\boldsymbol{x}) \leq y_{j-1}\}$, where $y_{j-1} = \min_{i=1,\cdots,j-1} f(\boldsymbol{x}_i)$ is the record value at the $(j-1)$th trial.
PBS (Population-Based Search). Similar to MGS but group (population) of points are probabilistically transformed into subsequent group rather than points to points in MGS.
RMS (Random Multi-Start). Local descents are performed from a number of random points in $\mathcal{X}$.

non-smoothness, discontinuity and presence of noise in the functions. Furthermore, some algorithms can guarantee theoretical convergence. On the other hand, practical efficiency of the algorithms often strongly depends on the dimension of problem and the theoretical convergence rate is known to be slow. Zhigljavsky and Žilinskas [2021] summarized principles and popular classes of the RS algorithms for minimization or maximization of a function $f(\cdot)$ as in Tables 2.1 and 2.2, respectively.

The RS algorithms assume that a sequence of random points $\boldsymbol{x}_1, \cdots, \boldsymbol{x}_n$ is generated, and for each $j \geq 1$ the point $\boldsymbol{x}_j$ has some probability distribution P_j, which is denoted by $\boldsymbol{x}_j \sim P_j$. For each $j \geq 2$, the distribution P_j may depend on the previous points $\boldsymbol{x}_1, \cdots, \boldsymbol{x}_{j-1}$ and the results of the function evaluations at these points. In order for an algorithm to be classified as an RS algorithm, at least one of the distributions P_j should be non-degenerate, i.e., at least one of $\boldsymbol{x}_j$ is a random point in $\mathcal{X}$, where $\mathcal{X}$ is a feasible domain for $\boldsymbol{x}_j$. Simplicity of PRS allows detailed investigation of the algorithm. However, in high-dimensional problems, the choice of the uniform distribution for generating points in $\mathcal{X}$ is not so efficient. The PBS including population-based evolutionary algorithms, e.g., genetic algorithms, where one group of points are probabilistically transformed to another group of points, have potentially the same drawback.

A general scheme of the RS algorithm is introduced by Zhigljavsky and Žilinskas [2008] as Algorithm 2.1. When the distributions P_j are the same for all j, Algorithm 2.1 becomes PRS. When P_j depends only on the previous point $\boldsymbol{x}_{j-1}$ and its function value $f(\boldsymbol{x}_{j-1})$, Algorithm 2.1 becomes MGS. In Algorithm 2.1, the distributions can be updated after a certain number of points have been generated, which allows the possibility of using more information about the function $f(\cdot)$. PBS and RMS are motivated to improve the performance by this approach. From this point of view, Algorithm 2.1 can be reformulated with an emphasis on probabilistic transformation of a group (population) of points into another group as Algorithm 2.2, which is useful for describing group-based methods. Interestingly, RMS can be regarded as a special case of Algorithm 2.2: the first population consists of independent and identically distributed random variables (iidrvs) in $\mathcal{X}$ and the second population consists of the local minimizers reached from the points of the first population [Zhigljavsky and Žilinskas, 2008]. Thus, PBS and RMS can be said to be mathematically equivalent. Of course, the distribution P_j strongly depends on the local descent methods and the theoretical analysis may be difficult or impossible.

2.3.2 CONVERGENCE OF PRS

We say that the algorithm converges, if for any $\delta >0$, the sequence of points $\boldsymbol{x}_j$ $(j = 1, 2, \cdots)$ arrives at the set

$$W(\delta) = \{\boldsymbol{x} \in \mathcal{X} : f(\boldsymbol{x}) - f^* \leq \delta\} \tag{2.5}$$

with probability one, where $f^* = \min_{\boldsymbol{x}\in\mathcal{X}} f(\boldsymbol{x})$. This obviously implies convergence of record values $y_j = \min_{i=1,\cdots,j} f(\boldsymbol{x}_i)$ to f^* as $j \to \infty$ with probability

---

Algorithm 2.1 Random search algorithms: General form.

1: Choose a probability distribution P_1 on $\mathcal{X}$. Set iteration counter $j = 1$.
2: Obtain a point $\boldsymbol{x}_j$ in $\mathcal{X}$ by sampling from the distribution P_j. Evaluate the objective function $f(\boldsymbol{x}_j)$.
3: Check the stopping criterion.
4: Using the points $\boldsymbol{x}_1, \cdots, \boldsymbol{x}_j$ and their objective function values $f(\boldsymbol{x}_1), \cdots, f(\boldsymbol{x}_j)$, construct a probability distribution $P_{j+1}(\cdot)$ on $\mathcal{X}$.
5: Substitute $j + 1$ for j and return to Step 2.

---

one. Conditions on the distributions P_j ($j = 1, 2, \cdots$) ensuring convergence of the RS algorithms have been well studied; see, e.g., Zhigljavsky and Žilinskas [2021]. Convergence of an RS algorithm is usually judged using the *zero-one law*; see e.g., Borel-Cantelli lemma in Papoulis and Pillai [2002]. It is easy to construct an RS algorithm that has the theoretical property of convergence. The following theorem provides a result of convergence of Algorithm 2.1.

Theorem 2.2 *Consider an RS algorithm with* $\boldsymbol{x}_j \sim P_j$ *for the minimization problem defined in Algorithm 2.1. Let* $\mathcal{X}$ *be a compact set,* $f(\cdot)$ *be a function on* $\mathcal{X}$ *that has finite number of global minimizers,* $\boldsymbol{x}^*$ *be a global minimizer of* $f(\cdot)$*, and* $f(\cdot)$ *be continuous in the neighborhood of* $\boldsymbol{x}^*$*. Define*

$$q_j(\varepsilon) = \inf P_j(B(\boldsymbol{x}^*, \varepsilon)), \tag{2.6}$$

where the infimum is taken over all previous points, and $B(\boldsymbol{x}^*, \varepsilon) = \{\boldsymbol{x} \in \mathcal{X} : \|\boldsymbol{x} - \boldsymbol{x}^*\| \leq \varepsilon\}$ *is a ball centered at* $\boldsymbol{x}^*$*. Assume that*

$$\sum_{j=1}^{\infty} q_j(\varepsilon) = \infty \tag{2.7}$$

for any $\varepsilon > 0$*. Then for any* $\delta > 0$*, the sequence of points* $\boldsymbol{x}_j$ *falls infinitely often into the set* $W(\delta) = \{\boldsymbol{x} \in \mathcal{X} : f(\boldsymbol{x}) - f^* \leq \delta\}$ *with probability one.*

Theorem 2.2 ensures that the corresponding RS algorithm converges to the global minimum. This implies that the sequence of records y_j converges to f^* as $j \to \infty$ with probability one. Since the location of $\boldsymbol{x}^*$ is not known in advance in the general situation, the following sufficient condition for (2.7) with (2.6) can be used:

Condition 2.1

$$\sum_{j=1}^{\infty} \inf P_j(B(\boldsymbol{x}, \varepsilon)) = \infty, \tag{2.8}$$

for all $\boldsymbol{x} \in \mathcal{X}$ *and* $\varepsilon > 0$.

Algorithm 2.2 Random search algorithms: Group-based general form.

1: Choose a probability distribution P_1 on $\mathcal{X}$. Set iteration number $j = 1$.
2: Obtain n_j points $\boldsymbol{x}_1^{(j)}, \cdots, \boldsymbol{x}_{n_j}^{(j)}$ in $\mathcal{X}$ by sampling from the distribution P_j. Evaluate the objective function $f(\cdot)$ at these points.
3: Check the stopping criterion.
4: Using the points $\boldsymbol{x}_1^{(i)}, \cdots, \boldsymbol{x}_{n_i}^{(i)}$ ($i = 1, \cdots, j$) and the objective function values at these points, construct a probability distribution $P_{j+1}(\cdot)$ on $\mathcal{X}$, where n_{j+1} is some integer that may depend on the function values at the jth iteration.
5: Substitute $j + 1$ for j and return to Step 2.

If we use RS with uniform distribution U on $\mathcal{X}$, we obtain $q_j(\varepsilon) = \text{const} > 0$ in (2.6), and therefore the condition (2.7) clearly holds. In practice, a usual choice of the distribution P_j is

$$P_j = \alpha_j U + (1 - \alpha_j) Q_j, \tag{2.9}$$

where $0 \leq \alpha_j \leq 1$ and Q_j is a specific probability measure on $\mathcal{X}$. Sampling from the distribution (2.9) corresponds to sampling from U with probability α_j and sampling from Q_j with probability $1 - \alpha_j$. If distribution (2.9) is chosen, then simple condition $\sum_{j=1}^{\infty} \alpha_j = \infty$ is sufficient for (2.7) and (2.8) to hold. Note that if $\sum_{j=1}^{\infty} \alpha_j$ converges to a finite value, then there is a non-zero probability that the algorithm will never reach the global minimizer.

2.3.3 CURSE OF DIMENSIONALITY

Consider a PRS algorithm with $\boldsymbol{x}_j \sim U$ on $\mathcal{X}$. Let S be the target set, and assume probability measure $P(S) > 0$ and the value of $P(S)$ is small. The target set S may be given by, e.g., $S = B(\boldsymbol{x}^*, \varepsilon) = \{\boldsymbol{x} \in \mathcal{X} : \|\boldsymbol{x} - \boldsymbol{x}^*\| \leq \varepsilon\}$. Also consider Bernoulli trials where the success in the trial j mean that $\boldsymbol{x}_j \in S$. PRS generates a sequence of independent Bernoulli trials with the same success probability $\Pr\{\boldsymbol{x}_j \in S\} = P(S)$. From the independence of $\boldsymbol{x}_j$, we have $\Pr\{\boldsymbol{x}_1 \in S, \cdots, \boldsymbol{x}_n \in S\} = (1 - P(S))^n$, and therefore the probability

$$\Pr\{\boldsymbol{x}_j \in S \text{ for at least one } j \ (1 \leq j \leq n)\} = 1 - (1 - P(S))^n$$

tends to one as $n \to \infty$. Let n_γ be the number of points which are required for PRS to reach the set S with probability at least $1 - \gamma$, where $0 < \gamma < 1$; that is, $n_\gamma = \min\{n : 1 - (1 - P(S))^n \geq 1 - \gamma\}$ which leads to

$$n_\gamma = \left\lceil \frac{\log \gamma}{\log(1 - P(S))} \right\rceil, \tag{2.10}$$

where $\lceil \cdot \rceil$ is the ceiling function that represents the smallest integer greater than the variable. Consider an m-dimensional hyper-cubic domain $\mathcal{X} = [0, 1]^m$ and a target set $S = B(x^*, \varepsilon)$. The upper bound of the volume of S is given by a d-dimensional hyper-ball and thus we have

$$P(S) \leq P^{\mathrm{U}}(S) = \frac{\pi^{\frac{m}{2}}}{\Gamma\left(\frac{m}{2} + 1\right)} \varepsilon^m, \tag{2.11}$$

where $P^{\mathrm{U}}(S)$ is the upper bound of $P(S)$, i.e., the volume of the hyper-ball and $\Gamma(\cdot)$ is the gamma function. When the hyper-ball is completely included in the hyper-cube, $P(S) = P^{\mathrm{U}}(S)$. Substituting $P^{\mathrm{U}}(S)$ in (2.11) into (2.10), we obtain the lower bound of the number of points required for PRS as

$$n_\gamma^{\mathrm{L}} = \left\lceil \frac{\log \gamma}{\log(1 - P^{\mathrm{U}}(S))} \right\rceil. \tag{2.12}$$

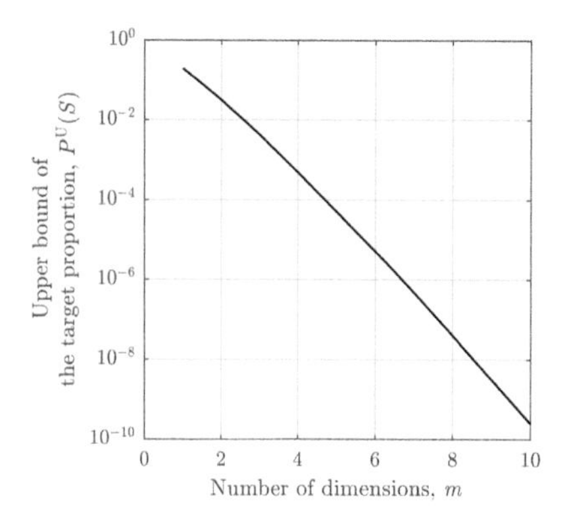

(a) Upper bound of the proportion of the set S to the domain $\mathcal{X}$

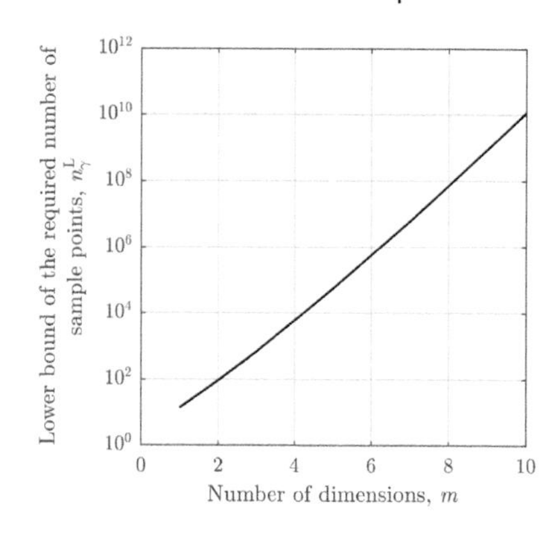

(b) Lower bound of the required number of sample points to reach the set S

Figure 2.1 Proportion of the target set S and required number of sample points to reach the set S with probability at least $1 - \gamma$ for $\varepsilon = 0.1$ and $\gamma = 0.05$, where S is a d-dimensional hyper-ball in hyper-cubic domain $\mathcal{X}$.

As the dimensionality d increases, the number n_{γ}^{L} exponentially increases. For example, for $\varepsilon = 0.1$, $\gamma = 0.05$ and $m = 1, \cdots, 10$, $P^{\mathrm{U}}(S)$ and n_{γ}^{L} are plotted in Fig. 2.1. This phenomenon is widely known as *curse of dimensionality* [Bellman, 1957]. Thus, even for moderate dimensions ($m \geq 10$), it is generally difficult to guarantee to find the global minimum (solution) by PRS in practically allowable time.

2.3.4 POPULATION-BASED SEARCH METHODS

Among various algorithms of improving a single solution by probabilistic search, simulated annealing [Kirkpatrick et al., 1983] and other MGS algorithms make use of only the last observation about the objective function gained during the process of search. Although use of the information is limited, many Markovian algorithms have proven to be more practically efficient than PRS algorithms. Thus, the possibility of using information even at one point in the current step leads to a visible improvement in efficiency. A natural next step in improving the efficiency of RS algorithms is to allow the possibility of using more information about the objective function keeping the structure of the algorithms relatively simple. Some algorithms utilize information of one group of points (current generation) for transforming it to another group of points (next generation) based on certain probabilistic rules. These types of algorithms are referred to as PBS algorithms. Note that the family of PBSs includes many evolutionary algorithms such as the very popular *genetic algorithms*. Algorithm 2.2 of group-based RS is extended to the general scheme of

Algorithm 2.3 Population-based algorithms: General form.

1: Choose a probability distribution $P_1(\cdot)$ on the m_1-fold product set $\mathcal{X} \times \cdots \times \mathcal{X}$, where $m_1 \geq 1$ is a given integer. Set iteration counter $j = 1$.
2: Obtain m_j points $\boldsymbol{x}_1^{(j)}, \cdots, \boldsymbol{x}_{m_j}^{(j)}$ in $\mathcal{X}$ by sampling from the distribution P_j. Evaluate the objective function $f(\cdot)$ at these points.
3: Check the stopping criterion.
4: Using the points $\boldsymbol{x}_l^{(j)}$ $(l = 1, \cdots, m_j)$ and the objective function values at these points, construct a probability distribution $P_{j+1}(\cdot)$ on the m_{j+1}–fold product set $\mathcal{X} \times \cdots \times \mathcal{X}$, where m_{j+1} is some integer that may depend on the search information.
5: Substitute $j + 1$ for j and return to Step 2.

global random search algorithms in Algorithm 2.3. We assume that the rule for constructing the probability distribution $P_{j+1}(\cdot)$ does not make use of the points $\boldsymbol{x}_{l_i}^{(i)}$ $(l_i = 1, \cdots, m_i;\ i = 1, \cdots, j - 1)$ and the results of the objective function evaluation at these points, where m_i is the number of points at the ith group or generation. That is, the probability distribution $P_{j+1}(\cdot)$ is constructed using only the points of the jth iteration $\boldsymbol{x}_{l_j}^{(j)}$ $(l_j = 1, \cdots, m_j)$ and the results of the objective function evaluation at these points.

We call the set of points of the jth iteration

$$\boldsymbol{x}_1^{(j)}, \cdots, \boldsymbol{x}_{m_j}^{(j)} \tag{2.13}$$

the *parent generation*, and the related set of points of the $(j + 1)$th iteration

$$\boldsymbol{x}_1^{(j+1)}, \cdots, \boldsymbol{x}_{m_{j+1}}^{(j+1)} \tag{2.14}$$

the *generation of descendants* or *children*. The choice of population sizes m_j is important and there are no restrictions for the values of m_j. When we use large values of m_j, the globality of the search increases. Conversely, when we have $m_j = 1$, the search becomes MGS.

The major difference between methods based on Algorithm 2.3 is related to the rules for obtaining the population of descendants (2.14) from the population of parents (2.13). For example, in genetic algorithm that is classified into one of PBSs, each descendant $x_l^{(j+1)}$ is obtained from two parents among $x_i^{(j)}$ $(i = 1, \cdots, m_j)$ by the operation called *crossover*. *Mutation* operation is also performed to randomly modify the variable. The mutation operation plays an important role because this operation can be regarded as a variant of PRS. Sampling from some probability distribution is referred to as the transition probability, which defines the way of choosing the point of the next generation in the neighborhood of a point (that may be in the sense of crossover of the genome) from the parent generation. The transition probabilities are often chosen so that one samples with uniform distribution on $\mathcal{X}$ with small probability $p_j \geq 0$ and some other distribution with probability $1 - p_j$. In this case, the condition $\sum_{j=1}^{\infty} p_j = \infty$ in the mutation operation guarantees the convergence of the algorithm. Therefore, the mutation operation with uniform distribution guarantees the convergence of genetic algorithms; see Zhigljavsky and Žilinskas [2021] for details.

2.4 STOPPING RULES FOR STOCHASTIC OPTIMIZATION

Optimization algorithms are roughly classified into two types, i.e., deterministic and stochastic, and each algorithm has its suitable type of problem to be applied. The no free lunch theorems state that no single optimization algorithm can consistently outperform all other algorithms when a wide class of problems is considered [Wolpert and Macready, 1997]. It is impossible to guarantee to find the global optimum if we cannot restrict classes of functions, e.g., Lipschitz continuous functions with known Lipschitz constants. In particular, it is difficult to deterministically guarantee the accuracy of the solution. Hence, deterministic measure is often relaxed into a confidence measure by introducing a stochastic element. A number of successful optimization algorithms belong to this class. When a stochastic algorithm is applied, it is important to use an appropriate stopping rule. Groenwold and Hindley [2002] developed a framework in which different algorithms compete in parallel for a contribution towards the same global stopping criterion, which is derived as a beta distribution by Bayesian approach. However, arguments for randomization in experimental design is difficult to reconcile with the Bayesian approach. Berry and Kadane [1997] suggested a reasonable Bayesian model in which randomization is optimal, thus removing the apparent conflict between Bayesian and randomized approaches. They pointed out that the amount of randomization depends on individual belief and different knowledge from the Bayesian viewpoint.

The structural optimization problems considered in this monograph are often formulated as combinatorial (discrete) optimization problems, which can be solved using some sophisticated metaheuristics that coordinate simple heuristics and rules to find good approximate solutions to computationally difficult problems. The main drawback of most metaheuristics is the absence of effective stopping criteria. Ribeiro et al. [2011] proposed a normal approximation-based stopping rule to estimate the probability of finding a solution at least as good as the currently known best solution at any iteration. This type of stopping rule for structural optimization will be introduced in Chap. 3.

On the other hand, RS approaches mainly deal with continuous optimization. Jezowski et al. [2005] addressed an RS-based optimization method for nonlinear problems with continuous variables. They observed that PBS methods can provide high reliability of finding the global optimum for highly nonlinear and multi-modal problems with sharp peaks. Zhigljavsky and Hamilton [2010] developed a methodology for defining stopping rules in a general class of RS algorithms that are based on the use of statistical procedures. They used the lowest k order statistics for constructing a confidence interval for the global minimizer and called it *k-adaptive global random search algorithm.* Order statistics can be linked to accuracy of RS, which will be introduced in Chap. 4.

The simplest RS algorithm is to pick some elements of S randomly and estimate the statistical value, which is referred to as *batch sampling.* We may be able to use more appropriate sample size for the current input data set. With this motivation, adaptive sampling techniques have been proposed [Watanabe, 2000; Lipton and Naughton, 1995]. A simplified stopping rule among them will be introduced below. Consider a sequence of points $\boldsymbol{x}$ in $\mathcal{X}$ by PRS in Sec. 2.3.2. Let S be the target set such that

$$S = \{\boldsymbol{x} \in \mathcal{X} : f(\boldsymbol{x}) \leq \bar{f} = Q(p)\},$$

where $Q(\cdot)$ and $\bar{f}$ are a quantile function and the pth quantile of f, receptively, and p is a given probability, i.e., $\Pr\{\boldsymbol{x} \in S\} = p$. Assume that we know only the value of $\bar{f}$ and want to estimate the value of p. Let $b(\cdot)$ be a Boolean function called indicator function defined on instances in $\mathcal{X}$ as follows:

$$b(\boldsymbol{x}) = \begin{cases} 1 & (\boldsymbol{x} \in S) \\ 0 & (\boldsymbol{x} \notin S). \end{cases}$$

Empirical probability $\tilde{p}$ by PRS is given by

$$\tilde{p} = \frac{1}{n} \sum_{j=1}^{n} b(\boldsymbol{x}_j),$$

where n is the sample size, and $\boldsymbol{x}_1, \cdots, \boldsymbol{x}_n$ are the points in $\mathcal{X}$ generated by PRS of Algorithm 2.1. Clearly, $\tilde{p}$ tends to p as n tends to infinity; however, we like to achieve a given accuracy with a sample size as small as possible, and consider the following approximation goal:

Approximation goal For given α $(0 < \alpha < 1)$ and ϵ $(0 < \epsilon < 1)$, the following accuracy is to be achieved:

$$\Pr\{|\tilde{p} - p| \leq \epsilon\} > \alpha. \tag{2.15}$$

We can also rewrite (2.15) to the following relations:

$$\begin{aligned} &1 - \Pr\{|\tilde{p} - p| \leq \epsilon\} < 1 - \alpha \\ &\Rightarrow \Pr\{|\tilde{p} - p| > \epsilon\} < 1 - \alpha. \end{aligned} \tag{2.16}$$

To achieve this approximation goal using PRS with a small sample size, Watanabe [2000] proposed some stopping rules based on the statistical bounds, among which we will introduce the two bounds. The first one is known as the Chernoff bound.

Theorem 2.3 (The Chernoff bound) *For any ε $(0 < \varepsilon < 1)$, we have the following relations:*

$$\Pr\{\tilde{p} > (1+\varepsilon)p\} \leq \exp\left(-\frac{pn\varepsilon^2}{3}\right),$$

$$\Pr\{\tilde{p} < (1-\varepsilon)p\} \leq \exp\left(-\frac{pn\varepsilon^2}{2}\right).$$

In Theorem 2.3, ε denotes relative error. Theorem 2.3 can be rewritten using absolute error $\epsilon = \varepsilon p$ as follows:

Theorem 2.4 (The Chernoff bound with absolute error) *For any ϵ $(0 < \epsilon < 1)$, we have the following relations:*

$$\Pr\{\tilde{p} > p + \epsilon\} \leq \exp\left(-\frac{n\epsilon^2}{3p}\right), \tag{2.17a}$$

$$\Pr\{\tilde{p} < p - \epsilon\} \leq \exp\left(-\frac{n\epsilon^2}{2p}\right). \tag{2.17b}$$

From Theorem 2.4, the following relations are satisfied:

$$\begin{aligned}\Pr\{|\tilde{p} - p| > \epsilon\} &= \Pr\{\tilde{p} > p + \epsilon\} + \Pr\{\tilde{p} < p - \epsilon\} \\ &\leq \exp\left(-\frac{n\epsilon^2}{3p}\right) + \exp\left(-\frac{n\epsilon^2}{2p}\right) \\ &\leq 2\exp\left(-\frac{n\epsilon^2}{3p}\right).\end{aligned} \tag{2.18}$$

From (2.16) and (2.18), we can derive the following inequality for the required number of samples n for satisfying the accuracy specified by ϵ and α:

$$\begin{aligned}2\exp\left(-\frac{n\epsilon^2}{3p}\right) &< 1 - \alpha, \\ \log\frac{2}{1-\alpha} &< \frac{n\epsilon^2}{3p}, \\ n &> \frac{3p}{\epsilon^2}\log\frac{2}{1-\alpha}.\end{aligned} \tag{2.19}$$

Another bound is the Hoeffding bound.

Theorem 2.5 (The Hoeffding bound) *For any ϵ $(0 < \epsilon < 1)$, we have the following relations:*

$$\Pr\{\tilde{p} < p - \epsilon\} \leq \exp\left(-2n\epsilon^2\right).$$

Table 2.3
Minimum Sample Size Requirement n by the Chernoff Bound and the Hoeffding Bound

(a) Chernoff bound

	$p = \epsilon = 0.1$	$p = \epsilon = 0.05$	$p = \epsilon = 0.01$
$\alpha = 0.9$	90	180	899
$\alpha = 0.95$	111	222	1107
$\alpha = 0.99$	159	318	1590

(b) Hoeffding Bound

	$\epsilon = 0.1$	$\epsilon = 0.05$	$\epsilon = 0.01$
$\alpha = 0.9$	150	600	14979
$\alpha = 0.95$	185	738	18445
$\alpha = 0.99$	265	1060	26492

From Theorem 2.5, the following inequality is obtained in a similar manner as (2.19):

$$n > \frac{1}{2\epsilon^2} \log \frac{2}{1-\alpha}. \tag{2.20}$$

Thus, we obtain required sample size n by the Chernoff bound and the Hoeffding bound, respectively, as follows:

$$\text{(Chernoff bound)} \qquad n_{\mathrm{C}} = \left\lceil \frac{3p}{\epsilon^2} \log \frac{2}{1-\alpha} \right\rceil, \tag{2.21}$$

$$\text{(Hoeffding bound)} \qquad n_{\mathrm{H}} = \left\lceil \frac{1}{2\epsilon^2} \log \frac{2}{1-\alpha} \right\rceil. \tag{2.22}$$

When $p < 1/6$, n_{C} is smaller than n_{H} for the same values of α and ϵ. However, we do not know the value of p in general. Although details are not explained here, Watanabe [2000] proposed use of estimated value of p by adaptive sampling. The minimum sample sizes for some given values of α and ϵ are shown in Table 2.3. It is seen from Table 2.3 that the required sample sizes by the Chernoff bound are smaller than those by the Hoeffding bound, which means the Chernoff bound is more appropriate than the Hoeffding bound in these cases.

Note that values in Table 2.3 are very conservative. The process of PRS can be expressed by the sum of independent Bernoulli trials. Let $Y_1, \cdots, Y_n$ be a sequence of independent Bernoulli trials such that

$$Y_j = \begin{cases} 1 & (\boldsymbol{x}_j \in S) \\ 0 & (\boldsymbol{x}_j \notin S), \end{cases}$$

Table 2.4
Values of $\Pr\{|Y/n - p| < \epsilon\}$ by Chernoff Bound Using the Corresponding Sample Size and Parameter Values in Table 2.3(a)

	$p = \epsilon = 0.1$	$p = \epsilon = 0.05$	$p = \epsilon = 0.01$
$\alpha = 0.9$	0.9986	0.9981	0.9976
$\alpha = 0.95$	0.9998	0.9996	0.9995
$\alpha = 0.99$	1.000	0.9999	0.9999

where $\Pr\{Y_j = 1\} = p$. Sum of the random variables is denoted by $Y = \sum_{j=1}^{n} Y_j$. Clearly, Y/n corresponds to $\tilde{p}$. We can easily compute following probability of Y

$$\begin{aligned} \Pr\left\{\left|\frac{Y}{n} - p\right| < \epsilon\right\} &= \Pr\{n(p-\epsilon) \leq Y \leq n(p+\epsilon)\} \\ &\leq \sum_{j=\lfloor n(p-\epsilon)\rfloor}^{\lceil n(p+\epsilon)\rceil} \binom{n}{j} p^j (1-p)^{1-j}, \end{aligned} \tag{2.23}$$

where $\lfloor\cdot\rfloor$ denotes the floor function, which is equal to the largest integer that is less than or equal to the variable. The probability (2.23) can be regarded as the true value of α in (2.15) when we know the true value of p. Some values of the probability (2.23) are computed and summarized in Table 2.4 using the minimum sample size by the Chernoff bound and corresponding parameter values in Table 2.3(a). As can be seen from Table 2.4, the probability values are substantially larger than those of α and close to one, which shows the Chernoff bound is conservative because the correct probability values should be close to α but not one.

2.5 SUMMARY

Overview of stochastic optimization has been presented in this chapter. Basics of random variable and random process have been presented, and random search methods for deterministic optimization problem have been classified. Pure random search is the simplest and most effective method for solving optimization problems with moderately small number of variables. However, due to the curse of dimensionality, some heuristic methods including population-based approaches and those based on local search should be used for a practical structural optimization problem with moderately large number of variables. Stopping criteria of random search can be defined by the Chernoff bound and Hoeffding bound; however, they are generally too conservative. Therefore, more strict rules are presented using attractors in Sec. 3.3 and order statistics in Sec. 4.2.

ACKNOWLEDGMENT

We acknowledge the permissions from publishers for use of the licensed contents as follows:

- Tables 2.1 and 2.2: Reprinted/adapted by permission from Springer Nature [Zhigljavsky and Žilinskas, 2021].

3 Random Search-based Optimization

In this chapter, two types of random search-based optimization methods are explained for problems with integer variables and parameters, namely, global random search and random multi-start local search. In Sec. 3.2, performances of random search with quantile-based approach is investigated using a mathematical problem and worst-case design of a building frame [Ohsaki and Katsura, 2012]. In Sec. 3.3, stopping rules for random multi-start local search are presented for application to structural optimization problems with moderately large number of local optima [Ohsaki and Yamakawa, 2018].

3.1 INTRODUCTION

In the conventional structural optimization methods, parameters representing the structural and material properties are given deterministically. However, in the practical design process, uncertainty in those parameters should be appropriately taken into account [Elishakoff and Ohsaki, 2010]. There are various approaches to such purpose; namely, reliability-based approach [Frangopol, 1995], probabilistic approach [Augusti et al., 1984], worst-case design [Rustem and Howe, 2002] and robust design [Gu et al., 2000a]. Among various sources of uncertainty in structural design, the parameters such as cross-sectional dimensions and material parameters often have lower and upper bounds, although their probabilistic properties are not known. In such situation, the concept of *unknown-but-bounded* [Elishakoff et al., 1994] is utilized to assume that the uncertain parameters exist in the specified bounded intervals, and the worst responses due to such uncertainty can be found using the standard approach of interval analysis [Moore, 1966]. However, this approach is not applicable to a problem with large number of uncertain parameters and/or highly nonlinear response functions.

Worst-case approach is a well-known robust design method [Parkinson et al., 1993; Zang et al., 2005; Ben-Tal et al., 2009], where the structures are designed considering the worst possible scenario. The process of finding the worst value for unknown-but-bounded uncertainty is referred to as worst-case analysis (WCA) [Ben-Haim and Elishakoff, 1990; Ben-Haim et al., 1996]. Thus, WCA can be regarded as a deterministic optimization process. Furthermore, the approach is often combined with structural optimization to formulate a worst-case design problem, in which objective function and/or constraints are assigned on the worst values of the structural responses, see e.g., Gu et al. [2000b]; Gurav et al. [2005]; Hashimoto and Kanno [2015]; ur Rehman and

DOI: 10.1201/9781003153160-3

Langelaar [2015]; Joseph [2017]; Yoo and Lee [2014]; Paly et al. [2013]. Therefore, the worst-case design problems are basically formulated as a two-level optimization problem of finding the worst responses with respect to the uncertain parameters and the problem of minimizing the objective function with respect to the design variables. This type of formulation is also referred to as bilevel optimization [Colson et al., 2007], robust counterpart approach [Beyer and Sendhoff, 2007] and optimization and anti-optimization framework [Elishakoff and Ohsaki, 2010]. The approach has significant methodological advantage, because the optimal design can be found by solving deterministic optimization problems; however, the worst-case-oriented decisions may be too conservative as stated by Ben-Tal et al. [2009].

For various problems of structural optimization, it is natural to expect that there exist many local optimal solutions, if constraints on response quantities such as stress, dynamic response, inelastic response, etc., are considered. Therefore, various methods of global optimization have been proposed [Riche and Haftka, 2012; Torn and Zhilinskas, 1989; Zabinsky et al., 2010]. Randomized algorithms can be successfully used for obtaining approximate optimal solutions [Ohsaki, 2001; Ohsaki and Katsura, 2012]. Especially for a worst-case design, the most difficult point is that the worst response value is very difficult to obtain. Therefore, the approach based on random search (RS) in Sec. 2.3.1 and quantile responses in Sec. 1.4 can be effectively used for obtaining the approximate worst response with a specified confidence level.

For problems with continuous functions and variables, gradient based nonlinear programming (NLP) approaches can be used to obtain a local optimal solution with good accuracy. However, in most of the design problems, the design variables have integer values, and the problem becomes a combinatorial optimization problem. For such problems, heuristic approaches have been developed for obtaining approximate optimal solutions within reasonable computational cost. They are classified into population-based search (PBS) approaches such as genetic algorithm (GA) and particle swarm optimization (PSO), and those based on local search (LS) such as simulated annealing (SA) and tabu search (TS). The PBSs demand substantial computational cost, because the function values of many solutions should be evaluated at each generation. By contrast, computational cost of the methods based on LS is rather small, because only the current solution and its neighborhood solutions are evaluated at each step. It is also important to note that the solution is always improved from the initial solution except for SA that accepts a non-improving solution with a specified probability. Note that the multi-start strategy is usually employed for the LS methods to improve possibility of finding the global solution.

Muselli [1997] compared the computational efficiency and accuracy between consecutively restarting searches and continuing searches with a single start for RS, random walk, grid search and various covering methods [Torn and Zhilinskas, 1989]. However, evaluation of objective and constraint functions requires much computational cost for structural optimization problems;

therefore, we cannot carry out function evaluation many times for optimization, although performances of algorithms are compared based on many function evaluations of mathematical test problems. Furthermore, obtaining the global optimal solution is not always necessary. Therefore, for a multi-start strategy, it is important to estimate the number of local optimal solutions including the global optimal solution, which is not known *a priori*, and the accuracy of the best solution obtained so far, at each step of generating another initial solution to judge when the multi-start steps can be terminated. Several stopping rules have been proposed for RMS and PRS [Dorea, 1983; Dorea and Gonçalves, 1993; Hart, 1998]. A simple rule may be defined based on the history of objective values; e.g., if the objective value is not improved during the prescribed number of trials, then the multi-start process is terminated, and the best solution obtained so far is conceived as an approximate global solution. Another strategy utilizes the attractor, which is the set of solutions leading to each local optimal solution by carrying out an LS. Some stopping rules using attractors are presented in Sec. 3.3.

3.2 RANDOM SEARCH FOR WORST-CASE DESIGN

3.2.1 TWO-LEVEL OPTIMIZATION PROBLEM

Consider a problem of optimizing integer variables such as the cross-sections of building frames that are selected from the pre-assigned list of standard sections. In the design problem of frames, as shown in the numerical example in Sec. 3.2.4, the members are classified into groups, each of which has the same section. Therefore, suppose the variables are classified into m groups that have the same value, respectively. The design variable vector is denoted by $\boldsymbol{J} = (J_1, \cdots, J_m)$, which has integer values. For example, if $J_i = k$, then the variables in the ith group have the kth value in the list. Let $f(\boldsymbol{J})$ denote the objective function representing, e.g., the total structural volume. The constraint functions defined by structural responses are denoted by $g_i(\boldsymbol{J})$ $(i = 1, \cdots, n)$, where n is the number of constraints. Then, the optimization problem is formulated as

$$\text{Minimize} \quad f(\boldsymbol{J}) \tag{3.1a}$$

$$\text{subject to} \quad g_i(\boldsymbol{J}) \leq \bar{g}_i \quad (i = 1, \cdots, n), \tag{3.1b}$$

$$J_i \in \{1, \cdots, s\} \quad (i = 1, \cdots, m), \tag{3.1c}$$

where $\bar{g}_i$ is the upper bound for g_i, and s is the number of sampling values in the list of variables.

The vector consisting of r uncertain parameters is denoted by $\boldsymbol{p} = (p_1, \cdots, p_r)$ representing the geometrical and material properties of the structure. We assign constraints on the worst values $g_i^{\mathrm{U}}(\boldsymbol{J})$ of functions $g_i(\boldsymbol{J}, \boldsymbol{p})$, and minimize the worst value $f^{\mathrm{U}}(\boldsymbol{J})$ of a function $f(\boldsymbol{J}, \boldsymbol{p})$. Hence, the optimization problem is formulated as

$$\text{Minimize} \quad f^{\mathrm{U}}(\boldsymbol{J}) \tag{3.2a}$$

$$\text{subject to} \quad g_i^{\mathrm{U}}(\boldsymbol{J}) \leq \bar{g}_i \quad (i = 1, \cdots, n), \tag{3.2b}$$

$$J_i \in \{1, \cdots, s\} \quad (i = 1, \cdots, m). \tag{3.2c}$$

The worst value $g_i^{\mathrm{U}}(\boldsymbol{J})$ is obtained by solving the following WCA problem, which is also called anti-optimization problem:

$$\text{Find} \qquad g_i^{\mathrm{U}}(\boldsymbol{J}) = \max_{\boldsymbol{p}} g_i(\boldsymbol{J}; \boldsymbol{p}) \tag{3.3a}$$

$$\text{subject to} \quad \boldsymbol{p}^{\mathrm{L}} \leq \boldsymbol{p} \leq \boldsymbol{p}^{\mathrm{U}}, \tag{3.3b}$$

where $\boldsymbol{p}^{\mathrm{U}} = (p_1^{\mathrm{U}}, \cdots, p_r^{\mathrm{U}})$ and $\boldsymbol{p}^{\mathrm{L}} = (p_1^{\mathrm{L}}, \cdots, p_r^{\mathrm{L}})$ are the upper and lower bounds for $\boldsymbol{p}$, respectively, which can be obtained from measurements, statistics and experiments. $f^{\mathrm{U}}(\boldsymbol{J})$ is found similarly. Hence, the optimal solution considering the worst function values can be found by solving a two-level optimization problem.

3.2.2 OPTIMIZATION METHODS

Here, we explain the optimization methods used in this section.

Tabu search (TS)

In the numerical examples, the performance of RS is compared with TS [Glover and Laguna, 1997] based on LS. TS is regarded as a deterministic approach, because the current solution at each step is replaced with the best solution among all the neighborhood solutions, although a randomness exists in the selection of initial solution. A tabu list is used to prevent a cyclic selection among a set of small number of solutions; see Appendix A.3 for details. However, for a problem with many variables, it is not desirable to carry out exhaustive local search at each step. Therefore, we can limit the number of neighborhood solutions in a similar manner as random LS [Ohsaki, 2001]. In the following examples, we apply TS to both upper- and lower-level problems for comparison purpose with the RS approach. The neighborhood solutions are generated using a random number $\tau \in [0, 1]$. Each variable is increased if $\tau \geq 2/3$, decreased if $\tau < 1/3$, and is not modified if $1/3 \leq \tau < 2/3$. Note that the length of tabu list is equal to the number of steps multiplied by the number of neighborhood solutions for all the following examples, which means that all the solutions generated throughout process are kept in the tabu list.

Random search (RS)

We first discuss applicability of RS, which is presented in Chap. 2, to the lower-level problem to maximize a function of r uncertain parameters. Based on the simple formulas of probability theory, the probability of obtaining the optimal solution using RS can easily be estimated as stated in Chap. 2. To compare the computational cost and accuracy of RS and TS, the parameters are assumed to take q different integer values, i.e., the total number of different parameter sets is $M = q^r$. When solving the lower-level problem, it often happens that the global worst solution is not necessary to be found, and only a good approximate worst solution is needed.

Accuracy of the worst solution may be evaluated using the objective function value; however, we cannot use this approach because the global worst value is unknown. Alternatively, the accuracy is defined based on the order of objective value for the specified number of samples, and an approximate worst value can be found as a quantile value of the response function. The randomly generated solutions are numbered in non-increasing order of the objective function value for discussing the accuracy of the worst solution in a similar manner as Sec. 1.4.2; i.e., the first solution has the worst (maximum) objective value.

For the lower-level problem, N worst solutions of parameters are assumed to be approximate worst solutions. Although probability distributions of the parameters are not known, suppose the r uncertain parameters p_i $(i = 1, \cdots, r)$ have the uniform probability to be sampled from q values $\{1, \cdots, q\}$. Then, the parameter vector $\boldsymbol{p}$ is replaced with an integer vector $\boldsymbol{I} = (I_1, \cdots, I_r)$. The possible value of I_i have the same probability $1/q$ to be sampled, and all parameter sets have the same probability to be selected. The objective function to be maximized in the lower-level problem is a function of $\boldsymbol{J}$ and $\boldsymbol{I}$ denoted as $g_i(\boldsymbol{J}, \boldsymbol{I})$. Therefore, the probability of failing to obtain an approximate worst solution is $1 - N/M$ for a randomly sampled solution from the set of M solutions. Hence, the probability that no approximate worst solution is found after t random selections is $(1 - N/M)^t$. Note that we utilize the procedure of *sampling with replacement*; i.e., a solution might be selected more than once, and the probability of selecting a particular solution is always $1/M$.

Figure 3.1 shows the probabilities of failing to obtain approximate worst solutions for $N/M = 0.1$, 0.05 and 0.01, i.e., the bounds of approximate worst values are 0.9, 0.95 and 0.99 quantiles of the response, respectively. Note that the vertical axis has a logarithmic scale. For example, the probability for $N/M = 0.05$ is less than 0.01 if random sampling is carried out 100 times. It should be noted here that the relation does not depend on N and M directly. The required number of analysis for the desired accuracy depends on the ratio N/M.

For the upper-level optimization problem, the objective function is written as $f(\boldsymbol{J}, \boldsymbol{I})$ and it is natural to define the accuracy of solution using the order of solutions, because it is an combinatorial problem. This way, the order of solutions can be utilized for both upper- and lower-level problems, and the two-level problem can be solved using an RS approach. Let M^{u} and M^{l} denote

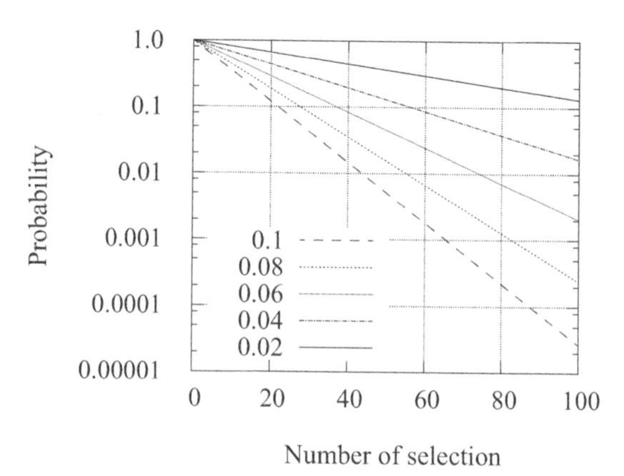

Figure 3.1 Probability of failing to obtain an approximate worst solution for various value of N/M.

the number of random sets to be generated, respectively, in the upper- and lower-level problems. The RS algorithm is summarized in Algorithm 3.1.

Algorithm 3.1 Algorithm of RS for a two-stage problem of optimization and WCA.

Require: Initialize objective function $f^{\rm opt} = \infty$.
1: **for** $i \leftarrow 1$ to $M^{\rm u}$ **do**
2: Randomly sample $\boldsymbol{J}$.
3: Initialize $g = -\infty$ and $f^{\rm U} = -\infty$.
4: **for** $j \leftarrow 1$ to $M^{\rm l}$ **do**
5: Randomly sample $\boldsymbol{I}$.
6: Assign $g \leftarrow \max_{i \in \{1,\cdots,n\}} (g_k(\boldsymbol{J};\boldsymbol{I}) - \bar{g}_k, g)$
7: Assign $f^{\rm U} \leftarrow \max\{f^{\rm U}, f(\boldsymbol{J};\boldsymbol{I})\}$
8: **if** $g > 0$ **then**
9: **break**
10: **end if**
11: Assign $f^{\rm opt} \leftarrow \min\{f^{\rm opt}, f^{\rm U}\}$
12: **end for**
13: **end for**
14: Output $f^{\rm opt}$ and corresponding sets of $\boldsymbol{J}$ and $\boldsymbol{I}$.

3.2.3 MATHEMATICAL EXAMPLE

Performances of TS and RS are first compared using a small mathematical optimization problem without constraints [Ohsaki and Katsura, 2012]. Let $-4 \leq x_i \leq 4$ $(i = 1, \cdots, 4)$ denote the variables. The uncertain parameters

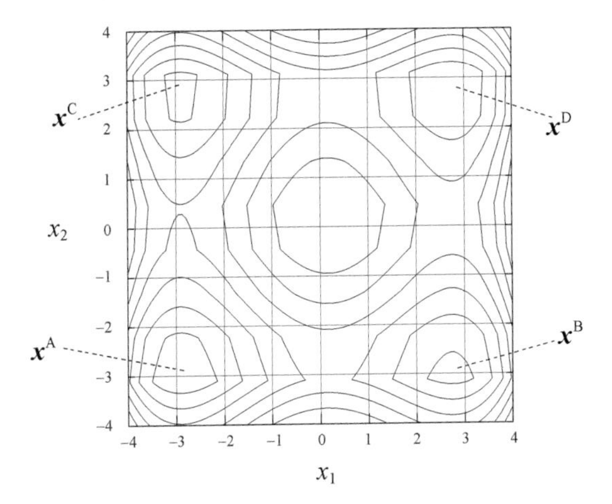

Figure 3.2 Contour lines of Test2N function for $n = 2$ without parameter uncertainty.

are denoted by $-1 \leq p_i \leq 1$ $(i = 1, \cdots, 4)$. The objective function is given as

$$\begin{aligned} f(\boldsymbol{x};\boldsymbol{p}) = &\sum_{i=1}^{4}(x_i^4 - 16x_i^2 + 5x_i) \\ &+ a[(p_1^2 + 0.1p_1)x_1 + (p_2^2 + 0.2p_2)x_2 + p_3x_3 + 0.9p_4x_4], \end{aligned} \tag{3.4}$$

where $a = 10.0$ is a specified parameter for the level of uncertainty, and the vectors of variables and parameters are denoted by $\boldsymbol{x} = (x_1, \cdots, x_4)$ and $\boldsymbol{p} = (p_1, \cdots, p_4)$, respectively. This function without parameter uncertainty $(a = 0)$ has 2^n local minima; hence, it is called Test2N function [Lagaris and Tsoulos, 2008]; see also Appendix A.1. Equation (3.4) is the case of $n = 4$, and Fig. 3.2 shows the contour lines for $n = 2$ without parameter uncertainty, i.e., $a = 0$. The optimal solution exists at $\boldsymbol{x}^{\mathrm{A}} = (-2.9035, -2.9035)$, and local minima are found at $\boldsymbol{x}^{\mathrm{B}} = (2.7468, -2.9035)$, $\boldsymbol{x}^{\mathrm{C}} = (-2.9035, 2.7468)$ and $\boldsymbol{x}^{\mathrm{D}} = (2.7468, 2.7468)$ with the function values $f(\boldsymbol{x}^{\mathrm{A}}) = -156.66$, $f(\boldsymbol{x}^{\mathrm{B}}) = -128.39$, $f(\boldsymbol{x}^{\mathrm{C}}) = -128.39$ and $f(\boldsymbol{x}^{\mathrm{D}}) = -100.12$.

The upper-level optimization problem is written as

$$\text{Minimize} \quad f^{\mathrm{U}}(\boldsymbol{x}) \tag{3.5a}$$

$$\text{subject to} \quad -4 \leq x_i \leq 4 \quad (i = 1, \cdots, 4), \tag{3.5b}$$

where the worst value $f^{\mathrm{U}}(\boldsymbol{x})$ is obtained by solving the following WCA problem:

$$\text{Find} \qquad f^{\mathrm{U}}(\boldsymbol{x}) = \max_{\boldsymbol{p}} f(\boldsymbol{x};\boldsymbol{p}) \tag{3.6a}$$

$$\text{subject to} \quad -1 \leq p_i \leq 1 \quad (i = 1, \cdots, 4), \tag{3.6b}$$

The variables and parameters are defined using integer values as follows:

$$x_i = -4 + \Delta x(J_i - 0.5) \quad (i = 1, \cdots, 4), \tag{3.7a}$$
$$p_i = -1 + \Delta p(I_i - 0.5) \quad (i = 1, \cdots, 4), \tag{3.7b}$$

where

$$\Delta x = 8/(s-1), \quad J_i \in \{1, \cdots, s\}, \tag{3.8a}$$
$$\Delta p = 2/(q-1), \quad I_i \in \{1, \cdots, q\}. \tag{3.8b}$$

Finally, the worst-case design problem with integer variables $\boldsymbol{J} = (J_1, \cdots, J_4)$ and integer parameters $\boldsymbol{I} = (I_1, \cdots, I_4)$ is formulated as

$$\text{Minimize} \quad f^{\text{U}}(\boldsymbol{J}) \tag{3.9a}$$
$$\text{subject to} \quad J_i \in \{1, \cdots, s\} \quad (i = 1, \cdots, 4), \tag{3.9b}$$
$$I_i \in \{1, \cdots, q\} \quad (i = 1, \cdots, 4), \tag{3.9c}$$

with the lower-level WCA problem

$$\text{Find} \quad f^{\text{U}}(\boldsymbol{J}) = \max_{\boldsymbol{I}} f(\boldsymbol{J}, \boldsymbol{I}) \tag{3.10a}$$
$$\text{subject to} \quad I_i \in \{1, \cdots, q\} \quad (i = 1, \cdots, 4). \tag{3.10b}$$

Performance of RS for WCA

The performance of RS for WCA problem is compared with that of TS [Ohsaki and Katsura, 2012]. The numbers of possible values of both variables and parameters are 16, i.e., $s = q = 16$, and accordingly, we have $16^4 = 65536$ solutions and 65536 parameter sets for each solution. Hence, the total number of different solution-parameter sets is $16^8 \simeq 4.2950 \times 10^9$. To evaluate the performance of RS for the lower-level WCA problem, the worst value is found from the 2000 sets, which is about 3.05% of the total 65536 sets of randomly sampled parameters, for each of 100 solutions. Suppose we define the approximate worst value by the 0.99th quantile. Then, the probability of failing to obtain the worst 1% value by an RS is $0.99^{2000} \simeq 1.864 \times 10^{-9}$ as discussed in Sec. 2.3.2.

Let R_{ij} $(i = 1, \cdots, 100; j = 1, \cdots, 2000)$ denote the order of the jth parameter set, which is assigned in non-increasing order of the objective function, among the 65536 parameter sets for the ith solution in the set of 100 randomly generated solutions. The smallest (worst) order of the parameter set for each solution is denoted by $R_i^{\min}$; i.e.,

$$R_i^{\min} = \min_j R_{ij} \quad (i = 1, \cdots, 100). \tag{3.11}$$

Table 3.1(a) shows the maximum, minimum and average values, as well as the standard deviation of $R_i^{\min}$ among the 100 solutions for five cases denoted by

Table 3.1
Values of $R_i^{\min}$ of 100 Solutions for Five Cases with Different Random Seeds by RS and TS

(a) RS

Case	1	2	3	4	5	Average
Maximum	175	146	133	318	128	180.0
Minimum	1	1	1	1	1	1
Average	35.900	31.610	32.180	34.730	28.860	32.656
Std. dev.	33.748	27.998	27.992	44.376	27.815	32.386

(b) TS

Case	1	2	3	4	5	Average
Maximum	1092	1480	664	1047	1092	1075.0
Minimum	1	1	1	1	1	1
Average	63.510	58.990	49.780	69.130	61.900	60.666
Std. dev.	176.94	199.50	117.015	169.641	174.126	167.450

Cases 1–5 with different initial random seeds. The average values among five cases are also listed in the last column. As seen from the table, the average value of $R_i^{\min}$ for RS is 32.656, which is sufficiently small compared with the total number 65536 of the parameter sets. Furthermore, the worst parameter set, i.e., $R_{ij} = 1$, is obtained in at least one of the 100 solutions for all five cases. The maximum value of R_{ij} is 318, which is 0.485% of the 65536 parameter sets.

The results of TS for neighborhood solutions $N^{\mathrm{b}} = 40$ and steps $N^{\mathrm{s}} = 50$ corresponding to 2000 function evaluations for each of 100 solutions are listed in Table 3.1(b). It is seen from the table that average performance of TS is worse than RS although the worst solution is obtained in one of the 100 solutions. It is also noted that the maximum values of R_{ij} for TS are larger than those of RS for all five cases. Therefore, RS is a very effective approach for optimization problems that have several local optima, because no problem-dependent parameter exists for RS.

The performances of RS and TS for smaller numbers of function evaluations, i.e., 600 and 1200 are investigated. Tables 3.2(a) and (b) for RS and TS, respectively, show the average values of five cases of maximum, minimum, average and standard deviation of $R_i^{\min}$ among 100 solutions; the results of 2000 function evaluations are also listed. As seen from the table, the average value and standard deviation for RS monotonically decrease as the number of selections is increased, although the worst value could not be obtained, i.e., $R_i^{\min} \neq 1$ for 600 function evaluations. By contrast, the average values of maximum, minimum, average and standard deviation for TS are almost

Table 3.2
Values of $R_i^{\min}$ of 100 Solutions with Different Numbers of Analyses by RS and TS

(a) RS

Number of selections	Maximum	Minimum	Average	Std. dev.
600	493.400	2.4	111.110	104.050
1200	293.6	1	55.180	54.599
2000	180.0	1	32.656	32.386

(b) TS

N^{b}	N^{s}	Maximum	Minimum	Average	Std. dev.
20	30	1075.0	1	60.008	165.720
30	40	1160.6	1	61.458	172.850
40	50	1075.0	1	60.666	167.450

the same, respectively, among 600, 1200 and 2000 function evaluations, which means that 600 function evaluations are enough for TS.

Finally, a quantitative evaluation is carried out for RS using a larger set of parameters with $q = 20$. The variables J_i $(i = 1, \cdots, 4)$ are fixed at 8 for $s = 16$, and only the second term in (3.4) corresponding to parameter uncertainty is considered; therefore, the total number of combination is $20^4 = 160000$. Random selection is carried out 10000 times from five different random seeds. The histories of order of worst solutions are plotted with respect to the number of selections in Fig. 3.3 in logarithmic scale. We can see from the figure that

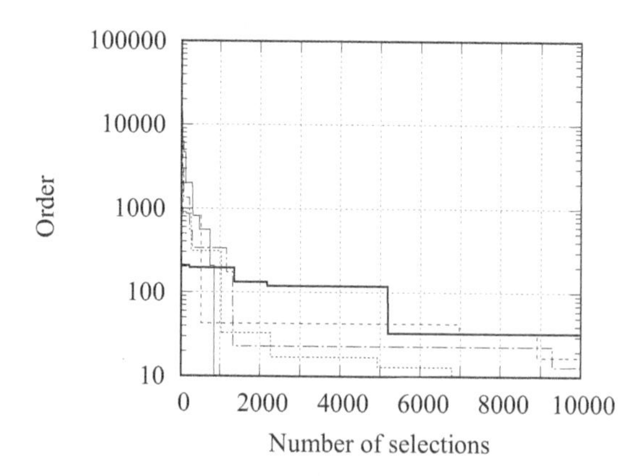

Figure 3.3 History of order of worst solution for five cases with $q = 20$.

Table 3.3
Approximate Optimal Solutions by RS and TS

(a) RS

	J				I				f^{A}	f^{W}	Order
Case 1	2	3	3	2	9	9	1	1	−250.063	−248.969	8
Case 2	2	2	2	2	7	6	1	1	−238.520	−236.488	13
Case 3	3	3	4	5	8	10	1	2	−238.152	−233.949	60
Case 4	3	4	4	3	7	8	1	1	−244.320	−243.805	5
Case 5	4	3	3	4	10	7	1	1	−244.742	−243.305	12

(b) TS

	J				I				f^{A}	f^{W}	Order
Case 1	3	14	3	3	4	16	1	2	−214.672	−204.359	58
Case 2	3	3	3	14	8	8	1	16	−233.449	−233.449	1
Case 3	3	3	3	3	8	8	1	1	−260.969	−260.969	1
Case 4	3	14	14	14	4	16	16	15	−259.672	−249.359	58
Case 5	14	14	13	3	1	1	16	1	−136.895	−121.426	51

the solution within 160th worst corresponding to the accuracy of 0.1% can be found with less than 1500 selections for all five cases.

Performance of RS for worst-case optimization problem

We compare the performances of RS and TS for the two-level optimization and WCA problem, where the same numbers $s = q = 16$ are used here for discretization of variables and parameters. Since good approximate worst solution is found with 2000 selections for the lower-problem, we fix the number of selections as 2000 for both of upper- and lower-level problems. The approximate optimal solutions found by RS are listed for five cases with different random seeds in Table 3.3(a), where *objective value* f^{A} denotes the value of objective function corresponding to the approximate worst parameter set $\boldsymbol{I}^{\mathrm{A}}$ of the approximate optimal solution $\boldsymbol{J}^{\mathrm{A}}$, and *worst value* f^{W} denotes the exact worst objective value of the approximate optimal solution; i.e.,

$$f^{\mathrm{A}} = f(\boldsymbol{J}^{\mathrm{A}}; \boldsymbol{I}^{\mathrm{A}}), \tag{3.12a}$$

$$f^{\mathrm{W}} = \max_{\boldsymbol{I}} f(\boldsymbol{J}^{\mathrm{A}}; \boldsymbol{I}). \tag{3.12b}$$

The order of f^{A} in the complete list of $16^4 = 65536$ parameter sets by enumeration for each solution is denoted by *order*. Note that the objective values cannot be compared among different cases, because they have different approximate solutions.

The optimal solution by enumeration of 16^8 variable-parameter sets is $\boldsymbol{J} = (3, 3, 3, 3)$ with the worst objective value -260.949 corresponding to the parameters $\boldsymbol{I} = (9, 9, 1, 1)$. Although the exact solution has not been found within five trials, Case 1 has the smallest approximate objective function value $f^{\mathrm{A}} = -250.063$ for the solution $\boldsymbol{J} = (2, 3, 3, 2)$ with the true objective function value $f^{\mathrm{W}} = -248.969$ which means that $\boldsymbol{I} = (9, 9, 1, 1)$ is not the worst parameter vector for the solution $\boldsymbol{J} = (2, 3, 3, 2)$.

The results of TS with numbers of neighborhood solutions $N^{\mathrm{b}} = 40$ and steps $N^{\mathrm{s}} = 50$ corresponding to 2000 function evaluations are listed in Table 3.3(b). As seen from the table, the exact optimal solution is found for Case 3 with the exact worst parameter value. However, a solution with a very large f^{A} is found for Case 5.

3.2.4 WORST-CASE DESIGN OF BUILDING FRAME

Optimal cross-sections are found for a 4-story single-span plane steel frame as shown in Fig. 3.4 subjected to seismic motions [Ohsaki and Katsura, 2012]. The same notations as Sec. 3.2.3 are used for the variables and parameters. The number of variables and parameters are both 4, and they are discretized into 5 and 9 values; i.e., $s = 5$ and $q = 9$, respectively. The objective function is the total structural volume. The steel sections of beams and columns are selected from the list of standard sections in Table 3.4, where H means wide-flange section, and HSS means tubular hollow square section in Japanese specification.

The design is designated by the section numbers in the list. Beams are classified into two groups denoted by Beam 1 in 2nd and 3rd floors and Beam 2 in 4th floor and roof, for which the sections are designated by the variables J_1 and J_2, respectively. Columns are also classified into two groups denoted by Columns 1 and 2, respectively, consisting of those in 1st and 2nd stories and those in 3rd and 4th stories, which are defined by the variables J_3

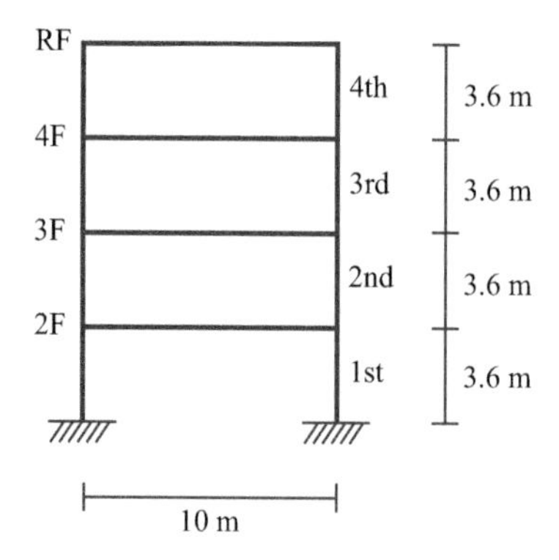

Figure 3.4 A 4-story plane frame.

Table 3.4
List of Standard Sections for Beams and Columns

	Beam 1 (2F, 3F)	Beam 2 (4F, RF)	Column
1	H-400×200×9×12	H-250×125×6×9	HSS-300×300×9
2	H-400×200×9×16	H-300×150×6.5×9	HSS-300×300×12
3	H-400×200×9×19	H-350×175×7×11	HSS-300×300×16
4	H-400×200×9×22	H-400×200×8×13	HSS-300×300×19
5	H-400×200×12×22	H-450×200×9×14	HSS-300×300×22

and J_4, respectively. Thus, the number of design variables is 4. The sections of all columns are selected from the same list, because the external size of all columns should be the same for this low-rise building frame. The standard model consisting of H-400×200×9×19 for Beam 1, H-350×175×7×11 for Beam 2 and HSS-300×300×16 for Columns 1 and 2 is designated by $\boldsymbol{J} = (3, 3, 3, 3)$.

The steel material has a bilinear stress-strain relation, where Young's modulus is $E = 205$ GPa, the lower-bound value of hardening ratio of linear kinematic hardening is $0.01E = 2.05$ GPa, and the lower-bound values of yield stresses of beams and columns are 235 MPa. The artificial seismic motions are generated using the standard superposition method of sinusoidal waves; see Ohsaki and Katsura [2012] for details. The seismic motion with duration 20 sec. is applied at the base of the frame in horizontal direction. A constraint is given for the maximum value among the mean-maximum interstory drifts of all stories against five artificial motions, which is simply denoted by *maximum interstory drift* $d(\boldsymbol{J}; \boldsymbol{I})$. A frame analysis software OpenSees [McKenna, 2011] is used for response history analysis. Each column is modeled by a beam-column element, whereas each beam is divided into two elements. The sections of elements are divided into fibers; see Ohsaki and Katsura [2012] for details. The stiffness-proportional damping is used with the damping ratio 0.02 for the first mode. The first natural period of the standard design with $\boldsymbol{J} = (3, 3, 3, 3)$ is 0.71 sec.

WCA of standard design

Based on the results of preliminary parametric study, the yield stress of columns and the hardening ratios of beams and columns are fixed at their lower-bound values 235 MPa and $0.01E$, respectively, because d is a monotonically decreasing function of these parameters and obviously the lower-bound values correspond to the worst values.

By contrast, d is not a monotonic function of the yield stress $\sigma_{\mathrm{y}}^{\mathrm{b}}$ of beams. Therefore, the integer parameters I_1 and I_2 are assigned for $\sigma_{\mathrm{y}}^{\mathrm{b}}$ of Beams 1 and 2, respectively, as $235.0 + 4.7(I_i - 0.5)$ MPa $(i = 1, 2)$ corresponding to

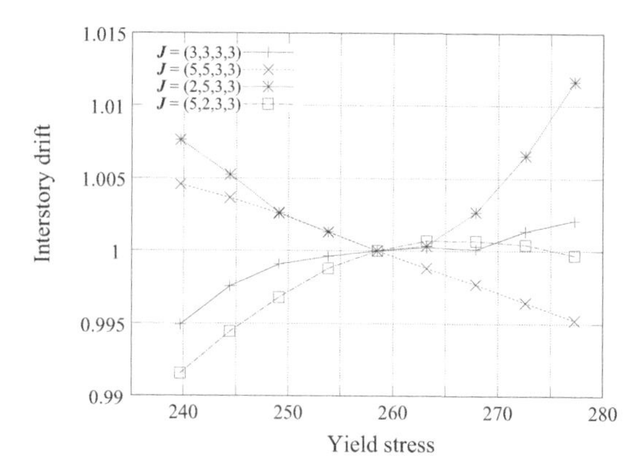

Figure 3.5 Relation between yield stress of beams and normalized values of maximum interstory drifts of various designs.

the integer value $I_i \in \{1, 2, \cdots, 9\}$. In addition to material parameters, the cross-sectional geometry also has uncertainty. We consider uncertainty in the thicknesses of flanges of Beams 1 and 2, defined by I_3 and I_4, respectively, which are discretized into nine equally spaced values. Hence, the possible values are determined by multiplying 0.96, 0.97, 0.98, 0.99, 1.0, 1.01, 1.02, 1.03 and 1.04 to the nominal value in Table 3.4.

Variation of d is investigated with respect to the values of $\sigma_{\mathrm{y}}^{\mathrm{b}}$ that is varied simultaneously for Beams 1 and 2, while the flange thickness has the nominal value, i.e., $I_3 = I_4 = 5$. Figure 3.5 shows the values of d normalized by those for $\sigma_{\mathrm{y}}^{\mathrm{b}} = 256.15$ MPa corresponding to $I_1 = I_2 = 5$ for four designs with different beam sections and the same column sections $J_3 = J_4 = 3$. For the standard design $\boldsymbol{J} = (3, 3, 3, 3)$, d is an increasing function of $\sigma_{\mathrm{y}}^{\mathrm{b}}$, because a strong beam leads to a column-collapse mechanism that has small energy dissipation and large local interstory drift. By contrast, for the design (5,5,3,3) with stronger beam, d is a decreasing function of $\sigma_{\mathrm{y}}^{\mathrm{b}}$, because, in this case, a larger $\sigma_{\mathrm{y}}^{\mathrm{b}}$ leads to larger energy dissipation under the same collapse mechanism. Finally, the values of d for intermediate designs (2,5,3,3) and (5,2,3,3) are not monotonic functions of $\sigma_{\mathrm{y}}^{\mathrm{b}}$.

The lower-level WCA problem is formulated as

$$\text{Find} \quad d^{\mathrm{U}}(\boldsymbol{J}) = \max_{\boldsymbol{I}} d(\boldsymbol{J}; \boldsymbol{I}) \tag{3.13a}$$

$$\text{subject to} \quad I_i \in \{1, \cdots, 9\} \quad (i = 1, \cdots, 4). \tag{3.13b}$$

Since we consider uncertainty in four parameters with nine possible values, the total number of combinations of parameters is $9^4 = 6561$. We assume the parameter sets corresponding to d up to the 200th maximum value are regarded as approximate worst parameter sets, and select 200 parameter sets randomly, which is about 3% of the 6561 samples. Then, the probability for

Table 3.5
Maximum and Minimum Values of d (m) and Their Orders by RS and TS for WCA of the Standard Design with Five Different Random Seeds

(a) RS

Case	**1**	**2**	**3**	**4**	**5**
Max.	0.0510992	0.0509448	0.0510325	0.0511712	0.0 509368
Order	31	96	55	11	105
Min.	0.0482927	0.0482812	0.0484463	0.0481460	0.0483789
Order	6541	6542	6507	6556	6527

(b) TS

Case	**1**	**2**	**3**	**4**	**5**
Max.	0.0511016	0.0512862	0.0512862	0.0512862	0.0512862
Order	29	1	1	1	1

missing an approximate worst parameter set through this random sampling with replacement is $(1 - 200/6561)^{200} = 0.0020468$, which is very small.

For TS, the number of neighborhood solutions N^{b} is 10, and the number of steps N^{s} is 20; hence, the number of analyses is also 200. We carry out RS and TS five times starting with different initial random seeds for comparison purpose. The maximum values of d obtained by RS and TS for the standard design $\boldsymbol{J} = (3, 3, 3, 3)$ are listed in Tables 3.5(a) and (b), respectively, where the second row is the order of the worst value (minimum order in the list of 6561 parameter sets). Note that the minimum value and its order are also shown for RS.

Although the global worst solution could not be found, RS can find a good approximate worst solution within the analyses of about 3% of the total solutions. For example, the largest minimum order among the five cases is 105 which is 1.5% of the total number of parameter sets. It is also important to note that the range of function values can also be found using RS. For TS, the worst parameter set has been found for four cases among the five cases; however, the 29th worst set has been found for Case 1. This way, TS mostly has a good performance through analyses of about 3% of the size of the original list, but sometimes fails to obtain a good approximate solution, although the 29th worst solution can be regarded as an approximate worst solution in our definition.

Worst-case design problem

In the upper-level optimization problem, we select member sections from the pre-assigned list of standard sections in Table 3.4. A constraint is given such

Table 3.6
Approximate Optimal Solutions by RS and TS

(a) RS

	Variables J				Parameters I				Objective value	Interstory drift	Order	Worst value
Case 1	3	4	1	1	2	7	1	4	0.52000	0.053095	82	0.053175
Case 2	3	4	1	1	2	7	1	4	0.52000	0.053095	82	0.053175
Case 3	3	5	1	1	1	9	1	6	0.53451	0.053176	47	0.053186
Case 4	3	4	1	1	2	7	1	4	0.52000	0.053095	82	0.053175
Case 5	3	3	1	2	2	8	1	4	0.53984	0.052404	82	0.052691

(b) TS

	Variables J				Parameters I				Objective value	Interstory drift	Order	Worst value
Case 1	3	3	1	1	9	6	1	1	0.49549	0.052678	1	0.052678
Case 2	3	3	1	1	9	6	1	1	0.49549	0.052678	1	0.052678
Case 3	2	3	1	1	9	7	9	1	0.49169	0.052823	2511	0.054365
Case 4	3	3	1	1	9	6	1	1	0.49549	0.052678	1	0.052678
Case 5	2	3	1	1	9	7	9	1	0.49169	0.052823	2511	0.054365

that the worst value of $d(\boldsymbol{J};\boldsymbol{I})$ is not more than $\bar{d} = 0.054$ m, which is equal to 1.5% of the story height. The structural optimization problem for minimizing the total structural volume $V(\boldsymbol{J})$ considering parameter uncertainty is formulated as

$$\text{Minimize} \quad V(\boldsymbol{J}) \tag{3.14a}$$

$$\text{subject to} \quad d^{\mathrm{U}}(\boldsymbol{J};\boldsymbol{I}) \leq \bar{d}, \tag{3.14b}$$

$$J_i \in \{1, \cdots, 5\} \quad (i = 1, \cdots, 4), \tag{3.14c}$$

$$I_i \in \{1, \cdots, 9\} \quad (i = 1, \cdots, 4). \tag{3.14d}$$

Note that uncertainty in the flange thickness is not considered when calculating the structural volume.

The performances of RS and TS are compared. For the upper-level problem, the number of trials for RS is 100, and accordingly, $N^{\mathrm{b}} = 10$, $N^{\mathrm{s}} = 10$ for TS. For the lower-level problem, the number of trials for RS is 200, and $N^{\mathrm{b}} = 10$, $N^{\mathrm{s}} = 20$ for TS. In the TS for the upper-level problem, the solutions violating the constraint is simply rejected. The results by RS and TS are listed in Tables 3.6(a) and (b), respectively. Among five trials of RS and TS, the best solution is $\boldsymbol{J} = (3, 3, 1, 1)$ with the worst parameter set $\boldsymbol{I} = (9, 6, 1, 1)$, which is regarded as an approximate optimal solution, and its details are shown in Table 3.7. For TS, the optimal solution has been found three times, whereas it could not be found by RS. However, it is seen from

Table 3.7
Approximate Optimal Solution $J = (3,3,1,1)$ with the Worst Parameter Set $I = (9,6,1,1)$

Beam 1 (2F, 3F)	H-400×200×9×19
Beam 2 (4F, RF)	H-350×175×7×11
Column 1 (1S, 2S)	HSS-300×300×9
Column 2 (3S, 4S)	HSS-300×300×9
Yield stress:	
Beam 1	274.95 MPa
Beam 2	260.85 MPa
Thickness of flange:	
Beam 1	18.24 mm
Beam 2	10.56 mm
Objective function	0.49549 m^3
Max. interstory drift	0.052678 m

Table 3.6(b) that TS may sometimes lead to very bad solution as Cases 3 and 5, which have smaller objective values than the optimal solution, but they actually violate the constraint. By contrast, all solutions obtained by RS actually satisfy the constraint.

3.3 STOPPING RULE OF RANDOM MULTI-START LOCAL SEARCH

3.3.1 LOCAL SEARCH

In this section, some stopping rules are presented for a process of randomly generating the initial solutions of random multi-start LS for improving a single solution to find a local optimal solution [Ohsaki and Yamakawa, 2018]. Combinatorial problem is considered for clear identification of the local optima. The vector of m integer variables is denoted by $\boldsymbol{x} = (x_1, \cdots, x_m)$. The stopping rules presented below can be used for any deterministic optimization method based on LS; e.g., grid search, tabu search and greedy method for local improvement. However, we consider the basic LS, which is described for a problem of minimizing the objective function $f(\boldsymbol{x})$ as shown in Algorithm 3.2.

There are several definitions of neighborhood solutions. Suppose x_j can take an integer value in the set $\{1, \cdots, q_j\}$, and, for simplicity, suppose the relation $1 < x_j^k < q_j$ is satisfied for the current value x_j^k of x_j at the kth step, i.e., x_j^k does not take its upper or lower bound value. Then the following definitions may be used for the neighborhood solutions $\boldsymbol{y}^i = (y_1^i, \cdots, y_m^i)$ $(i = 1, \cdots, N)$ of $\boldsymbol{x}^k$:

Algorithm 3.2 Algorithm of local search (LS).

1: Sample an initial random point $\boldsymbol{x}^0$ from a uniform probability distribution in the feasible domain. Set $k = 0$.
2: Enumerate all N neighborhood solutions of $\boldsymbol{x}^k$, denoted by $\boldsymbol{y}^i = (y^i_1, \cdots, y^i_m)$ $(i = 1, \cdots, N)$, and compute $f(\boldsymbol{y}^i)$.
3: Select the best solution $\boldsymbol{y}^{\min}$, which has the smallest value of $f(\boldsymbol{y}^i)$ among all neighborhood solutions.
4: If $f(\boldsymbol{x}^k) > f(\boldsymbol{y}^{\min})$, let $\boldsymbol{x}^{k+1} \leftarrow \boldsymbol{y}^{\min}$, $k \leftarrow k + 1$, and go to Step 2; otherwise, output $\boldsymbol{x}^k$ as a local optimal solution and terminate the process.

Moore neighborhood:

$$\begin{aligned} &y^i_j \in \{x^k_j - 1, x^k_j, x^k_j + 1\} \quad (j = 1, \cdots, m), \ \boldsymbol{y}^i \neq \boldsymbol{x}^k \\ &N = 3^m - 1. \end{aligned} \tag{3.15}$$

Neumann neighborhood:

$$\begin{aligned} &\begin{cases} y^i_j \in \{x^k_j - 1, x^k_j + 1\} & \text{for } j = \hat{j} \\ y^i_j = x^k_j & \text{for } j \neq \hat{j} \end{cases} \qquad (\hat{j} = 1, \cdots, m), \\ &N = 2m. \end{aligned} \tag{3.16}$$

Neighborhood for grid search:

$$\begin{aligned} &\begin{cases} y^i_j \in \{1, \cdots, x^k_j - 1, x^k_j + 1, \cdots, q_j\} & \text{for } j = j^* \\ y^i_j = x^k_j & \text{for } j \neq j^* \end{cases} \qquad (j^* \in \{1, \cdots, m\}), \\ &N = q_{j^*} - 1. \end{aligned} \tag{3.17}$$

The neighborhood solutions for the case $m = 2$, $q_1 = 5$ and $j^* = 1$ are illustrated in Fig. 3.6. Note that the grid search finds the best solution among the neighborhood solutions in a fixed direction.

3.3.2 STOPPING RULE USING ATTRACTORS

Stopping rules without using intermediate solutions

In the process of random multi-start LS, a local optimal solution is found for each specified initial solution, and generally there exist multiple solutions or a region of solutions that leads to the same local optimal solution. The *attractor* or *region of attraction* is defined as a set of initial solutions, denoted by X_i, leading to a particular local optimal solution $\boldsymbol{x}^*_i$ by carrying out LS [Zieliński, 1981; Lagaris and Tsoulos, 2008; Ohsaki and Yamakawa, 2018]. Suppose we obtain w local optimal solutions $\boldsymbol{x}^*_1, \cdots, \boldsymbol{x}^*_w$ by carrying out LS t

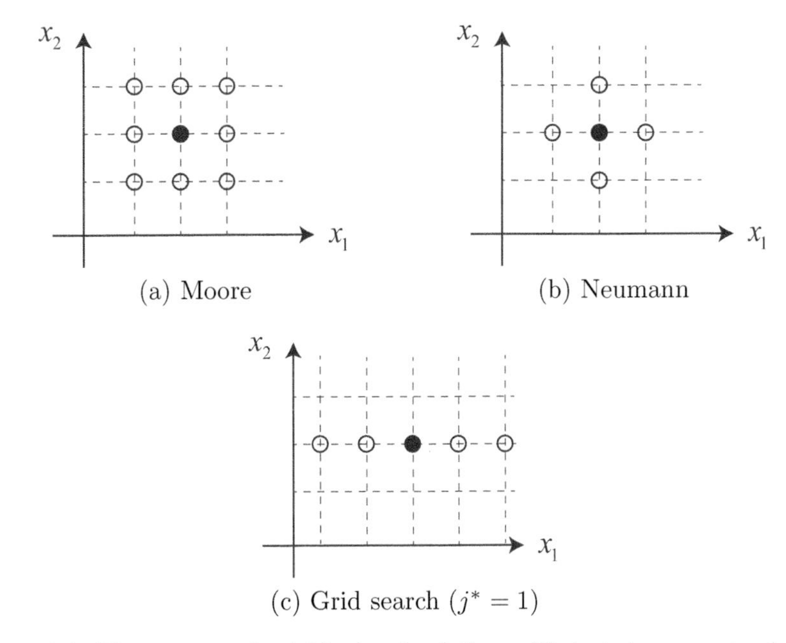

Figure 3.6 Three types of neighborhood solutions; filled circle: current solution, blank circle: neighborhood solution.

times from randomly generated initial solutions. The number of LSs that find $\boldsymbol{x}_i^*$ is denoted by n_i, i.e., $n_1 + \cdots + n_w = t$.

Although the size of attractor is difficult to obtain for a problem with continuous variables, the size of attractor is represented by the number of solutions in the attractor for problems with integer variables. If a sufficient amount of attractors are obtained, then the multi-start process can be terminated judging that the required number of local optimal solutions have been obtained and an approximate optimal solution with good accuracy has been found. Lagaris and Tsoulos [2008] used a variance of the number of expected optimal solutions as well as the expected ratio of the region covered by the attractors to define a stopping rule. Zabinsky et al. [2010] proposed a criterion based on trade-off between the computational cost and probability of obtaining the global optimal solution.

The number of attractors is usually counted by the number of initial solutions. If we use a deterministic LS, the same solution $\boldsymbol{x}_i^*$ is found starting from any intermediate solution between the initial solution and $\boldsymbol{x}_i^*$. Therefore, all intermediate solutions along the path to $\boldsymbol{x}_i^*$ can be included in X_i. The total number of solutions visited during the t trials of LS, including initial, optimal and intermediate solutions along the path, is denoted by $\hat{t}$, which is counted without duplication. Figure 3.7 illustrates the process of LSs for a problem with two variables $\boldsymbol{x} = (x_1, x_2)$. Three solutions $\boldsymbol{x}_1^*$, $\boldsymbol{x}_2^*$ and

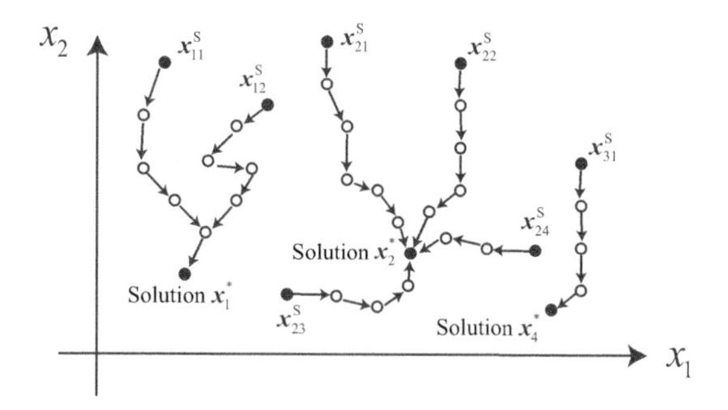

Figure 3.7 Process of local search and definition of attractors ($w = 3$, $n_1 = 2$, $n_2 = 4$, $n_3 = 1$, $t = 7$, $\hat{t} = 35$).

$\boldsymbol{x}_3^*$ are found, i.e., $w = 3$ from two, four and one initial solutions $\{\boldsymbol{x}_{11}^{\mathrm{S}}, \boldsymbol{x}_{12}^{\mathrm{S}}\}$, $\{\boldsymbol{x}_{21}^{\mathrm{S}}, \boldsymbol{x}_{22}^{\mathrm{S}}, \boldsymbol{x}_{23}^{\mathrm{S}}, \boldsymbol{x}_{24}^{\mathrm{S}}\}$ and $\{\boldsymbol{x}_{31}^{\mathrm{S}}\}$, respectively; therefore, $n_1 = 2$, $n_2 = 4$, $n_3 = 1$ and $t = n_1 + n_2 + n_3 = 7$. The two, four and one paths leading to $\boldsymbol{x}_1^*$, $\boldsymbol{x}_2^*$ and $\boldsymbol{x}_3^*$ have 11, 19 and 5 solutions, respectively; therefore, $\hat{t} = 11 + 19 + 5 = 35$.

Let s_i denote the size (number of solutions) of attractor X_i. The magnitude of X_i relative to other attractors can be measured by the *share* c_i of the attractor, which is the ratio of s_i to the number of all feasible solutions satisfying the constraints [Zieliński, 2009]. Several Bayesian approaches have been proposed assuming that no prior information is available for the number of local optimal solutions or the sizes of attractors. Zieliński [1981] proposed a method for minimizing the Bayesian risk of the estimated number of remaining attractors utilizing the information of the known attractors. Boender and Rinnooy Kan [1987] used a Bayesian approach to estimation of the total number of local optimal solutions and derivation of a stopping rule.

Let h denote the number of local optimal solutions, which is equal to the number of attractors. Suppose the share of attractor is a uniform random variable between 0 and 1 satisfying $c_1 + \cdots + c_h = 1$, and the intermediate solutions are not included in the attractors. When w local optimal solutions have been found during t trials, i.e., $n_1 + \cdots + n_w = t$, the number of local optimal solutions is estimated as the mean value of posterior estimate of h, denoted by $\tilde{h}$, as follows:

$$\tilde{h} = \frac{w(t-1)}{t-w-2}. \tag{3.18}$$

Using $\tilde{h}$, the following rule is proposed:

Rule 3.1 *Terminate random multi-start LS if w is sufficiently close to $\tilde{h}$ [Boender and Rinnooy Kan, 1987].*

However, Boender and Rinnooy Kan [1987] noted that the convergence of $\tilde{h}$ is very slow. The ratio $\tilde{\Omega}$ of sum of sizes of attractors $X_1, \cdots, X_w$ to the total feasible region is also estimated as

$$\tilde{\Omega} = \frac{(t-w-1)(t+w)}{t(t-1)}. \tag{3.19}$$

Using $\tilde{\Omega}$, another stopping rule is proposed as follows:

Rule 3.2 *Terminate random multi-start LS if $\tilde{\Omega} \geq e_1$ is satisfied, where e_1 is a prescribed value slightly less than one [Boender and Rinnooy Kan, 1987].*

Stopping rule using intermediate solutions

Note again that the share c_i of attractor i is assumed to be uniformly distributed between 0 and 1 in the Bayesian approaches for Rules 3.1 and 3.2, which means that property of the specific problem to be solved is not fully incorporated in those stopping rules. Therefore, c_i may be alternatively estimated based on the history of finding the local optimal solutions in random multi-start LS. If only the initial solutions leading to $\boldsymbol{x}_i^*$ is considered for s_i, then $s_i = n_i$, $s_1 + \cdots + s_w = t$, and c_i is estimated as $c_i = s_i/t$. By contrast, if s_i includes all of the initial, intermediate and local optimal solutions, then $s_1 + \cdots + s_w = \hat{t}$ and c_i is estimated by the ratio $c_i = s_i/\hat{t}$. In this case, in the example of Fig. 3.7, $s_1 = 11$, $s_2 = 19$ and $s_3 = 5$; therefore, $c_1 = 11/35$, $c_2 = 19/35$ and $c_3 = 5/35$.

In the following, we summarize the rules by Ohsaki and Yamakawa [2018] including intermediate solutions in the attractors; i.e., the total number of solutions in attractors is $\hat{t}$, while t is used as the total number of trials. After visiting $\hat{t}$ solutions, suppose another solution x_{w+1}^* that has not been found is expected to exist. Assuming $s_{w+1} = k$, the value of c_i corresponding to $s_{w+1} = k$, denoted by $c_i^{(k)}$, is computed as

$$\begin{aligned} c_i^{(k)} &= \frac{s_i}{\hat{t}+k} \quad (i = 1, \cdots, w), \\ c_{w+1}^{(k)} &= \frac{k}{\hat{t}+k}. \end{aligned} \tag{3.20}$$

Because $s_1 + \cdots + s_w = \hat{t}$ holds, $c_1^{(k)} + \cdots + c_{w+1}^{(k)} = 1$ is satisfied from (3.20).

Based on the Bayesian approach, we can formulate the likelihood of observing this event. Since the number of trials n_i leading to the solution $\boldsymbol{x}_i^*$ is a random variable, it is written as N_i. Then the probability of finding solutions $\boldsymbol{x}_1^*, \cdots, \boldsymbol{x}_{w+1}^*$, respectively, $N_1, \cdots, N_{w+1}$ times in t trials is obtained as

$$P_w^{(k)}(N_1, \cdots, N_{w+1} | c_1^{(k)}, \cdots, c_{w+1}^{(k)}, w, t) = \frac{t!}{\prod_{i=1}^{w+1} N_i!} \prod_{i=1}^{w+1} (c_i^{(k)})^{N_i}. \tag{3.21}$$

After n_i times finding $\boldsymbol{x}_i^*$ $(i = 1, \cdots, w)$, respectively, and 0 time $\boldsymbol{x}_{w+1}^*$ in t trials; i.e., $n_1 + \cdots + n_w = t$ and $n_{w+1} = 0$, the likelihood $L_w^{(k)}$ for observing this event is computed as

$$\begin{aligned} L_w^{(k)} &= P_w^{(k)}(N_1 = n_1, \cdots, N_w = n_w, N_{w+1} = 0 \,|\, c_1^{(k)}, \cdots, c_{w+1}^{(k)}, w, t) \\ &= \frac{t!}{\prod_{i=1}^{w+1} n_i!} \prod_{i=1}^{w+1} (c_i^{(k)})^{n_i} \\ &= \frac{t!}{\prod_{i=1}^{w} n_i!} \prod_{i=1}^{w} (c_i^{(k)})^{n_i}. \end{aligned} \tag{3.22}$$

The size s_{w+1} of X_{w+1} corresponding to the solution $\boldsymbol{x}_{w+1}^*$ which has not been found may be estimated using the sizes of attractors that have already been found. However, the smallest value s^{L} among $s_1, \cdots, s_w$ is too small as the estimate of s_{w+1}, while the largest value s^{U} is too large. Therefore, an intermediate value is to be expected. Suppose the sizes of attractors of unknown solutions are uniformly distributed between s^{L} and s^{U}. Then, the likelihood of missing another solution, denoted by $\bar{L}_w$, is obtained as the average value of $L_w^{(k)}$ for $k = s^{\mathrm{L}}, \cdots, s^{\mathrm{U}}$ as

$$\bar{L}_w = \frac{1}{s^{\mathrm{U}} - s^{\mathrm{L}} + 1} \sum_{k=s^{\mathrm{L}}}^{s^{\mathrm{U}}} L_w^{(k)}. \tag{3.23}$$

The value of $L_w^{(k)}$ when no solution is left, i.e., $w = h$, is denoted by $L_w^{(0)}$. We can stop the random multi-start LS process if the likelihood of missing a solution $\bar{L}_w$ is sufficiently smaller than $L_w^{(0)}$, and accordingly, a stopping rule can be defined as follows:

Rule 3.3 *Terminate random multi-start LS if $\bar{L}_w / L_w^{(0)}$ is smaller than a specified small positive value e_2 [Ohsaki and Yamakawa, 2018].*

This way, the history of trials of LS is fully incorporated to $\bar{L}_w$ and $L_w^{(0)}$ that are updated at each trial using (3.22) based on the updated ratios of attractors in (3.20), from which we obtain $c_i^{(0)} = c_i$ $(i = 1, \cdots, w)$. Therefore, using (3.20), (3.22) and $s_i = c_i \hat{t}$, $L_w^{(k)}/L_w^{(0)}$ is reformulated as

$$\begin{aligned} \frac{L_w^{(k)}}{L_w^{(0)}} &= \prod_{i=1}^{w} \left(\frac{c_i^{(k)}}{c_i^{(0)}} \right)^{n_i} \\ &= \left(\frac{\hat{t}}{\hat{t} + k} \right)^{t} \\ &= (1 - c_{w+1}^{(k)})^t, \end{aligned} \tag{3.24}$$

which is equal to the probability of missing the $(w+1)$th solution within t trials.

Using (3.23) and (3.24), we obtain

$$\frac{\bar{L}_w}{L_w^{(0)}} = \frac{1}{s^{\mathrm{U}} - s^{\mathrm{L}} + 1} \sum_{k=s^{\mathrm{L}}}^{s^{\mathrm{U}}} (1 - c_{w+1}^{(k)})^t. \tag{3.25}$$

In the example in Fig. 3.7 that has three local optima, $s^{\mathrm{L}} = 5$, $s^{\mathrm{U}} = 19$, and $\bar{L}_w / L_w^{(0)}$ is computed using (3.24) as

$$\frac{\bar{L}_w}{L_w^{(0)}} = \frac{1}{19 - 5 + 1} \sum_{k=5}^{19} \left(1 - \frac{k}{35 + k}\right)^7 = 0.1607. \tag{3.26}$$

It should be noted that $\bar{L}_w$ strongly depends on the size of extremely small and/or large attractor. For example, suppose we have found five solutions and $(n_1, \cdots, n_5) = (1, 5, 6, 7, 12)$. In this case, $s^{\mathrm{L}} = 1$ and $s^{\mathrm{U}} = 12$, and the likelihood for $n_i = 1, \cdots, 12$ including $n_i = 2, 3, 4, 8, 9, 10, 11$ should be summed for computing $\bar{L}_w$ in (3.23). Therefore, probability of missing another solution is overestimated if one solution has a very large/small attractor and $s^{\mathrm{U}} - s^{\mathrm{L}}$ has a large value. Hence, we next assume w possible choices of s_{w+1} from the already obtained values as $s_{w+1} = s_j$ $(j = 1, \cdots, w)$, and define c_i^{j*} as

$$c_i^{j*} = \frac{s_i}{\hat{t} + s_j} \quad (i = 1, \cdots, w). \tag{3.27}$$

Using (3.27), L_w^{j*} is computed as

$$L_w^{j*} = \frac{t!}{\prod_{i=1}^{w} n_i!} \prod_{i=1}^{w} (c_i^{j*})^{n_i} \quad (j = 1, \cdots, w). \tag{3.28}$$

and $\bar{L}_w^*$ is obtained as the mean value of L_w^{j*} as

$$\bar{L}_w^* = \frac{1}{w} \sum_{j=1}^{w} L_w^{j*}. \tag{3.29}$$

In a similar manner as (3.24) and (3.25), the likelihood ratio is obtained as

$$\frac{\bar{L}_w^*}{L_w^{(0)}} = \frac{1}{w} \sum_{j=1}^{w} (1 - c_{w+1}^{(s_j)})^t. \tag{3.30}$$

Using $\bar{L}_w^*$, the following rule may be used:

Rule 3.4 *Terminate random multi-start LS if $\bar{L}_w^* / L_w^{(0)}$ is smaller than a specified small positive value e_3 [Ohsaki and Yamakawa, 2018].*

In the example in Fig. 3.7, $\bar{L}^*_w/L^{(0)}_w$ is computed using (3.30) as

$$\frac{\bar{L}^*_w}{L^{(0)}_w} = \frac{1}{2}\left[\left(1-\frac{11}{35+11}\right)^7 + \left(1-\frac{19}{35+19}\right)^7 + \left(1-\frac{5}{35+5}\right)^7\right] = 0.2942. \tag{3.31}$$

3.3.3 NUMERICAL EXAMPLES

Mathematical problems

Rules 3.2, 3.3 and 3.4 are compared using the mathematical problems in Voglis and Lagaris [2009] and the Test2N function [Lagaris and Tsoulos, 2008] for the cases with moderately large number of solutions that may exist in structural optimization problems. The test functions are listed in Appendix A.1. We use the Moore neighborhood in the following examples. The variable x_i is discretized to 101 equally spaced values using the integer variables $J_i \in \{1, \cdots, 101\}$ as

$$x_i = x^{\mathrm{L}} + (J_i - 1)\frac{x^{\mathrm{U}} - x^{\mathrm{L}}}{100}, \tag{3.32}$$

where x^{U} and x^{L} are the upper and lower bounds of x_i specified for each problem in Appendix A.1. Since the number of local optima is known for each problem as shown in the first column N^{opt} in Table 3.8, ten sets of trials of random multi-start LS are carried out until all local optima are found. Let S denote the number of trials of LS needed to find all local optima. The results for six test functions are shown in Table 3.8. Although Voglis and Lagaris [2009] and Lagaris and Tsoulos [2008] use continuous variables, we confirmed that the same numbers of solutions are obtained after discretization by (3.32). It is seen from Table 3.8 that the value of $\tilde{\Omega}$ in Rule 3.2, when all local optima have been found, is close to one for the functions Guillin Hill's, Piccioni's and M0, but not for Helder and Test2N; $\bar{L}_w/L^{(0)}_w$ in Rule 3.3 has small values for Guillin Hill's and M0, but not for other four problems; $\bar{L}^*_w/L^{(0)}_w$ in Rule 3.4

Table 3.8
Values When All Solutions are Found

	N^{opt}	S	$\tilde{\Omega}$	$\bar{L}_w/L^{(0)}_w$	$\bar{L}^*_w/L^{(0)}_w$
Ackley's	121	969.7	0.9806	0.02391	0.02003
Guillin Hill's	25	3589.4	0.9999	0.00044	0.06901
Helder	85	443.1	0.9595	0.05069	0.03346
Piccioni's	28	453.2	0.9942	0.01245	0.07872
M0	66	2923.1	0.9987	0.00533	0.01635
Test2N	32	125.1	0.9110	0.06020	0.09360

Table 3.9
Distribution of Ratios of Attractors

	Max.	Min.	Mean	Std. dev.
Ackley's	0.0201	0.00275	0.00826	0.0030
Guillin Hill's	0.5468	0.00012	0.04000	0.1084
Helder	0.0147	0.00844	0.01176	0.0014
Piccioni's	0.1043	0.00431	0.03571	0.0344
M0	0.0484	0.00028	0.01562	0.0135
Test2N	0.0406	0.02439	0.03125	0.0038

has the same order for all the six problems, which suggests that Rule 3.4 may be effective for all types of problem.

Distribution of the ratios c_i of attractors is listed in Table 3.9, which shows that the minimum values are very small for Guillin Hill's and M0, and the standard deviations are large for Guillin Hill's, Piccioni's and M0. Table 3.10 compares the results for the parameter values $e_1 = 0.95$, 0.98 and 0.995 for Rule 3.2, $e_2 = 0.02$ and 0.05 for Rule 3.3, and $e_3 = 0.02$ and 0.05 for Rule 3.4. In the table, N_{95}^{opt} is the average value of N^{opt} when $\tilde{\Omega} > 0.95$ (Rule 3.2) is satisfied, and $\bar{N}_{02}^{\mathrm{opt}}$ and $\bar{N}_{02}^{\mathrm{opt}*}$ are the average values of N^{opt} when $\bar{L}_w/L_w^{(0)} < 0.02$ (Rule 3.3) and $\bar{L}_w^*/L_w^{(0)} < 0.02$ (Rule 3.4), respectively,

Table 3.10
Values When Stopping Criteria are Satisfied

	$N_{95}^{\mathrm{opt}}(S)$	$N_{98}^{\mathrm{opt}}(S)$	$N_{995}^{\mathrm{opt}}(S)$	
Ackley's	116.4 (523.9)	119.9 (852.3)	121.0 (1719.0)	
Guillin Hill's	10.7 (50.8)	13.3 (98.5)	18.7 (272.6)	
Helder	84.0 (378.7)	84.8 (604.4)	85.0 (1210.0)	
Piccioni's	21.5 (99.5)	25.0 (181.0)	27.2 (392.8)	
M0	54.1 (245.1)	58.5 (418.5)	61.7 (880.8)	
Test2N	31.7 (144.7)	32.0 (231.0)	32.0 (461.0)	
	$\bar{N}_{02}^{\mathrm{opt}}(S)$	$\bar{N}_{05}^{\mathrm{opt}}(S)$	$\bar{N}_{02}^{\mathrm{opt}*}(S)$	$\bar{N}_{05}^{\mathrm{opt}*}(S)$
Ackley's	119.9 (930.7)	116.4 (508.0)	119.5 (836.8)	117.8 (568.2)
Guillin Hill's	11.6 (64.6)	7.5 (27.2)	24.7 (4926.6)	23.9 (2815.5)
Helder	84.9 (558.3)	84.0 (442.7)	84.7 (484.8)	83.1 (353.7)
Piccioni's	27.1 (287.5)	23.2 (124.5)	27.6 (671.7)	27.8 (518.2)
M0	59.3 (603.9)	53.9 (250.2)	63.4 (1817.3)	59.5 (553.5)
Test2N	32.0 (186.3)	31.4 (125.9)	31.9 (192.6)	31.8 (145.9)

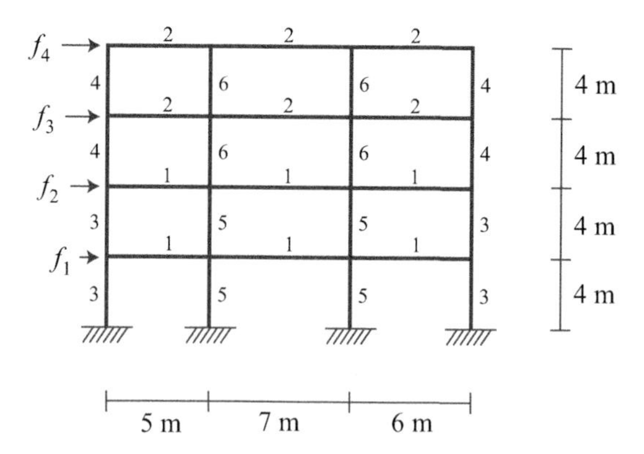

Figure 3.8 A 3-span 4-story plane frame.

are satisfied. Other parameters are defined similarly. The value in parentheses is the average number of trials to satisfy the stopping rule.

It is seen from the results that a large portion of local optima is missed for Guillin Hill's and M0, if Rule 3.2 or 3.3 is used. Even when e_1 is increased to 0.995, which is close to 1, for Rule 3.2, the value of N_{995}^{opt} is still less than $\bar{N}_{02}^{\text{opt}*}$ for Guillin Hill's and M0, while large number of unnecessary trials should be carried out for Ackley's, Helder and Test2N. By contrast, $\bar{N}_{02}^{\text{opt}*}$ is close enough to N^{opt} for all the six problems; i.e., Rule 3.4 gives better estimate of the number of local optimal solutions than Rules 3.2 and 3.3. It is also seen from Table 3.10 that $\bar{N}_{05}^{\text{opt}*}$ is slightly smaller than $\bar{N}_{02}^{\text{opt}*}$ for all problems except Piccioni's, for which $\bar{N}_{05}^{\text{opt}*} > \bar{N}_{02}^{\text{opt}*}$ is satisfied. This is because all results depend on the initial random seed, and almost all solutions are found except for Piccioni's before satisfying $\bar{L}_w^*/L_w^{(0)} < 0.05$.

Optimization of plane steel frame

Random multi-start LS is applied to optimization of a 3-span 4-story plane steel frame as shown in Fig. 3.8, where the numbers beside the beams and columns are group numbers; i.e., there are six groups ($n^{\text{g}} = 6$). The horizontal loads are given as $f_1, \cdots, f_4 = 180, 210, 240, 270$ (kN). Concentrated downward vertical load of 70 kN representing the self-weight is applied at the beam-column joint and the center of each beam.

Each member has sandwich section with the height H. The cross-sectional areas $A^{(i)}$ (m^2) of the members in the ith group are selected from the list of n^{s} different sections as

$$A_j^{(i)} = A_0^{(i)} + 0.002J_i \quad (i = 1, \cdots, n^{\text{g}}; J_i \in \{1, \cdots, n^{\text{s}}\}). \tag{3.33}$$

The values of $A_0^{(i)}$ $(i = 1, \cdots, n^{\text{g}})$ are given as 0.019, 0.007, 0.017, 0.007, 0.021, 0.007 (m^2), and $n^{\text{s}} = 10$. The second moment of area $I_j^{(i)}$ (m^4) and the

Table 3.11
Values When Stopping Criteria are Satisfied

	N^{opt}	S	$N_{95}^{\mathrm{opt}}(S)$	$N_{98}^{\mathrm{opt}}(S)$	$N_{995}^{\mathrm{opt}}(S)$
$\theta = 0.010$	12	1847.2	8.6 (41.4)	11.2 (70.8)	9.8 (146.2)
$\theta = 0.011$	15	2158.6	7.4 (36.0)	8.6 (65.2)	10.8 (160.4)
$\theta = 0.012$	10	696.2	6.0 (29.8)	7.0 (54.0)	7.4 (112.6)
	$\bar{N}_{02}^{\mathrm{opt}}(S)$	$\bar{N}_{05}^{\mathrm{opt}}(S)$	$\bar{N}_{02}^{\mathrm{opt}*}(S)$	$\bar{N}_{05}^{\mathrm{opt}*}(S)$	
$\theta = 0.010$	9.6 (72.2)	8.6 (42.2)	10.6 (1101.4)	10.2 (268.4)	
$\theta = 0.011$	7.6 (47.6)	5.2 (18.8)	13.8 (2409.0)	10.0 (242.2)	
$\theta = 0.012$	6.6 (42.4)	5.4 (19.2)	8.4 (376.2)	7.0 (89.8)	

section modulus $Z_j^{(i)}$ (m^3) of the jth section of the members in the ith group are functions of $A_j^{(i)}$ as $I_j^{(i)} = (H/2)^2 A_j^{(i)}$ and $Z_j^{(i)} = (H/2)A_j^{(i)}$, respectively, where $H = 0.25$ (m) in the following example.

The design variables are the cross-sectional areas A_i $(i = 1, \cdots, n^{\mathrm{g}})$. Let $L^{(i)}$ denote the total length of members in the ith group. The objective function is the total structural volume defined as

$$f(\boldsymbol{x}) = \sum_{i=1}^{n^{\mathrm{g}}} A^{(i)} L^{(i)}, \tag{3.34}$$

which is minimized under constraints on the interstory drift angle and the stress. The upper-bound stress is 235 MPa, and the upper bound θ of interstory drift angle is varied parametrically as 0.010, 0.011 and 0.012. Moore neighborhood is used; i.e., there are at most $3^6 - 1 = 728$ neighborhood solutions at each step of LS.

Five sets of random multi-start LS are carried out from different initial solutions, where each set consists of 5000 trials (restarts) of LS to find most of the local optimal solutions. Although the exact number of local optimal solutions is not known, N^{opt} has the same value for five sets, respectively, for three cases of θ, and the average number of trials S for obtaining the last local optimal solution for three cases are sufficiently smaller than the total number of trials 5000; therefore, we assume all solutions have been found.

Table 3.11 shows the mean values among five sets when Rules 3.2 ($e_1 =$ 0.95, 0.98, 0.995), Rule 3.3 ($e_2 = 0.02$, 0.05) and Rule 3.4 ($e_3 = 0.02$, 0.05) are satisfied. Note that Rule 3.2 with $e_1 = 0.95$ is not appropriate, because N_{95}^{opt} in Table 3.11 is too small for all cases. Even for $e_1 = 0.995$ that is close to 1, the ratio of missing optimal solutions is not small enough especially for $\theta = 0.011$. By contrast, Rule 3.4 with $e_2 = 0.02$ may be appropriate as the termination rule for this frame, because $\bar{N}_{02}^{\mathrm{opt}*}$ is only slightly smaller than N^{opt}.

Table 3.12
Three Best Solutions for $\theta = 0.010, 0.011$ and 0.012

	J_1	J_2	J_3	J_4	J_5	J_6	$F(x)$	n_i	s_i
$\theta = 0.010$	7	3	4	2	5	6	2.864	1410	3558
	8	3	2	3	5	4	2.872	492	1040
	8	3	3	3	4	4	2.872	504	1109
$\theta = 0.011$	6	3	3	2	3	4	2.632	2140	4546
	4	3	3	2	7	5	2.648	699	1508
	5	3	3	3	5	3	2.656	142	290
$\theta = 0.012$	3	3	4	2	4	4	2.480	2282	4949
	3	3	3	2	5	4	2.480	193	673
	4	3	5	2	2	3	2.488	1521	3133

Table 3.12 shows the three best solutions for each case. As seen from the table, there are several solutions that has the same or almost the same objective values. Furthermore, the best solution has large number s_i of attractor, and the large number n_i of the trials to reach the solution. Table 3.13 shows 12 solutions obtained from a set of 5000 trials for $\theta = 0.010$. As seen from the table, the best solution is found at the 8th trial, and it is the most frequently obtained solution with largest size of attractor. Therefore, for this frame optimization problem, effect of the existence of local optimal solution with small attractor may be negligible.

Table 3.13
All 12 Solutions Obtained by a Set of 5000 Trials for $\theta = 0.010$

Rank	J_1	J_2	J_3	J_4	J_5	J_6	$f(x)$	n_i	s_i	S
1	7	3	4	2	5	6	2.864	1410	3558	8
2	8	3	2	3	5	4	2.872	492	2040	1
2	8	3	3	3	4	4	2.872	504	1109	3
2	8	3	2	2	5	6	2.872	7	10	1351
5	6	2	4	2	7	5	2.888	1158	2949	2
5	6	3	3	2	8	7	2.888	14	23	19
5	6	3	3	3	8	5	2.888	86	161	76
5	6	3	4	2	7	7	2.888	30	65	160
9	7	3	7	2	3	6	2.896	364	839	5
9	7	3	6	2	4	6	2.896	582	1476	10
11	8	3	5	3	3	4	2.904	268	565	4
12	8	3	7	3	2	4	2.936	85	185	59

3.4 SUMMARY

Some results have been presented for WCA and design problems using RS. Although convergence to an optimal solution is slow, both global and local RS methods are promising for complex worst-case structural design problems, because there is no hyper-parameter to be assigned, and accuracy can be estimated using simple theoretical formula. Effectiveness of RS method has been demonstrated for an optimization problem of a plane frame subjected to seismic excitations. Note that finding exact worst values, which demands substantial computational cost, is not generally necessary in structural analysis and design problems. Stopping rules have also been presented for random start deterministic local search. The number of missing local optimal solutions can be estimated using the total number and sizes of the attractors, each of which consists of a set of solutions leading to the specific local optimal solution. Effectiveness of some stopping rules for problems with moderately large number of local optimal solutions has been demonstrated using mathematical problems and a frame design problem.

ACKNOWLEDGMENT

We acknowledge the permissions from publishers for use of the licensed contents as follows:

- Figure 3.8, and Tables 3.8, 3.9, 3.10, 3.11, 3.12 and 3.13: Reproduced with permission from Springer Nature [Ohsaki and Yamakawa, 2018].

4 Order Statistics-based Robust Design Optimization

In this chapter, robust design optimization problem is regarded as a two-level optimization problem, and is linked to pure random search (PRS) with order statistics in the lower-level problem, i.e., worst-case analysis. A stopping rule of PRS is described on the basis of order statistics theory. Through mathematical examples, it is shown that the order statistics-based approach can find approximately worst value with pre-specified accuracy. Some sampling methods with linear constraints are also introduced. Clarifying the relation between the accuracy of the solution and the stopping rules in the framework of order statistics is the main subject. Numerical examples of design problems with dynamic analysis under uncertainty are presented, where lack of taking uncertainties of the seismic motions into consideration may cause severe damage in structures. Approximate worst-case designs considering uncertainty in earthquake motions can be found using the method based on order statistics.

4.1 INTRODUCTION

Robust design is known as an effective design method under uncertainty, where a design is found to be insensitive to environment and other uncontrollable factors. Among several formulations of robust design, worst-case design is a popular approach, where the objective function and/or the constraints are defined as the worst values of the structural responses under uncertainty. Finding such worst values is often referred to as worst-case analysis (WCA) as discussed in Chap. 3. WCA can be interpreted as an optimization process to maximize the structural responses within the pre-specified set of uncertain parameters. Hence, finding or predicting the exact worst (extreme) value is a kind of global optimization, which is the most difficult field in computer science. In such situation, use of the kth worst value instead of the exact worst value is effective [Ohsaki et al., 2019]. Key concept of the approach is estimation of the worst value by random search (RS) with the kth order statistics described in Sec. 1.4.2, which is hereinafter referred to as order statistics-based RS. This approach is based on the formulation that is also known as distribution-free one-sided tolerance interval. We can predict and control accuracy of the approximation by the order statistics-based RS.

In Chap. 3, we formulated the robust optimization problem based on WCA as a two-level optimization problem of finding the worst responses with respect

DOI: 10.1201/9781003153160-4

to the uncertain parameters and the problem of minimizing the objective function with respect to the design variables. To overcome the difficulties for solving the two-level problem within the reasonable computational cost, a pure random search (PRS) approach is introduced. PRS is basically the simplest method that can be carried out without tuning any parameters. Furthermore, genetic algorithm (GA) and simulated annealing (SA) can also be regarded as special cases of RS with some heuristics [Zhigljavsky and Žilinskas, 2008] as mentioned in Sec. 2.3.1. To say the least, PRS is obviously simpler than GA and SA to find a near-optimal solution.

There are still several important issues to deal with PRS. One of them is an appropriate selection of the stopping rule as discussed in Chap. 3. It has also been pointed out that PRS may be impractical or less efficient due to the difficulty of predicting the exact extreme values from only finite small samples. However, in practical situations, we may not need such exact extreme values. It may be preferred that approximately worst solutions are obtained quickly and accuracy of the solutions are quantified. For this purpose, the worst value may be relaxed to the kth worst value to improve the efficiency of the PRS-based approach. This approach enables us to predict and control the behavior of PRS more accurately. The theoretical basis of such relaxation is clarified and rigorous theoretical background is provided by order statistics [David and Nagaraja, 2003; Krishnamoorthy and Mathew, 2008], which have been successfully applied to reliability-based design [Rosowsky and Bulleit, 2002; Makkonen, 2008; Okasha, 2016]. Traditional probabilistic models lead to reliable evaluation of structural responses if the idealized model exactly characterizes the uncertainty of the parameter; however, the probabilistic model is limited to be mathematically simple and may not be applicable to realistic problems [Augustin and Hable, 2010]. The order statistics-based approach is distribution-free, which is to say that we do not need to consider the exactness of the assumed probabilistic models, which is an advantage in application to engineering problems.

4.2 ORDER STATISTICS-BASED ROBUSTNESS EVALUATION

4.2.1 TWO-LEVEL OPTIMIZATION PROBLEM

We first define a robust design optimization (RDO) problem, which can be written as a two-level problem. In particular, the lower-level optimization problem has continuous uncertain parameters, and the upper-level optimization problem may have real (continuous) or integer (discrete) design variables. Similar formulation can be found in Elishakoff and Ohsaki [2010], where the lower-level problem is called anti-optimization problem. In general, the worst-case design is regarded as one of the robust design methods as stated in Chap. 3 [Beck et al., 2015; Beyer and Sendhoff, 2007; Ben-Tal et al., 2009].

Consider a problem of optimizing some variables of structures, e.g., the cross-sectional dimensions, the material parameters, etc., which may be selected from the pre-assigned list of standard specifications. They are assumed to have real values, and the design variable vector is denoted by $\boldsymbol{x} = (x_1, \cdots, x_m) \in \mathbb{R}^m$, where m is the number of design variables. The vector consisting of uncertain parameters is denoted by $\boldsymbol{\theta} = (\theta_1, \cdots, \theta_r) \in \Omega \subset \mathbb{R}^r$, where r is the number of uncertain parameters and Ω is a pre-specified uncertainty set. We assume that the uncertain parameters are continuous and bounded. The uncertainty is incorporated into objective function $g_0(\boldsymbol{x}; \boldsymbol{\theta})$ and constraint functions $g_i(\boldsymbol{x}; \boldsymbol{\theta})$ $(i = 1, \cdots, l)$, which may represent structural volume and responses, and l is the number of constraints with uncertainty. Other constraints are simply denoted by $\boldsymbol{x} \in \mathcal{X}$, where $\mathcal{X}$ is the feasible set or region of $\boldsymbol{x}$. Thus, an RDO problem we are interested in may be described as

$$\underset{\boldsymbol{x}}{\text{Minimize}} \quad g_0(\boldsymbol{x}; \boldsymbol{\theta}) \quad \text{for all } \boldsymbol{\theta} \in \Omega \tag{4.1a}$$

$$\text{subject to} \quad g_i(\boldsymbol{x}; \boldsymbol{\theta}) \leq \bar{g}_i \quad \text{for all } \boldsymbol{\theta} \in \Omega \quad (i = 1, \cdots, l), \tag{4.1b}$$

$$\boldsymbol{x} \in \mathcal{X}, \tag{4.1c}$$

where $\bar{g}_i$ $(i = 1, \cdots, l)$ are the specified upper bounds of g_i.

The original problem (4.1) is converted to a two-level problem. The lower-level optimization problems are formulated for finding the worst values of the structural volume and the structural response, respectively, as follows:

$$\text{Find} \quad g_i^{\mathrm{U}}(\boldsymbol{x}) = \max_{\boldsymbol{\theta} \in \Omega} g_i(\boldsymbol{x}; \boldsymbol{\theta}) \quad (i = 0, 1, \cdots, l). \tag{4.2}$$

By using (4.2), we can formulate the upper-level optimization problem as

$$\text{Minimize} \quad g_0^{\mathrm{U}}(\boldsymbol{x}) \tag{4.3a}$$

$$\text{subject to} \quad g_i^{\mathrm{U}}(\boldsymbol{x}) \leq \bar{g}_i \quad (i = 1, \cdots, l), \tag{4.3b}$$

$$\boldsymbol{x} \in \mathcal{X}. \tag{4.3c}$$

This framework is essentially the same as the formulation in Chap. 3 with the exception of dealing with continuous design variables. To solve the lower-level optimization problem, the RS approach can be successfully applied to obtain a near-optimal solution, as demonstrated in Chap. 3, which is interpreted as *approximate worst-case approach*. Considering the relation between the accuracy of the solution and the stopping rules in the framework of order statistics is the main subject in this section.

4.2.2 RANDOM SAMPLING AND PROBABILISTIC CONSTRAINTS UTILIZING ORDER STATISTICS

To make the description short, we fix the design variables $\boldsymbol{x}$ and index i, and omit the dependence of the function g_i on $\boldsymbol{x}$ to focus on the lower-level optimization problem, which is to be given as

$$\text{Find} \quad g^{\mathrm{U}} = \max_{\boldsymbol{\theta} \in \Omega} g(\boldsymbol{\theta}). \tag{4.4}$$

We apply PRS to solve problem (4.4) or to determine whether constraint (4.3b) is satisfied in some sense. Hereafter, $\boldsymbol{\theta}$ is treated as a random variable vector in Ω. Following Algorithm 2.1 in Sec. 2.3, a sequence of random points $\boldsymbol{\theta}_1, \cdots, \boldsymbol{\theta}_n$ is generated in the uncertain parameter space so that the point $\boldsymbol{\theta}_j$ has prescribed probability distribution P_j for each j $(1 \leq j \leq n)$. We will refer to this general scheme as general random search (GRS); see, e.g., Zhigljavsky and Žilinskas [2008]. GRS becomes PRS when all the distributions are the same, i.e., $P_j = P$ for all j, and the random variables $\boldsymbol{\theta}_j$ are identically distributed. As a result of application of PRS, we obtain independent samples $\{\boldsymbol{\theta}_1, \cdots, \boldsymbol{\theta}_n\}$ from a distribution P on Ω. Accordingly, we obtain independent samples $\{G_1 = g(\boldsymbol{\theta}_1), \cdots, G_n = g(\boldsymbol{\theta}_n)\}$ of the function values at these points. The random variables G_j $(j = 1, \cdots, n)$ form a set of independent identically distributed random variables (iidrv) with the continuous cumulative distribution function (cdf) of a random variable $G = g(\boldsymbol{\theta})$, which is given by

$$\begin{aligned} F_G(t) &= \Pr\{\boldsymbol{\theta} \in \Omega : G = g(\boldsymbol{\theta}) \leq t\} \\ &= \int_{G \leq t} P(dG) = \int_{-\infty}^{t} f_G(g) dg, \end{aligned} \tag{4.5}$$

where f_G is the probability density function (pdf). The iidrv $G_1, \cdots, G_n$ are arranged in decreasing order, and for $1 \leq k \leq n$ the kth value is denoted by

$$G_{k:n} = G_{k:n}(G_1, \cdots, G_n)$$

such that

$$G_{1:n} \geq G_{2:n} \geq \cdots \geq G_{n:n}. \tag{4.6}$$

In the upper-level of the RDO problem, the constraint on g^{U} of (4.4) is given as

$$g^{\mathrm{U}} \leq \bar{g}, \tag{4.7}$$

and an approach using the following quantile constraint will be considered:

$$G_{k:n} \leq \bar{g}. \tag{4.8}$$

If we set $k = 1$ and $n \to \infty$, (4.8) is the same as the worst-case constraint (4.7) with probability one. In Sec. 4.2.3, we will show that (4.8) has a close relation to

$$\Pr\{F_G(\bar{g}) \geq \gamma\} \geq \alpha, \tag{4.9}$$

where both α and γ are the pre-assigned constants $(0 \leq \alpha \leq 1,\ 0 \leq \gamma \leq 1)$. Implications of (4.9) are clearer than (4.8). It should be noted that (4.9) is a generalization of the chance constraint and the worst-case constraint. When we set $\gamma = 1$, (4.9) corresponds to the chance constraint:

$$\Pr\{F_G(\bar{g}) = 1\} \geq \alpha \Rightarrow \Pr\{g(\boldsymbol{\theta}) \leq \bar{g}\} \geq \alpha. \tag{4.10}$$

When we set $\alpha = \gamma = 1$, (4.9) corresponds to the worst-case constraint with probability one:

$$\Pr\{F_G(\bar{g}) = 1\} = 1 \Rightarrow \Pr\{g(\boldsymbol{\theta}) \leq \bar{g}\} = 1. \tag{4.11}$$

Note that the constraint (4.11) is not exactly the same as the worst-case constraint (4.7); however, both are the same in the sense of probability. In Sec. 4.2.3, we will show that the number of samples of PRS and the kth value of the samples are related to the parameters α and γ. This means that the accuracy of the solution is specified by the parameters α and γ, by which the stopping rule of PRS is presented.

4.2.3 STOPPING RULES OF ORDER STATISTICS-BASED RANDOM SEARCH

We first summarize the basic theory of the kth order statistics. From the theory, distribution-free tolerance intervals are derived, and stopping rules of PRS are presented.

The iidrv $Y_1, \cdots, Y_n$ with the common cdf F_Y are arranged in decreasing order of magnitude as $Y_{1:n} \geq \cdots \geq Y_{n:n}$, where $Y_{k:n}$ is called the kth order statistics as stated in Sec. 1.4.2. The cdf F_Y is assumed to be continuous but unknown. Then we have

$$\Pr\{Y_{k:n} \leq y\} = I_{F_Y(y)}(n-k+1, k), \tag{4.12}$$

where $I_p(a,b)$ is the incomplete beta function:

$$I_p(a,b) = \frac{\int_0^p t^{a-1}(1-t)^{b-1}dt}{\int_0^1 t^{a-1}(1-t)^{b-1}dt}; \tag{4.13}$$

see Sec. 1.4.2 and monographs, e.g., David and Nagaraja [2003]; Krishnamoorthy and Mathew [2008] for further details. For $0 < \gamma < 1$, define the γth quantile of Y with the cdf F_Y as

$$\xi_\gamma = F_Y^{-1}(\gamma) = \inf\{y : F_Y(y) \geq \gamma\}. \tag{4.14}$$

From the continuity of F_Y, we have

$$\begin{aligned}\Pr\{F_Y(Y_{k:n}) \geq \gamma\} &= 1 - \Pr\{F_Y(Y_{k:n}) < \gamma\} \\ &= 1 - \Pr\{Y_{k:n} < \xi_\gamma\} \\ &= 1 - I_{F_Y(\xi_\gamma)}(n-k+1, k) \\ &= 1 - I_\gamma(n-k+1, k).\end{aligned} \tag{4.15}$$

Note that $\Pr\{Y_{k:n} < \xi_\gamma\} = \Pr\{Y_{k:n} \leq \xi_\gamma\}$ and $F_Y(\xi_\gamma) = \gamma$ are satisfied from the assumption of continuity of F_Y. Thus the following theorem can be derived:

Theorem 4.1 *If we choose the values of n and k such that*

$$1 - I_\gamma(n-k+1, k) \geq \alpha, \tag{4.16}$$

then the following relation holds:

$$\Pr\{F_Y(Y_{k:n}) \geq \gamma\} \geq \alpha. \tag{4.17}$$

Interval $(-\infty, Y_{k:n})$ is known as *one-sided tolerance interval* [David and Nagaraja, 2003; Krishnamoorthy and Mathew, 2008]. The interval $(-\infty, Y_{k:n})$ is not explicitly dependent on the cdf, and hence it is called *distribution-free interval.* The observed value of $Y_{k:n}$ is denoted by $y_{k:n}$. Theorem 4.1 implies that

Proposition 4.1 *If condition* (4.16) *is satisfied, probability of "100γ% of yet unobserved samples are less than $y_{k:n}$ in values" is more than or equal to* 100α%.

Roughly speaking, $y_{k:n}$ is the top $(1-\gamma) \times 100\%$ worst value with probability of 100α% if condition (4.16) is satisfied. When $y_{k:n} \leq \bar{y}$ is observed under condition (4.16), we can say that

$$\Pr\{F_Y(\bar{y}) \geq \gamma\} \geq \alpha, \tag{4.18}$$

i.e., at least a proportion γ of the entire population is less than $\bar{y}$ with at least 100α% confidence. To apply this approach, we do not need to know the cdf.

Note that $F_Y(Y_{k:n})$ is also a random variable, which will be referred to as *content ratio*. In fact, if the cdf F_Y is continuous, it is known that

$$F_Y(Y_{k:n}) \overset{d}{=} U_{k:n}, \tag{4.19}$$

where $U_{k:n}$ is the kth order statistics for the n samples from a standard uniform distribution and $\overset{d}{=}$ stands for equality in distribution; see Appendix A.2 for details. That is, $F_Y(Y_{k:n})$ is distributed as $U_{k:n}$. It is also known that $U_{k:n}$ follows a beta distribution with the pdf

$$f_{k:n}(u) = \frac{1}{B(n-k+1, k)} u^{n-k}(1-u)^{k-1} \quad (0 < u < 1), \tag{4.20}$$

where $B(\cdot,\cdot)$ is the beta function defined in (1.38). Thus, random variable $F_Y(Y_{k:n})$ has the pdf (4.20). The mean value and variance of the content ratio in the tolerance interval are respectively denoted by

$$\mu_{k:n} = \mathbb{E}[F_Y(Y_{k:n})] = \frac{n-k+1}{n+1}, \tag{4.21}$$

$$\sigma^2_{k:n} = \mathrm{Var}[F_Y(Y_{k:n})] = \frac{k(n-k+1)}{(n+1)^2(n+2)}. \tag{4.22}$$

Accordingly, the stopping rule of PRS becomes the selection of k and n. We select k and n such that condition (4.16) is satisfied by the left-hand-side value in excess of α by an amount as small as possible. For given k, α and γ, the following problem is formulated:

$$\underset{n \in \mathbb{N}}{\text{Minimize}} \quad n \tag{4.23a}$$

$$\text{subject to} \quad 1 - I_\gamma(n-k+1, k) \geq \alpha. \tag{4.23b}$$

We can numerically solve problem (4.23) by enumeration method to obtain the minimum sample size requirement $n = n(k)$ as shown in Table 4.1 for some

Table 4.1
Minimum Sample Size Requirements and Corresponding Content Ratios

(a) $\alpha = \gamma = 0.9$

k	1	2	3	4	5	6	7	8	9	10
$n(k)$	22	38	52	65	78	91	104	116	128	140
$\mu_{k:n}$	95.7%	94.9%	94.3%	93.9%	93.7%	93.5%	93.3%	93.2%	93.0%	92.9%
$\sigma_{k:n}$	4.16%	3.49%	3.14%	2.92%	2.72%	2.56%	2.42%	2.32%	2.23%	2.15%

k	11	12	13	14	15	16	17	18	19	20
$n(k)$	152	164	175	187	199	210	222	233	245	256
$\mu_{k:n}$	92.8%	92.7%	92.6%	92.6%	92.5%	92.4%	92.4%	92.3%	92.3%	92.2%
$\sigma_{k:n}$	2.08%	2.02%	1.97%	1.91%	1.86%	1.82%	1.77%	1.74%	1.70%	1.67%

(b) $\alpha = 0.99$, $\gamma = 0.9$

k	1	2	3	4	5	6	7	8	9	10
$n(k)$	44	64	81	97	113	127	142	156	170	183
$\mu_{k:n}$	97.8%	96.9%	96.3%	95.9%	95.6%	95.3%	95.1%	94.9%	94.7%	94.6%
$\sigma_{k:n}$	2.17%	2.13%	2.06%	1.99%	1.91%	1.86%	1.80%	1.75%	1.70%	1.67%

k	11	12	13	14	15	16	17	18	19	20
$n(k)$	197	210	223	236	249	262	275	287	300	312
$\mu_{k:n}$	94.4%	94.3%	94.2%	94.1%	94.0%	93.9%	93.8%	93.8%	93.7%	93.6%
$\sigma_{k:n}$	1.62%	1.59%	1.56%	1.53%	1.50%	1.47%	1.44%	1.42%	1.40%	1.38%

(c) $\alpha = \gamma = 0.99$

k	1	2	3	4	5	6	7	8	9	10
$n(k)$	459	662	838	1001	1157	1307	1453	1596	1736	1874
$\mu_{k:n}$	99.8%	99.7%	99.6%	99.6%	99.6%	99.5%	99.5%	99.5%	99.5%	99.5%
$\sigma_{k:n}$	0.22%	0.21%	0.21%	0.20%	0.19%	0.19%	0.18%	0.18%	0.17%	0.17%

k	11	12	13	14	15	16	17	18	19	20
$n(k)$	2010	2144	2277	2409	2539	2669	2798	2925	3052	3179
$\mu_{k:n}$	99.5%	99.4%	99.4%	99.4%	99.4%	99.4%	99.4%	99.4%	99.4%	99.4%
$\sigma_{k:n}$	0.16%	0.16%	0.16%	0.15%	0.15%	0.15%	0.15%	0.14%	0.14%	0.14%

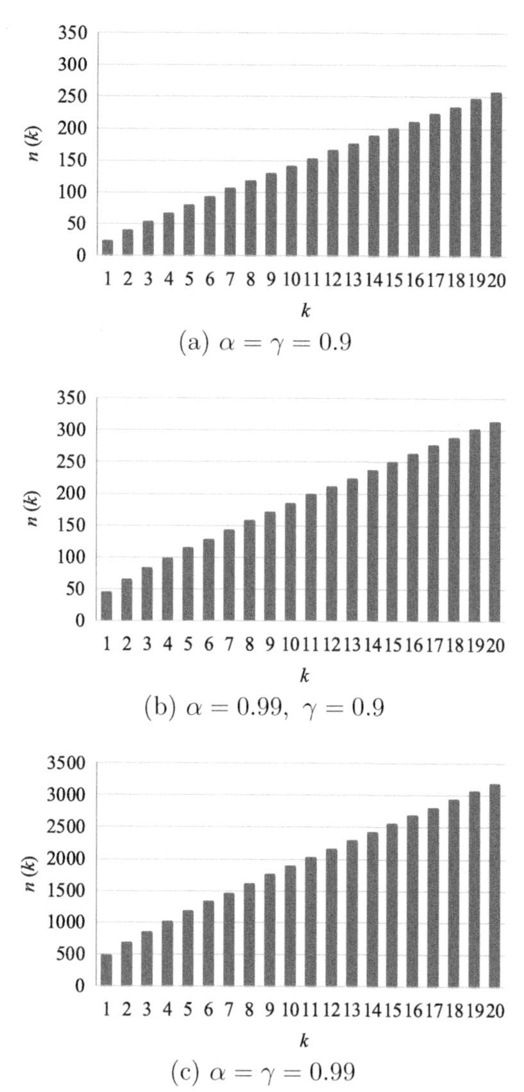

Figure 4.1 Comparison of sample size requirements.

given α and γ. Comparison of $n(k)$ for some given α and γ is also illustrated in Fig. 4.1, which clearly shows that a larger $n(k)$ is needed for a larger α, γ and k. This means that high confidence requires a large sample size. From Proposition 4.1, the following proposition is satisfied:

Proposition 4.2 *If we select $n(k)$ for given α and γ as shown in Table 4.1, probability of $\{100\gamma\%$ of yet unobserved samples are less than $Y_{k:n(k)}$ in values$\}$ is more than or equal to $100\alpha\%$.*

It is obviously easier to compute the order statistics than to find the exact worst value, and the meaning of the use of the order statistics is clear in the sense of (4.9). Thus, the order statistics can be effectively utilized for probabilistic WCA. The mean value and standard deviation of the content ratio $F_Y(Y_{k:n(k)})$ are easily computed using (4.21) and (4.22), and the parameter values corresponding to the minimum sample size requirements are to be selected from Table 4.1, or by solving problem (4.23).

4.2.4 SIMPLE MATHEMATICAL EXAMPLE

We first investigate the properties of function values generated by random sampling approach based on the kth order statistics using Test2N function given as:

$$y(\boldsymbol{x};\boldsymbol{\theta}) = \sum_{i=1}^{d}\{-(x_i+\theta_i)^4 + 16(x_i+\theta_i)^2 - 5(x_i+\theta_i)\}, \tag{4.24}$$

where $\boldsymbol{x} = (x_1, \cdots, x_d) \in \mathbb{R}^d$ and $\boldsymbol{\theta} = (\theta_1, \cdots, \theta_d) \in \Omega_d \subset \mathbb{R}^d$ denote the variables and the uncertain parameters, respectively; see also Appendix A.1. Hereinafter, the prescribed d-dimensional uncertainty set is denoted by Ω_d. Test2N function in (4.24) has 2^d local maxima at $x_i + \theta_i = -2.9035$ and 2.7468 $(i = 1, \cdots, d)$; see Lagaris and Tsoulos [2008]. The case of $d = 4$ was investigated in Sec. 3.2.3 with a different definition of uncertainty. Note that the sign of the function is reversed to investigate a maximization problem instead of a minimization problem.

First, let us consider Test2N function for $d = 2$ and $x_1 = x_2 = 0$. Here again, we abbreviate $y(\boldsymbol{x};\boldsymbol{\theta})$ by $y(\boldsymbol{\theta})$. The function has $2^2 = 4$ local maxima, and the global maximum value is 156.66 at $\theta_1 = \theta_2 = -2.9035$. As already mentioned, the uncertain parameters $\boldsymbol{\theta} = (\theta_1, \theta_2)$ are treated as random variables in Ω_2. Let us set

$$\Omega_2 = \{(\theta_1, \theta_2) | -4 \le \theta_1 \le 4, -4 \le \theta_2 \le 4\}. \tag{4.25}$$

A sequence of random points $\boldsymbol{\theta}_1, \cdots, \boldsymbol{\theta}_n$ generated by PRS is assumed to follow the uniform distribution $U(\Omega_2)$. In Fig. 4.2, the filled circles represent a set of randomly generated $n = 300$ points in Ω_2, and the solid lines represent the contour lines of Test2N function.

Next, let us consider Test2N function for $d = 100$ and $x_i = 0$ $(i = 1, \cdots, 100)$. Here again, Ω_{100} is given by

$$\Omega_{100} = \{(\theta_1, \cdots, \theta_{100}) | -4 \le \theta_i \le 4,\ i = 1, \cdots, 100\}. \tag{4.26}$$

The global maximum value of the function in Ω_{100} is 7833.2. It is extremely difficult to find this exact maximum value if we do not have any prior information, because this function has $2^{100} \approx 1.27 \times 10^{30}$ local maxima in Ω_{100}. However, we are not so interested in the exact worst value, which corresponds

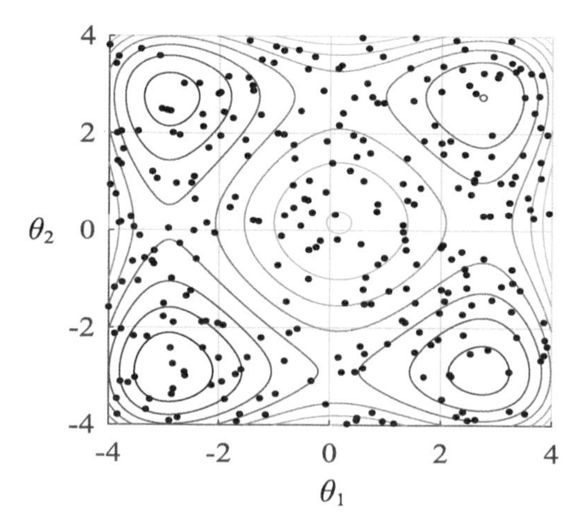

Figure 4.2 An example of sample points of Test2N function with contour lines for $d = 2$ and $x_1 = x_2 = 0$.

to a very rare event within a probabilistic framework. Our interest focuses on finding approximately worst value with pre-specified accuracy.

When the kth order statistics is computed using $n(k)$ in Table 4.1, the value is expected to correspond to the $100\gamma\%$ worst value with a probability of $100\alpha\%$. Let us consider $\alpha = 0.99$ and $\gamma = 0.9$, i.e., the case of Table 4.1(b). From $M = 10^4$ different sample sets $\{\boldsymbol{\theta}_1, \cdots, \boldsymbol{\theta}_{n(k)}\}$, the following M sets of function values are generated:

$$\mathcal{Y}_{n(k)}^{(j)} = \left\{ y_1^{(j)} = y(\boldsymbol{\theta}_1^{(j)}), \cdots, y_{n(k)}^{(j)} = y(\boldsymbol{\theta}_{n(k)}^{(j)}) \right\} \quad (j = 1, \cdots, M). \tag{4.27}$$

For each sample set, the observed order statistics are denoted by $y_{k:n(k)}^{(j)}$. We have different $y_{k:n(k)}^{(j)}$ for $j = 1, \cdots, M$, which can be denoted by

$$y_{k:n(k)}^{(j)} \in \mathcal{Y}_{n(k)}^{(j)} \quad (j = 1, \cdots, M). \tag{4.28}$$

The mean value $\overline{y}_{k:n(k)}$ and the standard deviation $s_{k:n(k)}$ of the observed order statistics $y_{k:n(k)}^{(j)}$ of the M sets are computed as

$$\overline{y}_{k:n(k)} = \frac{1}{M} \sum_{j=1}^{M} y_{k:n(k)}^{(j)}, \tag{4.29}$$

$$s_{k:n(k)} = \sqrt{\frac{1}{M-1} \sum_{j=1}^{M} (y_{k:n(k)}^{(j)} - \overline{y}_{k:n(k)})^2}, \tag{4.30}$$

and are summarized in Table 4.2.

Table 4.2
Summary of the $M = 10^4$ Observed Values of Order Statistics and Validation with $MN = 10^8$ Samples of Test2N Function for $\alpha = 0.99$ and $\gamma = 0.90$

k	1	3	5	7	9	10
$\overline{y}_{k:n(k)}$	3969	3881	3853	3839	3828	3823
$s_{k:n(k)}$	124.4	70.3	56.5	47.3	42.0	40.0
$\bar{F}_Y(y_{k:n(k)})$	97.7%	96.3%	95.6%	95.1%	94.8%	94.6%
$s(\bar{F}_Y(y_{k:n(k)}))$	2.26%	2.17%	2.05%	1.84%	1.74%	1.71%
$\bar{F}(\bar{F}_Y(y_{k:n(k)}) \geq \gamma)$	99.0%	98.4%	98.6%	98.7%	98.8%	98.8%
k	11	13	15	17	19	20
$\overline{y}_{k:n(k)}$	3820	3814	3809	3805	3803	3800
$s_{k:n(k)}$	38.3	35.0	33.3	31.7	30.5	29.9
$\bar{F}_Y(y_{k:n(k)})$	94.5%	94.2%	94.0%	93.9%	93.8%	93.6%
$s(\bar{F}_Y(y_{k:n(k)}))$	1.67%	1.57%	1.54%	1.52%	1.47%	1.47%
$\bar{F}(\bar{F}_Y(y_{k:n(k)}) \geq \gamma)$	98.9%	99.0%	99.3%	98.5%	98.4%	98.3%

To test the validity of prediction by the order statistics, samples are further generated until the number of samples reaches a sufficiently large value N ($\gg n(k)$). This operation is repeated M times to obtain the sets

$$\left\{\boldsymbol{\theta}_1^{(j)}, \cdots, \boldsymbol{\theta}_N^{(j)}\right\} \quad (j = 1, \cdots, M), \tag{4.31}$$

and the corresponding function values denoted by

$$\mathcal{Y}_N^{(j)} = \left\{y_1^{(j)} = g(\boldsymbol{\theta}_1^{(j)}), \cdots, y_N^{(j)} = g(\boldsymbol{\theta}_N^{(j)})\right\} \supset \mathcal{Y}_{n(k)}^{(j)} \quad (j = 1, \cdots, M). \tag{4.32}$$

Thus, for $M = N = 10^4$, the following $MN = 10^8$ samples in total are used for this example:

$$\mathcal{Y}_{MN} = \{y_1^{(1)}, \cdots, y_N^{(1)}, \cdots, y_1^{(M)}, \cdots, y_N^{(M)}\}. \tag{4.33}$$

By using the sample sets, we compute the following empirical cdf:

$$\begin{aligned} &\bar{F}_Y(y_{k:n(k)}^{(j)}) \\ &= \frac{\text{number of elements that are equal to or less than } y_{k:n(k)}^{(j)} \text{ in } \mathcal{Y}_N^{(j)}}{N} \\ &= \frac{1}{N}\sum_{i=1}^{N} I(y_i^{(j)} \leq y_{k:n(k)}^{(j)}), \end{aligned} \tag{4.34}$$

where $I(E)$ is the indicator function of the event E, which in (4.34) means

$$I(y_i^{(j)} \leq y_{k:n(k)}^{(j)}) = \begin{cases} 1 & \text{if } y_i^{(j)} \leq y_{k:n(k)}^{(j)}, \\ 0 & \text{otherwise.} \end{cases} \tag{4.35}$$

Note that

$$\bar{F}_Y(y^{(j)}_{k:n(k)}) \to F_Y(y^{(j)}_{k:n(k)}) \quad \text{as} \quad N \to \infty. \tag{4.36}$$

The mean value and the standard deviation of the empirical cdf $\bar{F}_Y(y^{(j)}_{k:n(k)})$ of $M = 10^4$ sets of $N = 10^4$ samples in (4.34) are computed as

$$\bar{F}_Y(y_{k:n(k)}) = \frac{1}{M}\sum_{j=1}^{M} \bar{F}_Y(y^{(j)}_{k:n(k)}), \tag{4.37}$$

$$s(\bar{F}_Y(y_{k:n(k)})) = \sqrt{\frac{1}{M-1}\sum_{j=1}^{M}\{\bar{F}_Y(y^{(j)}_{k:n(k)}) - \bar{F}_Y(y_{k:n(k)})\}^2}, \tag{4.38}$$

which are listed in Table 4.2 for $k = 1, \cdots, 20$. As already mentioned, the mean value (expectation) and standard deviation of the content ratio can be computed without assumption on the distribution function. In fact, for uniform distribution, the values of $\bar{F}_Y(y_{k:n(k)})$ and $s(\bar{F}_Y(y_{k:n(k)}))$ in Table 4.2 are almost the same as those of $\mu_{k:n}$ and $\sigma_{k:n}$ in Table 4.1(a).

Furthermore, an empirical proportion is defined using the indicator function as

$$\bar{F}(\bar{F}_Y(y_{k:n(k)}) \geq \gamma) = \frac{1}{M}\sum_{j=1}^{M} I(\bar{F}_Y(y^{(j)}_{k:n(k)}) \geq \gamma), \tag{4.39}$$

which is also shown in Table 4.2 for $M = N = 10^4$. Note again that

$$\bar{F}(\bar{F}_Y(y_{k:n(k)}) \geq \gamma) \to \Pr\{F_Y(Y_{k:n(k)}) \geq \gamma\} \quad \text{as} \quad N, M \to \infty. \tag{4.40}$$

Thus, $\bar{F}(\bar{F}_Y(y_{k:n(k)}) \geq \gamma)$ approaches α for sufficiently large M and N if the Proposition 4.2 is true. Indeed, $\bar{F}(\bar{F}_Y(y_{k:n(k)}) \geq \gamma)$ shown in Table 4.2 approximately coincides with $\alpha = 0.99$ for $k = 1, \cdots, 20$; accordingly, the validity of Proposition 4.2 has been verified.

For any k, the values of $\bar{F}(\bar{F}_Y(y_{k:n(k)}) \geq \gamma)$ shown in Table 4.2 are almost the same. If we choose a smaller value for k, corresponding $n(k)$ also becomes smaller. This is an advantage in view of obtaining an accurate maximum value; however, smaller k has also disadvantages. In Table 4.2, the cases of smaller k have larger mean value and larger standard deviation in both of $y_{k:n(k)}$ and $\bar{F}_Y(y^{(j)}_{k:n(k)})$. This fact can be interpreted that a smaller k have larger uncertainty that leads to more conservative expectation, which may cause too conservative and expensive design when applied to RDO based on WCA. How should we select an appropriate value of k? We can only say that it depends on the type of problem and the purpose of design, and it is better to set a large value for k in view of the allowable computational cost.

4.2.5 ROBUST DESIGN OPTIMIZATION OF BUILDING FRAME

9-story 2-span braced steel frame

Let us consider a 9-story 2-span X-braced steel building frame as shown in Fig. 4.3. The braces are pin-jointed to the frame, and the two braces at each story are not connected at their intersection point. The steel sections of beams and columns are summarized in Table 4.3. I-shaped cross-sections are used for the beams, and their height (mm), flange width (mm), web thickness (mm) and flange thickness (mm) are shown in Table 4.3 using Japanese specification. Square hollow structural sections (HSSs) are used for the columns, and their height (mm) and thickness (mm) are also shown in Table 4.3. Furthermore, steel bars are used for braces, which are designed to resist only tensile force under axial deformation. We arrange two sets of the bundled four steel bars at each story. The diameter (mm) of each bar is also shown in Table 4.3. The steel materials of the beams, columns and braces are assumed to have bilinear stress–strain relations, where Young's modulus is $E = 205\,\mathrm{kN/mm^2}$, and the kinematic hardening ratio is $0.01E$. The yield stress is given as $\sigma_\mathrm{y} = 325\,\mathrm{N/mm^2}$ for beams and columns, and $\sigma_\mathrm{y} = 1080\,\mathrm{N/mm^2}$ for braces. The floor mass of 7.2×10^4 kg is divided and concentrated at the nodes of each floor. The floor is assumed to be rigid allowing bending deformation, i.e., the

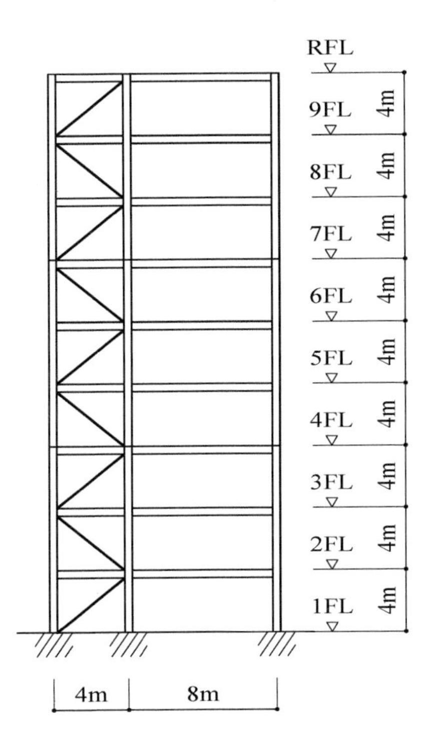

Figure 4.3 9-story 2-span braced steel frame.

Table 4.3
List of Cross-sections

	Beam	
G1	(2F to 4F)	H-500×250×12×25
G2	(5F to 7F)	H-500×250×12×22
G3	(8F to RF)	H-500×250×9×19
	Column	
C1	(1S to 3S)	HSS-400×25
C2	(4S to 6S)	HSS-400×22
C3	(7S to 9S)	HSS-400×19
	Bar diameter of brace (mm)	
B1	(1Sto 3S)	40
B2	(4S to 6S)	36
B3	(7S to 9S)	32

horizontal displacements of the nodes on the same floor have the same value. Considering composite action of the steel beam and the concrete slab, we multiply the factor 1.5 to the flexural rigidity of each beam. The first, second and third natural periods are 1.214, 0.340 and 0.175 (sec.), respectively.

Seismic motion

The artificial seismic motions are generated using the standard superposition method of sinusoidal waves [Iyengar and Rao, 1979]. The design acceleration response spectrum for 5% damping at the engineering bedrock surface specified by Japanese design code is given as

$$S_a(T) = \begin{cases} 3.2 + 30T & \text{if } T < 0.16, \\ 8 & \text{if } 0.16 \leq T < 0.64, \\ 5.12/T & \text{if } 0.64 \leq T, \end{cases} \tag{4.41}$$

where T (sec.) is the natural period of the frame. For the sake of simplicity, we regard $S_a(T)$ in (4.41) as the design acceleration response spectrum without considering the ground surface amplification. The seismic motion with duration $T_d = 60\,\text{sec.}$ and sampling time interval $\Delta t = 0.01\,\text{sec.}$ is applied. The acceleration time history waveform at sample points is given as

$$a_j = \sum_{k=0}^{n_d/2} A_k \cos(j\omega_k \Delta t + \theta_k) \quad (j = 0, \cdots, n_d - 1), \tag{4.42}$$

where $n_d = T_d/\Delta t = 6000$ is the number of sample points, $\omega_k = 2\pi k/T_d$ is the kth circular frequency, A_k and θ_k $(k = 0, \cdots, n_d/2)$ with $\theta_0 = \theta_{n_d/2} = 0$ respectively denote amplitude and phase of the discrete Fourier spectrum of the seismic motion. The phase spectrum is often linked to stochastic models for practical seismic design, see, e.g., Ohsaki [1979]. One of the popular phase generation methods is the uniformly distributed random phase. Thus, we treat $\boldsymbol{\theta} = (\theta_1, \cdots, \theta_{n_d/2-1})$ as the uncertain parameters uniformly distributed in $\Omega_{n_d/2-1}$, i.e.,

$$\boldsymbol{\theta} \in \Omega_{n_d/2-1}, \tag{4.43}$$

where

$$\Omega_{n_d/2-1} = \left\{ (\theta_1, \cdots, \theta_{n_d/2-1}) \;\middle|\; -\pi \leq \theta_k \leq \pi,\ k = 1, \cdots, \frac{n_d}{2} - 1 \right\}. \tag{4.44}$$

The amplitude A_k is determined to fit the target spectrum under the given $\boldsymbol{\theta}$. Hence, the acceleration time history waveform can be regarded as a function of the phase spectrum, i.e., $(a_0(\boldsymbol{\theta}), \cdots, a_{n_d-1}(\boldsymbol{\theta}))$. Acceleration response spectrum S_a for 5% damping of a generated artificial seismic motion and the target spectrum are shown in Fig. 4.4(a), and the corresponding acceleration time history waveform is also shown in Fig. 4.4(b).

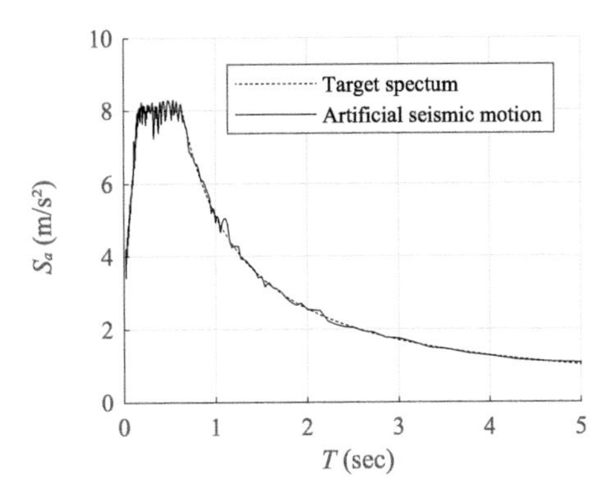

(a) Acceleration response spectrum S_a for 5% damping and the target spectrum

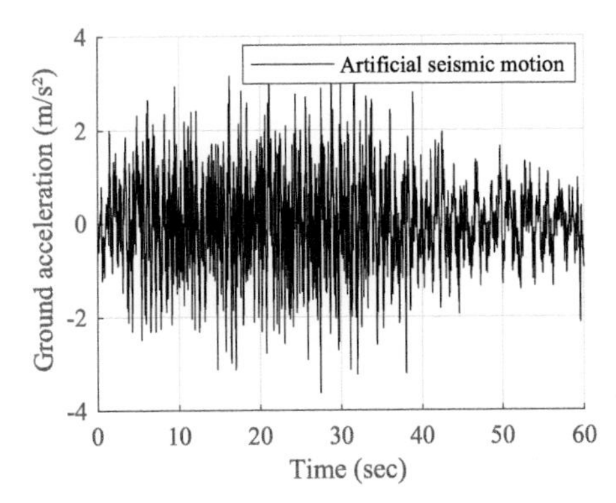

(b) Acceleration time history (input earthquake motions)

Figure 4.4 An example of generated artificial seismic motion.

Seismic response analysis

A software framework OpenSees [McKenna, 2011] for simulating the performance of structural and geotechnical systems subjected to earthquakes is used for seismic response analysis of the frame. Each beam and column is modeled by a force-based beam-column element, the section of which is divided into fibers. The flange and web of the beam are discretized into 4 and 16 fibers, respectively. Integration along the element is based on Gauss-Lobatto quadrature rule. The number of integration points is 8. The standard Newmark-β method ($\beta = 0.25, \gamma = 0.5$) is used for transient analysis with increment of 0.005 sec. The initial stiffness-proportional damping is used with the damping ratio 0.02 for the first mode.

In this example, we do not change the design of frame, e.g., shape of sections, material properties and geometry of frame. The seismic response can be regarded as a function of the uncertain parameters in the seismic motion. The ith interstory drift at time $t_m = m\Delta t$ is denoted by

$$\delta_i(\boldsymbol{\theta}, t_m) \quad (i = 1, \cdots, 9;\ m = 0, \cdots, n_d - 1) \tag{4.45}$$

and the maximum value of the ith interstory drift angle (rad) is obtained as

$$d_i(\boldsymbol{\theta}) = \max_m \frac{|\delta_i(\boldsymbol{\theta}, t_m)|}{h_i} \quad (i = 1, \cdots, 9), \tag{4.46}$$

where h_i is the ith story height. Furthermore, the representative response function $g(\boldsymbol{\theta})$ is computed as the maximum interstory drift angle (rad) among all the stories as

$$g(\boldsymbol{\theta}) = \max_i d_i(\boldsymbol{\theta}) = \max_{i,m} \frac{|\delta_i(\boldsymbol{\theta}, t_m)|}{h_i}. \tag{4.47}$$

Note again that the maximum interstory drift angle, which is chosen as the objective function of the WCA problem, is a function of only uncertain parameters in the seismic motion. For example, the worst artificial seismic motion among the randomly generated 44 observations is shown in Fig. 4.4(b). Corresponding maximum value of the ith interstory drift angle $d_i(\boldsymbol{\theta})$ and its maximum value among all stories $g(\boldsymbol{\theta}) = 0.01300\,\text{rad}$ are plotted in Fig. 4.5 using the square symbol and the solid line, respectively.

Probabilistic worst-case analysis

To test the validity of prediction by the order statistics, we first generate $N = 10^4$ samples of the phase set $\boldsymbol{\theta}$, which are denoted by $\{\boldsymbol{\theta}_1, \cdots, \boldsymbol{\theta}_N\}$. As defined in (4.43) and (4.44), $\boldsymbol{\theta}_i \in \Omega_{n_d/2-1}$ with $n_d/2 - 1 = 2999$ for $i = 1, \cdots, N$ is used, and a set of corresponding function values are denoted by

$$\mathcal{G}_N = \{g_1 = g(\boldsymbol{\theta}_1), \cdots, g_N = g(\boldsymbol{\theta}_N)\}. \tag{4.48}$$

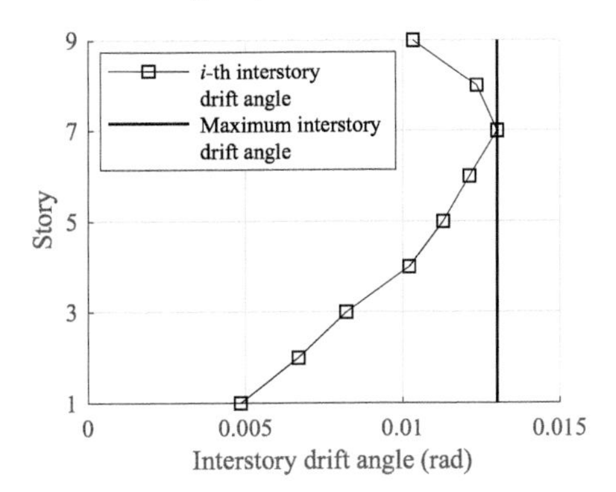

Figure 4.5 Maximum interstory drift angle to the artificial seismic motion in Fig. 4.4.

Here, only one set of the sample values is used, i.e., $M = 1$ in (4.32) because the transient analysis is time-consuming. To compensate this simplification, we randomly select $L = 10^4$ sets of $n(k)$ sample values from $\mathcal{G}_N$, which are denoted by

$$\mathcal{G}^{(j)}_{n(k)} = \left\{ g^{(j)}_1, \cdots, g^{(j)}_{n(k)} \right\} \subset \mathcal{G}_N \quad (j = 1, \cdots, L). \tag{4.49}$$

By using the same formulation as (4.32), (4.49) can be expressed as

$$\mathcal{G}^{(j)}_N = \mathcal{G}_N \quad (j = 1, \cdots, L). \tag{4.50}$$

We observe the kth order statistics $g^{(j)}_{k:n(k)}$ from $\mathcal{G}^{(j)}_{n(k)}$. Similarly to Table 4.2, the mean value $\overline{g}_{k:n(k)}$ and the standard deviation $s_{k:n(k)}$ of the observed order statistics for the building frame are computed and summarized in Table 4.4, which also shows the mean value $\overline{F}_G(g_{k:n(k)})$, the standard deviation $s_N(g_{k:n(k)})$ of the empirical indices and the empirical proportion $\bar{F}(\bar{F}_G(g_{k:n(k)}) \geq \gamma)$ that is similarly defined as in (4.39). Note that we use $N = 10^4$ samples for this example, while $MN = 10^8$ samples are used for the simple mathematical example in Sec. 4.2.4. To avoid misunderstandings, the number of observed order statistics for this example is denoted by L instead of M.

Let us compare the results for $k = 1$ with those of $k = 20$ for the same α and γ. The empirical proportion $\bar{F}(\bar{F}_G(g_{k:n(k)}) \geq \gamma)$ for $k = 1$ and 20 are 99.0% and 98.9%, respectively. They approximately coincide with the value of $\alpha = 0.99$. This fact indicates the validity of Proposition 4.2 again; however, the mean values of the observed order statistics are different between $k = 1$ and 20, i.e., $\overline{g}_{1:44} = 0.01346$ and $\overline{g}_{20:312} = 0.01284$ (rad). The filled circles in Fig. 4.6 refer to the observed sample values of the maximum interstory drift angle of each story. The observed order statistics for $k = 1$ and 20, i.e., $g_{1:44}$ and $g_{20:312}$, are indicated by the solid lines in Figs. 4.6(a) and (b), respectively.

Table 4.4
Summary of the $L = 10^4$ Observed Values of Order Statistics and Validation with $N = 10^4$ Samples of the Building Frame for $\alpha = 0.99$ and $\gamma = 0.90$, where $\bar{g}_{k:n(k)}$ and $s_{k:n(k)}$ are in Units of $1/1000$ rad

k	1	3	5	7	9	10
$\bar{g}_{k:n(k)}$	13.46	13.15	13.05	12.99	12.95	12.93
$s_{k:n(k)}$	0.41	0.26	0.21	0.18	0.16	0.15
$\bar{F}_G(g_{k:n(k)})$	97.8%	96.3%	95.6%	95.1%	94.7%	94.6%
$s(\bar{F}_G(g_{k:n(k)}))$	2.18%	2.07%	1.95%	1.83%	1.72%	1.68%
$\bar{F}(\bar{F}_G(g_{k:n(k)}) \geq \gamma)$	99.0%	99.0%	99.0%	99.0%	99.0%	98.9%
k	11	13	15	17	19	20
$\bar{g}_{k:n(k)}$	12.92	12.89	12.87	12.86	12.85	12.84
$s_{k:n(k)}$	0.14	0.13	0.12	0.11	0.11	0.10
$\bar{F}_G(g_{k:n(k)})$	94.4%	94.2%	94.0%	93.8%	93.7%	93.6%
$s(\bar{F}_G(g_{k:n(k)}))$	1.63%	1.57%	1.52%	1.45%	1.40%	1.38%
$\bar{F}(\bar{F}_G(g_{k:n(k)}) \geq \gamma)$	98.9%	98.9%	98.9%	99.0%	99.0%	98.9%

As can be seen from Fig. 4.6, the observed values of $g_{1:44}$ more widely spread than $g_{20:312}$ as $s_{1:44} = 0.41 > s_{20:312} = 0.10$; see Table 4.4. However, the lower bounds of them are 0.01231 and 0.01244 (rad), which are almost the same. Empirical cdfs of the observed maximum interstory drift angle of samples and order statistics are shown in Fig. 4.7. The filled circles in Figs. 4.7(a) and (b) represent the observed kth order statistics for $k = 1$ and 20, respectively. Although some circles are seen to be below 90% of the cdf, they are only 1.03% in Fig. 4.7(a) and 1.10% in Fig. 4.7(b) of the total number of samples.

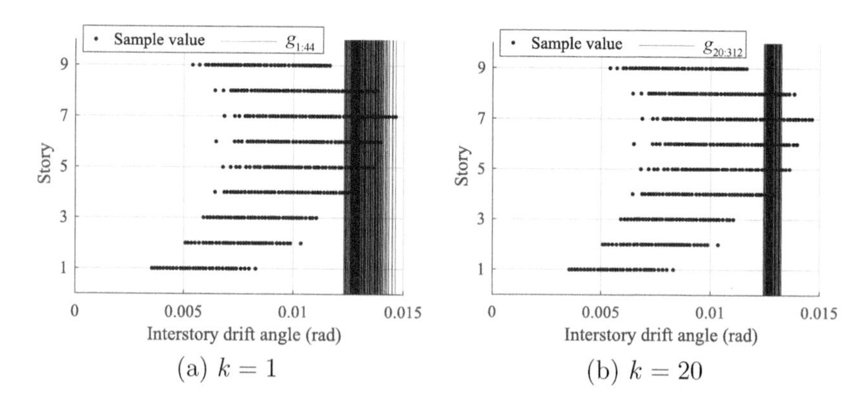

Figure 4.6 Observed maximum interstory drift angles of each story of samples and corresponding order statistics.

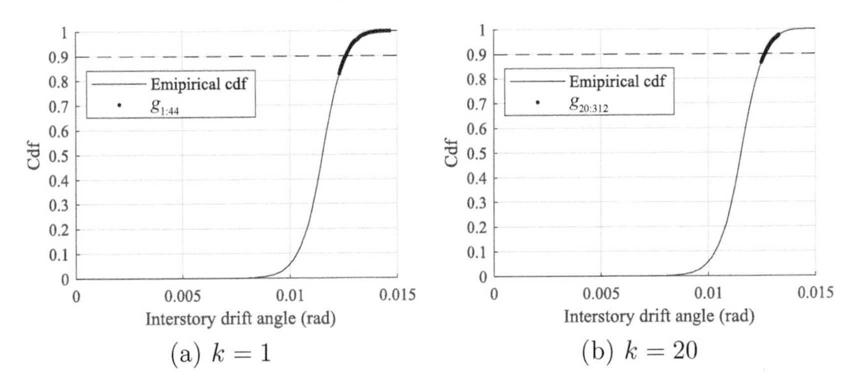

Figure 4.7 Empirical cdf of observed maximum interstory drift angle of samples and corresponding order statistics.

Thus, by using only $n(1) = 44$ samples, we could predict the top 10% worst value with probability of almost 99% even for such a complex problem with transient dynamic analysis.

4.3 ORDER STATISTICS-BASED ROBUST DESIGN OPTIMIZATION

4.3.1 FORMULATION OF A SIMPLIFIED PROBLEM

Let us first define an RDO problem, which is a simplified version of problem (4.1). The design variable vector is denoted by $\boldsymbol{x} = (x_1, \cdots, x_m)$ and a vector consisting of uncertain parameters is denoted by $\boldsymbol{\theta} = (\theta_1, \cdots, \theta_r) \in \Omega$. For simplicity, only the lower and upper bounds of the variables, denoted by x^{L} and x^{U}, respectively, are given as constraints, denoted by $x^{\mathrm{L}} \leq x_i \leq x^{\mathrm{U}}$ $(i = 1, \cdots, m)$. The objective function is denoted by $g = g(\boldsymbol{x}, \boldsymbol{\theta})$, which is a structural response function corresponding to the design variables $\boldsymbol{x}$ and uncertain parameters $\boldsymbol{\theta}$. Thus, an RDO problem we are interested in can be described as

$$\begin{aligned} &\text{Minimize} && g(\boldsymbol{x}; \boldsymbol{\theta}) && \text{for all } \boldsymbol{\theta} \in \Omega && (4.51\text{a}) \\ &\text{subject to} && x^{\mathrm{L}} \leq x_i \leq x^{\mathrm{U}} && (i = 1, \cdots, m). && (4.51\text{b}) \end{aligned}$$

It is difficult to directly solve problem (4.51) and hence an order statistics-based RS approach is utilized. The response function in (4.51a) is regarded as a random variable depending on $\boldsymbol{x}$, which is denoted by $G(\boldsymbol{x}) = g(\boldsymbol{x}; \boldsymbol{\Theta})$, where $\boldsymbol{\Theta} \in \Omega$ is a random variable vector and $\boldsymbol{\theta}$ can be regarded as a realization of $\boldsymbol{\Theta}$. Thus, $g(\boldsymbol{x}; \boldsymbol{\theta})$ is also regarded as a realization of $g(\boldsymbol{x}; \boldsymbol{\Theta})$ but not as a random variable. We generate iidrvs $\boldsymbol{\Theta}_1, \cdots, \boldsymbol{\Theta}_n \in \Omega$ and corresponding function values are denoted by

$$G_1 = g(\boldsymbol{\Theta}_1), \cdots, G_n = g(\boldsymbol{\Theta}_n), \qquad (4.52)$$

where $G_1, \cdots, G_n$ can also be regarded as random variables. The random variables are arranged in order of magnitude as

$$G_{1:n} \geq \cdots \geq G_{n:n}. \tag{4.53}$$

The random variable $G_{k:n}$ is the kth order statistics as stated before. For given parameters α and γ, we choose the number of sample points $n = n(k)$ as shown in Table 4.1, then the following relation is satisfied:

$$\Pr\{F_G(G_{k:n}) \geq \gamma\} \geq \alpha. \tag{4.54}$$

For example, for $\alpha = \gamma = 0.9$ and $k = 2$, we have

$$\begin{aligned}&\text{probability of \{more than 90\% of the entire population}\\ &\qquad\qquad\qquad\text{are less than } G_{2:38} \text{ in values\}} \geq 0.9\end{aligned} \tag{4.55}$$

A set of samples of $\{\boldsymbol{\Theta}_1, \cdots, \boldsymbol{\Theta}_n\}$ is randomly generated and denoted by $\{\boldsymbol{\theta}_1, \cdots, \boldsymbol{\theta}_n\}$, and the corresponding function values are denoted by $g_1(\boldsymbol{x}) = g(\boldsymbol{x}; \boldsymbol{\theta}_1), \cdots, g_n(\boldsymbol{x}) = g(\boldsymbol{x}; \boldsymbol{\theta}_n)$. The values are arranged in decreasing order of magnitude as $g_{1:n}(\boldsymbol{x}) \geq \cdots \geq g_{n:n}(\boldsymbol{x})$, which are regarded as realizations of order statistics $G_{1:n}(\boldsymbol{x}), \cdots, G_{n:n}(\boldsymbol{x})$. In more details, it is expressed as

$$\begin{aligned}&g_{k:n}(\boldsymbol{x}) = g_{k:n}(\boldsymbol{x}; \boldsymbol{\theta}_1, \cdots, \boldsymbol{\theta}_n)\\ &\qquad\qquad \text{as a realization of } G_{k:n}(\boldsymbol{x}) = g_{k:n}(\boldsymbol{x}; \boldsymbol{\Theta}_1, \cdots, \boldsymbol{\Theta}_n)\\ &\qquad\qquad\qquad\qquad\qquad\qquad\qquad\qquad (k = 1, \cdots, n),\end{aligned}$$

The sample size requirement $n = n(k)$ is shown in Table 4.1. Thus, an RDO problem using the order statistics-based RS is formulated as

$$\begin{aligned}&\text{Minimize} && g_{k:n}(\boldsymbol{x}) && \text{(4.56a)}\\ &\text{subject to} && x^{\mathrm{L}} \leq x_i \leq x^{\mathrm{U}} \quad (i = 1, \cdots, m). && \text{(4.56b)}\end{aligned}$$

Thus, we solve problem (4.56) instead of problem (4.51). This approach is justified by relation (4.54), e.g., as stated in (4.55).

In a similar way, the approach is applicable to a more general problem with uncertainty in both objective and constraint functions. For example, we can consider the following problem using the order statistics-based RS to problem (4.1) as

$$\begin{aligned}&\text{Minimize} && g_{0,k:n}(\boldsymbol{x}) && \text{(4.57a)}\\ &\text{subject to} && g_{i,k:n}(\boldsymbol{x}) \leq \bar{g}_i \quad (i = 1, \cdots, l), && \text{(4.57b)}\\ & && \boldsymbol{x} \in \mathcal{X}, && \text{(4.57c)}\end{aligned}$$

where $g_{i,k:n}(\boldsymbol{x})$ is regarded as a realization of the kth order statistic to the set of iidrvs $\{g_i(\boldsymbol{x}; \boldsymbol{\Theta}_1), \cdots, g_i(\boldsymbol{x}; \boldsymbol{\Theta}_n)\}$.

4.3.2 SEISMIC MOTION CONSIDERING UNCERTAIN PROPERTIES OF SURFACE GROUND

We apply the order statistics-based approach to a seismic design problem [Yamakawa et al., 2018]. Design seismic motions are generated by simple superposition of sinusoidal waves [Iyengar and Rao, 1979]. Their amplitudes of acceleration response spectra at engineering bedrock surface are specified by a design target spectrum; however, their phase spectra are supposed to contain uncertainty. Furthermore, amplification of seismic motions on surface ground is taken into account by incorporating uncertainty in its properties. Thus, the seismic motions to a structure should be influenced by uncertainty in the phase spectra and the surface-ground properties; see also Yamakawa and Ohsaki [2021].

Amplitude spectrum and phase spectrum

The discretized acceleration time history waveform at the engineering bedrock surface can be denoted by

$$\boldsymbol{a}^{\mathrm{B}} = \left(a_0^{\mathrm{B}}, a_1^{\mathrm{B}}, \cdots, a_{n_d-1}^{\mathrm{B}}\right), \tag{4.58a}$$

$$a_j^{\mathrm{B}} = \sum_{k=0}^{n_d/2} \xi_k \cos(j\omega_k \Delta t + \psi_k) \quad (j = 0, \cdots, n_d - 1), \tag{4.58b}$$

where n_d is the number of time points, ω_k is the kth circular frequency, $\boldsymbol{\xi} = (\xi_0, \xi_1, \cdots, \xi_{n_d-1})$ and $\boldsymbol{\psi} = (\psi_0, \psi_1, \cdots, \psi_{n_d-1})$ respectively denote the amplitude and phase of the discrete Fourier spectrum of the seismic motion. The design acceleration response spectrum for 5% damping at the engineering bedrock surface specified by Japanese design code is given as (4.41). The design seismic motions at the engineering bedrock surface are generated as follows: (i) a set of c phase spectra $\boldsymbol{\psi}_1, \cdots, \boldsymbol{\psi}_c$ is given and one of them is randomly selected, which is denoted by $\boldsymbol{\psi}(\theta_1) = \boldsymbol{\psi}_{\theta_1}$, where $\theta_1 \in \{1, \cdots, c\}$ is an uncertain parameter of index of the phase spectrum, and (ii) an amplitude spectrum denoted by $\boldsymbol{\xi}^{\mathrm{B}}$ is generated by simple superposition of sinusoidal waves to fit the design acceleration response spectrum specified by (4.41). Thus, the acceleration time history at the engineering bedrock surface is calculated by the inverse Fourier transform of $\boldsymbol{\xi}^{\mathrm{B}}$ and $\boldsymbol{\psi}(\theta_1)$, which can be regarded as a vector-valued function denoted by $\boldsymbol{a}^{\mathrm{B}} = \boldsymbol{a}^{\mathrm{B}}(\theta_1) = \boldsymbol{a}^{\mathrm{B}}(\boldsymbol{\psi}(\theta_1); \boldsymbol{\xi}^{\mathrm{B}})$.

Seismic motion considering the uncertain amplification of surface ground

Let us consider an N-layer soil model, as shown in Fig. 4.8. Numbering of soil layers starts at the ground surface and the engineering bedrock is given as the Nth layer. The layer thickness and the shear wave velocity of the ith soil layer are denoted by H_i and V_i, respectively. The mass density, the shear modulus and the damping ratio of the ith layer are respectively denoted by

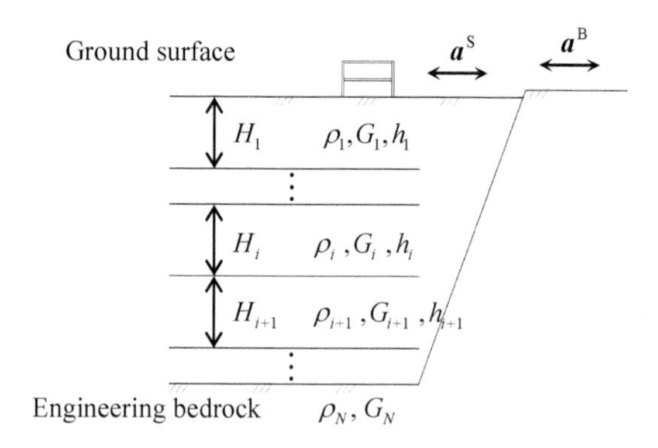

Figure 4.8 N-layer soil model.

ρ_i, G_i and h_i. Nonlinear relation between shear modulus G and damping ratio h at a given shear strain γ is provided by the Hardin-Drnevich model [Hardin and Drnevich, 1972] denoted by

$$G/G_0 = \frac{1}{1+\gamma/\gamma_{0.5}}, \qquad h = h_{\max}\left(1 - G/G_0\right), \tag{4.59}$$

where G_0 is the small strain shear modulus at initial loading, $h_{\max}$ is the maximum damping ratio at maximum strain and $\gamma_{0.5}$ is the reference shear strain, in the following examples, which are given as $\gamma_{0.5} = 0.15\%$ and $h_{\max} = 15\%$, respectively.

We use an equivalent linear method to obtain input seismic motions to a structure [Schnabel et al., 1972], where shear modulus and damping ratio are assumed to be given, and $\gamma_{\max}$ is evaluated by some linear analyses. In a linear analysis, the incident wave and the reflected wave in ith soil layer are denoted by E_i and F_i. The recursive formula expresses the amplitudes E_{i+1} and F_{i+1} in the $(i+1)$th layer in terms of the amplitudes in the ith layer as

$$\begin{pmatrix} E_{i+1} \\ F_{i+1} \end{pmatrix} = \frac{1}{2} \begin{pmatrix} (1+\alpha_i)e^{i(\omega H_i/V_i)} & (1-\alpha_i)e^{-i(\omega H_i/V_i)} \\ (1-\alpha_i)e^{i(\omega H_i/V_i)} & (1+\alpha_i)e^{-i(\omega H_i/V_i)} \end{pmatrix} \begin{pmatrix} E_i \\ F_i \end{pmatrix},$$

where $\alpha_i = \rho_i V_i/(\rho_{i+1}V_{i+1})$ is the impedance ratio, and shear wave velocity is obtained by $V_i = \sqrt{G_i/\rho_i}$, and ω is the frequency of the harmonic displacement. The effective strain γ_{eq} is computed from the maximum strain $\gamma_{\max}$ by the linear analysis as

$$\gamma_{\mathrm{eq}} = 0.65\gamma_{\max}, \tag{4.60}$$

where 0.65 in (4.60) is a constant of effective strain conversion. We evaluate $\gamma_{\max}$ recursively until a certain convergence criterion is satisfied. Thus, the amplitude spectrum at the surface ground denoted by $\boldsymbol{\xi}^{\mathrm{S}}$ is calculated from $\boldsymbol{\xi}^{\mathrm{B}}$ at the engineering bedrock and the surface ground properties. Some

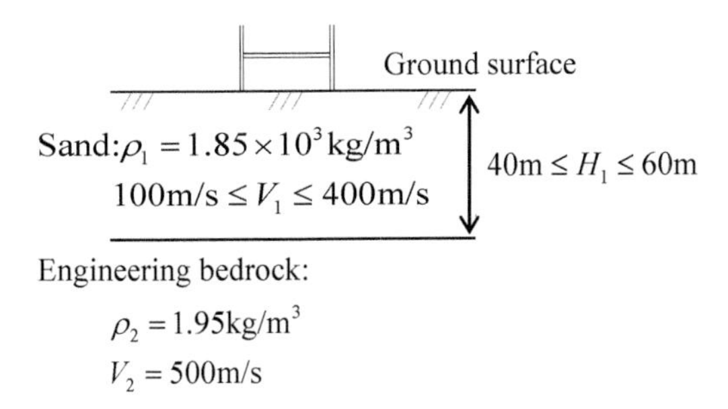

Figure 4.9 Two-layer ground model.

ground properties are linked with uncertain parameters $\theta_2, \cdots, \theta_r$ as described in the following numerical examples, and the amplitude spectrum at the surface ground is regarded as a vector-valued function denoted by $\boldsymbol{\xi}^{\mathrm{S}} = \boldsymbol{\xi}^{\mathrm{S}}(\theta_2, \cdots, \theta_r; \boldsymbol{\xi}^{\mathrm{B}})$. The acceleration time-history at the surface ground is obtained by the inverse Fourier transform of $\boldsymbol{\xi}^{\mathrm{S}}$ and $\boldsymbol{\psi}$, which is denoted by $\boldsymbol{\xi}^{\mathrm{S}} = \boldsymbol{\xi}^{\mathrm{S}}(\boldsymbol{\theta}) = \boldsymbol{\xi}^{\mathrm{S}}(\boldsymbol{\psi}(\theta_1), \theta_2, \cdots, \theta_r; \boldsymbol{\xi}^{\mathrm{B}})$. The response of the building depending on the design variable vector $\boldsymbol{x}$ subjected to the seismic motion of the surface ground with uncertainty is denoted by $g(\boldsymbol{x}) = g(\boldsymbol{x}; \boldsymbol{\xi}^{\mathrm{S}}(\boldsymbol{\theta}))$. Thus, we formulate an RDO problem with amplification of seismic ground motion by uncertain surface ground properties as expressed in problem (4.56) based on order statistics.

Numerical example

We assume a phase spectrum of a design seismic motion, layer thickness and shear wave velocity of the surface ground have uncertainties. Thus, the number of uncertain parameters is set to $r = 3$. The phase spectra are given by Hachinohe EW of the Tokachi-oki Earthquake in 1968, Tohoku University NS of the Miyagi-oki Earthquake in 1978 and Kobe NS of the Hyogoken-Nambu Earthquake in 1995, which are denoted by $\boldsymbol{\psi}(1), \boldsymbol{\psi}(2)$ and $\boldsymbol{\psi}(3)$, respectively, i.e., $c = 3$. One of them is randomly selected as $\boldsymbol{\phi}(\theta_1)$ for $\theta_1 \in \{1, 2, 3\}$.

We consider a two-layer ground model as shown in Fig. 4.9, i.e., $N = 2$. Layer thickness and shear wave velocity of the surface ground are taken as the uncertain parameters denoted by $\theta_2 = H_1(\mathrm{m})$ and $\theta_3 = V_1(\mathrm{m/s})$. Finally, the uncertain set Ω is given as

$$\Omega = \{(\theta_1, \theta_2, \theta_3) \,|\, \theta_1 \in \{1, 2, 3\},\ 40 \leq \theta_2 \leq 60,\ 100 \leq \theta_3 \leq 400\}.$$

Building model and seismic response analysis

Let us consider a 9-story 2-bay braced steel frame as shown in Fig. 4.10. The beams are rigidly connected to the columns, and the ends of the braces are pinned. The steel sections of beams and columns are summarized in Table 4.5.

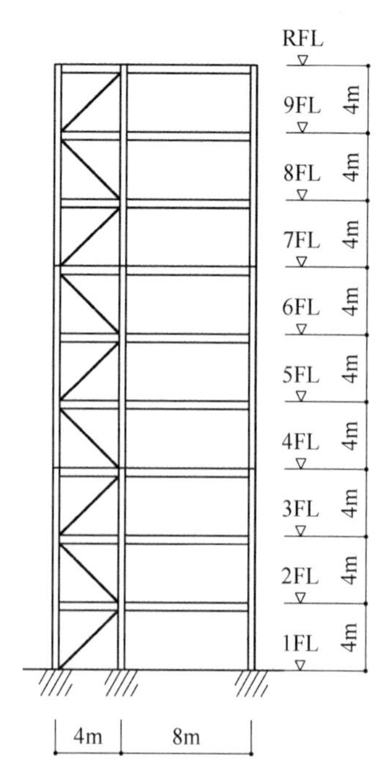

Figure 4.10 9-story 2-span braced steel frame.

Table 4.5
List of Cross-sections

Beam		
G1	(2F to 4F)	H-500×250×12×25
G2	(5F to 7F)	H-500×250×12×22
G3	(8F to RF)	H-500×250×9×19
Column		
C1	(1S to 3S)	HSS-400×25
C2	(4S to 6S)	HSS-400×22
C3	(7S to 9S)	HSS-400×19
Cross-sectional area of brace (cm^2)		
	(1S to 9S)	$0 \leq x_i \leq 200$ $(i = 1, \cdots, 9)$

Wide flange cross-sections are used for the beams, and their height (mm), flange width (mm), web thickness (mm) and flange thickness (mm) are shown in Table 4.5. Square hollow structural sections are used for the columns, and their height (mm) and thickness (mm) are also shown in Table 4.5. We use buckling-restrained braces, and its cross-sectional area of central steel material (cm^2) is taken as the design variable as shown in Table 4.5, i.e., the number of design variables is set as $m = 9$. The upper and lower bounds of the design variables are given as $x^{\mathrm{U}} = 200\,\mathrm{cm}^2$ and $x^{\mathrm{L}} = 0$, respectively. The steel material of the elements is assumed to have bilinear stress-strain relation, where Young's modulus is $E = 205\,\mathrm{kN/mm}^2$, and the kinematic hardening ratio is $0.01E$. The yield stress is given as $\sigma_{\mathrm{y}} = 325\,\mathrm{N/mm}^2$ for beams and columns and $\sigma_{\mathrm{y}} = 100\,\mathrm{N/mm}^2$ for braces. The floor mass of $7.2 \times 10^4\,\mathrm{kg}$ is distributed to the nodes of each floor. The floor diaphragm is assumed to be rigid, i.e., the horizontal displacements of the nodes on the same floor have the same value. Considering composite action of the steel beam and the concrete slab, we multiply the factor 1.5 to the flexural rigidity of each beam.

A software framework to simulate the performance of structural and geotechnical systems called OpenSees [McKenna, 2011] is used for seismic response analysis of the frame, where analysis conditions are same as Sec. 4.2.5.

Results

We choose the maximum interstory drift angle as the response function of this example, which is denoted by $g(\boldsymbol{x}) = g(\boldsymbol{x}; \boldsymbol{\xi}^{\mathrm{S}}(\boldsymbol{\theta}))$. As an RDO problem, we find $\boldsymbol{x}$ that minimizes $g_{k:n}(\boldsymbol{x})$ for $(k, n) = (2, 38)$ corresponding to $\alpha = \gamma = 0.9$, which is referred to as robust optimal solution (ROS) obtained by solving problem (4.56). For comparison, we also find the solution that minimizes the maximum interstory drift angle without considering uncertain amplification of surface ground, where the problem is defined as

$$\underset{\boldsymbol{x}}{\text{Minimize}} \quad \max_{\theta_1 \in \{1,2,3\}} g(\boldsymbol{x}; \boldsymbol{\xi}^{\mathrm{B}}(\theta_1)) \tag{4.61a}$$

$$\text{subject to} \quad x^{\mathrm{L}} \leq x_i \leq x^{\mathrm{U}} \qquad (i = 1, \cdots, m). \tag{4.61b}$$

The solution of problem (4.61) is referred to as nominal optimal solution (NOS). Each optimal solution is shown in Table 4.6.

The average cross-sectional area of 1 to 3 stories of ROS is 1.3 times as large as those of NOS, and the average cross-sectional area of 7 to 9 stories of ROS is half of that of NOS. To test the validity of the results, we apply Monte Carlo simulation to ROS and NOS with random samples of size 1000. The maximum value, mean value and standard deviation of the observed maximum interstory drift angles are summarized in Table 4.7.

The 10% worst values (0.9th quantiles) are also shown in Table 4.7. Moreover, the value of the order statistic $g_{2:38}$ and the ratio of the number of samples smaller than $g_{2:38}$ are also summarized in Table 4.7. The standard deviation in ROS decreases by 41% from NOS. The numerical result indicates

Table 4.6
Optimal Solutions (Cross-sectional Area of Brace, Unit:cm^2)

	x_1	x_2	x_3	x_4	x_5	x_6	x_7	x_8	x_9
ROS	149.5	198.1	144.5	142.4	107.9	74.3	60.3	199.6	24.5
NOS	102.0	150.0	132.5	102.0	126.0	94.0	200.0	200.0	114.7

Table 4.7
Summary of Observed Maximum Interstory Drift Angles

	Maximum (rad)	Mean (rad)	Std. dev. (rad)	Worst 10% (rad)	$g_{2:38}$ (rad)	Ratio (%)
ROS	0.023	0.014	0.0025	0.017	0.020	98
NOS	0.030	0.014	0.0042	0.021	0.027	98

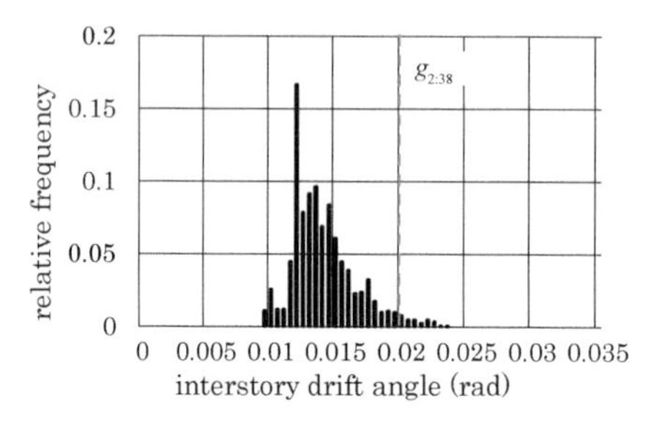

(a) Robust optimal solution

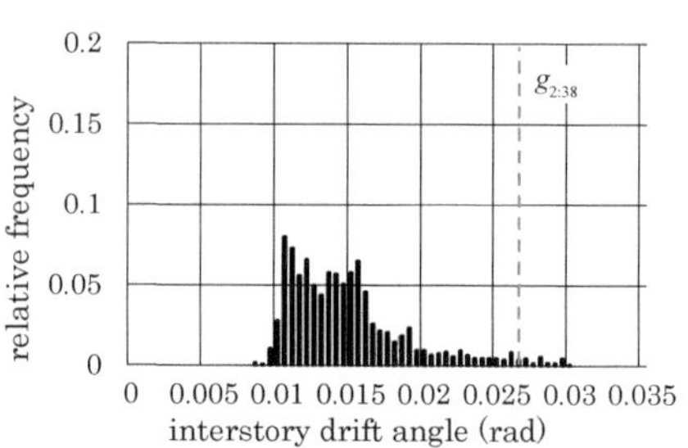

(b) Nominal optimal solution

Figure 4.11 Histogram of observed maximum interstory drift angles.

the use of several tens of samples enables to predict a larger value than the 10% worst value as shown in *Ratio* in Table 4.7. Histogram of the observed maximum interstory drift angles is shown in Fig. 4.11. It is confirmed that the variation of maximum interstory drift angles of ROS is smaller than that of NOS.

4.4 SAMPLING METHOD WITH LINEAR CONSTRAINTS

In this section, our interest focuses on the WCA in which variation of uncertain parameters is given within a set of linear constraints. To apply the order statistics-based RS to the linearly constrained problem, we deal with sampling a point from the uniform distribution over a bounded volume in

r-dimensional space defined by a set of linear constraints, i.e., a convex polytope, defined as

$$K = \left\{ \boldsymbol{\theta} \in \mathbb{R}^r \mid \boldsymbol{a}_i^\top \boldsymbol{\theta} \leq b_i,\ i = 1, \cdots, q \right\}, \tag{4.62}$$

where $\boldsymbol{a}_i \in \mathbb{R}^r$ is a column vector and $b_i \in \mathbb{R}$ is a real number. In this section, all vectors are assumed to be column vectors unless explicitly defined as row vector. Uniform sampling on the convex polytope is not a trivial task. Such sampling can be done by acceptance-rejection methods; however, it can be quite inefficient due to a low acceptance rate for a high-dimensional problem. The common approach is to use Markov chain Monte Carlo (MCMC) algorithms. Sampling by MCMC is the only known way for a polynomial-time algorithm. It has been recently reported that MCMC by Dikin walk (DW) based on interior point methods generates *asymptotically* uniform distribution, and run-time of the algorithm is strongly polynomial [Kannan and Narayanan, 2012]. We present an order statistics-based RS with MCMC for linearly constrained WCA. The method is applicable to the problem that has linear constraints and non-smooth, non-convex and noisy response functions.

4.4.1 FORMULATION OF A ROBUST DESIGN PROBLEM WITH POLYTOPE

Consider a optimization problem that is a specialized version of the problem defined in Sec. 4.2.1, where the uncertain parameter set is restricted to the polytope K. Thus, the problem is described as

$$\underset{\boldsymbol{x}}{\text{Minimize}} \quad g(\boldsymbol{x}; \boldsymbol{\theta}) \text{ for all } \boldsymbol{\theta} \in K \tag{4.63a}$$

$$\text{subject to} \quad \boldsymbol{x} \in \mathcal{X}. \tag{4.63b}$$

The order statistics-based RS method stated in Sec. 4.3.1 is utilized to problem (4.63). Thus, another RDO problem is formulated as

$$\text{Minimize} \quad g_{k:n}(\boldsymbol{x}) \tag{4.64a}$$

$$\text{subject to} \quad x^{\mathrm{L}} \leq x_i \leq x^{\mathrm{U}} \quad (i = 1, \cdots, m), \tag{4.64b}$$

where n samples of uncertain parameter vectors denoted by $\boldsymbol{\theta}_1, \cdots, \boldsymbol{\theta}_n \in K$ are randomly generated in the polytope, and $g_{k:n}(\boldsymbol{x})$ is the kth value among $g(\boldsymbol{x}; \boldsymbol{\theta}_1), \cdots, g_i(\boldsymbol{x}; \boldsymbol{\theta}_n)$ which is regarded as a realization of the kth order statistic as stated in Sec. 4.2.1.

4.4.2 BARRIER MONTE CARLO METHOD

To apply the order statistics-based RS to the linearly constrained problem, we deal with sampling a point from the uniform distribution over a convex polytope K defined in (4.62). MCMC methods begin at a starting point in the polytope and then randomly wander through the space. Several methods for generating random walks are known, e.g., the grid walk, the ball walk, the hit-and-run walk and the DW. In these random walks, the points generated after a sufficient number of steps are shown to follow a nearly uniform stationary distribution in the polytope K. The number of walk steps required to achieve this stationary distribution is called the mixing time of the walk. The first polynomial time algorithm appeared in Dyer et al. [1991]. The dependence of its mixing time on the dimension r was $O(r^{23})$. Kannan and Narayanan [2012] proposed the use of DW in the polytope, in which the interior-point method for nonlinear optimization is connected with the random walk. They proved that the mixing time of the DW is $O(qr)$ utilizing a warm start in the polytope. Another efficient approach is known as hit-and-run random walk [Smith, 1984], which was also combined with the technique of interior-point methods to propose the barrier hit-and-run walk (BHRW) by Polyak and Gryazina [2010]. Among these algorithms, we attempt to use the two algorithms of MCMC to WCA. One is the DW proposed by Kannan and Narayanan [2012] and the other is the BHRW. The two methods similarly exploit directions uniformly distributed in Dikin ellipsoids; however, the step size strategy is different. Before presenting details of these methods, we introduce an approach to exploit Dikin ellipsoid and barrier function.

Dikin ellipsoid

For polytope K, the log-barrier function is given by

$$\phi(\boldsymbol{\theta}) = -\sum_{i=1}^{q} \log\left(b_i - \boldsymbol{a}_i^\top \boldsymbol{\theta}\right),$$

which measures closeness of a point $\boldsymbol{\theta} \in \operatorname{int}(K)$ to the boundary. As $\boldsymbol{\theta}$ approaches the boundary of K, the value of $\phi(\boldsymbol{\theta})$ approaches infinity. Such barrier function is widely used in interior-point methods for nonlinear optimization [Nesterov and Nemirovskii, 1994]. Hessian of the log-barrier function is given by

$$\boldsymbol{H}_{\boldsymbol{\theta}} = \nabla^2 \phi(\boldsymbol{\theta}) = \sum_{i=1}^{q} \frac{\boldsymbol{a}_i \boldsymbol{a}_i^\top}{(b_i - \boldsymbol{a}_i^\top \boldsymbol{\theta})^2} = \boldsymbol{A}^\top \boldsymbol{D} \boldsymbol{A},$$

where

$$\boldsymbol{A} = \begin{pmatrix} \boldsymbol{a}_1 & \boldsymbol{a}_2 & \cdots & \boldsymbol{a}_q \end{pmatrix}^\top,$$

$$\boldsymbol{D} = \operatorname{diag}\left[\frac{1}{(b_1 - \boldsymbol{a}_1^\top \boldsymbol{\theta})^2}, \ldots, \frac{1}{(b_q - \boldsymbol{a}_q^\top \boldsymbol{\theta})^2}\right].$$

An ellipsoid specified by $\boldsymbol{H}_\theta$ and centered at $\boldsymbol{\theta} \in \operatorname{int}(K)$ with radius ρ such that

$$D_{\boldsymbol{\theta}} = \left\{\boldsymbol{\xi} \in \mathbb{R}^r \mid (\boldsymbol{\xi} - \boldsymbol{\theta})^\top \boldsymbol{H}_\theta (\boldsymbol{\xi} - \boldsymbol{\theta}) \leq \rho^2\right\}$$

is called Dikin ellipsoid. It can be proved that $D_\theta \subset K$ for $\rho \leq 1$, that is, the Dikin ellipsoid with radius 1 is the collections of all points around $\boldsymbol{\theta}$ whose

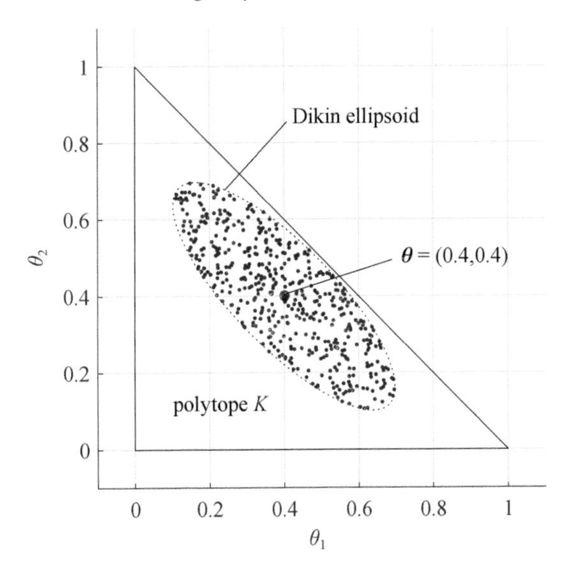

Figure 4.12 Dikin ellipsoid for a polytope K.

distance from $\boldsymbol{\theta}$ is within the unit threshold with respect to the Hessian norm [Nesterov, 2004].

We set radius $\rho = 1$ hereinafter. For example, Dikin ellipsoid centered at $\boldsymbol{\theta} = (0.4, 0.4)^\top$ and random uniform points in the ellipsoid for polytope

$$K = \{(\theta_1, \theta_2) \mid \theta_1 \geq 0,\ \theta_2 \geq 0,\ \theta_1 + \theta_2 \leq 1\}$$

are illustrated in Fig. 4.12.

Dikin walk

Kannan and Narayanan [2012] proposed DW as a new random walk using Dikin ellipsoids, and proved that the mixing time for DW is polynomial. DW is a *Metropolis* type walk which picks a move and then decides whether to *accept* the move or *reject* and stay. The algorithm of DW is shown in Algorithm 4.1, where vol$(\cdot)$ is the volume of the set.

BHRW is another type of random walk using Dikin ellipsoids [Polyak and Gryazina, 2010]. BHRW also assures that the distribution of sampled points tends to be uniform but lacks estimation of the rate of convergence to a uniform distribution. In theory, DW has advantage of strongly polynomial mixing time, however in practice, random walk of hit-and-run type is not inferior to DW and often shows better results than DW [Polyak and Gryazina, 2010; Huang and Mehrotra, 2015]. The BHRW method is described in Algorithm 4.2. By the both Algorithms 4.1 and 4.2, for number of desired points n, we obtain the feasible sample set

$$\{\boldsymbol{\theta}_1 = \boldsymbol{\eta}_1, \cdots, \boldsymbol{\theta}_n = \boldsymbol{\eta}_n\}$$

on polytope K.

Algorithm 4.1 Dikin walk (DW).

1: Assign the starting point $\boldsymbol{\theta}_0 \in \text{int}(K)$; set $i = 0$.
2: Flip a fair coin. If head, stay at $\boldsymbol{\theta}_i$.
3: Otherwise, pick a random point $\boldsymbol{\eta}$ from the ellipsoid $D_{\boldsymbol{\theta}_i}$.
4: If $\boldsymbol{\theta}_i \notin D_{\eta}$, then reject $\boldsymbol{\eta}$ (stay at $\boldsymbol{\theta}_i$); otherwise, accept $\boldsymbol{\eta}$ with probability

$$\alpha_{\text{DW}} = \min\left\{1, \frac{\text{vol}(D_{\boldsymbol{\theta}_i})}{\text{vol}(D_{\boldsymbol{\eta}})}\right\} = \min\left\{1, \sqrt{\frac{\det(\boldsymbol{H}_{\boldsymbol{\eta}})}{\det(\boldsymbol{H}_{\boldsymbol{\theta}_i})}}\right\}.$$

5: If $\boldsymbol{\eta}$ is accepted, then substitute i with $i+1$ and set $\boldsymbol{\theta}_i = \boldsymbol{\eta}$.
6: Go back to Step 2 until number of obtained points reaches the desired value n.

Algorithm 4.2 Barrier hit-and-run walk (BHRW).

1: Assign the starting point $\boldsymbol{\theta}_0 \in \text{int}(K) \subset \mathbb{R}^r$; set $i = 0$.
2: Pick a random direction $\boldsymbol{d}$ uniformly in $D_{\boldsymbol{\theta}_i}$.
3: Pick a point t uniformly distributed in $T = \{t \in \mathbb{R} \mid \boldsymbol{\theta}_i + t\boldsymbol{d} \in K\}$.
4: Generate a point $\boldsymbol{\eta} = \boldsymbol{\theta}_i + t\boldsymbol{d}$.
5: Calculate

$$\rho_{\boldsymbol{\theta}_i} = \frac{\|\boldsymbol{d}\|}{\sqrt{\boldsymbol{d}^\top \boldsymbol{H}_{\boldsymbol{\theta}_i}\boldsymbol{d}}}, \quad \rho_{\boldsymbol{\eta}} = \frac{\|\boldsymbol{d}\|}{\sqrt{\boldsymbol{d}^\top \boldsymbol{H}_{\boldsymbol{\eta}}\boldsymbol{d}}}.$$

6: Accept $\boldsymbol{\eta}$ with probability

$$\alpha_{\text{BHRW}} = \min\left\{1, \left(\frac{\rho_{\boldsymbol{\theta}_i}}{\rho_{\boldsymbol{\eta}}}\right)^r \sqrt{\frac{\det(\boldsymbol{H}_{\boldsymbol{\eta}})}{\det(\boldsymbol{H}_{\boldsymbol{\theta}_i})}}\right\}.$$

7: If $\boldsymbol{\eta}$ is accepted, then substitute i with $i+1$ and set $\boldsymbol{\theta}_i = \boldsymbol{\eta}$.
8: Go back to Step 2 until the number of obtained points reaches the desired value n.

Comparison of uniformity of sampled points

For comparison, the acceptance-rejection method (ARM) is also used in the test. The algorithm is summarized in Algorithm 4.3. ARM is one of the most useful general methods for sampling from uniform distribution. However, it is known that its efficiency rapidly deteriorates as the number of dimensions increases. To compare the uniformity of sampling by DW, BHRW and ARM, consider a polytope

$$K = \left\{\boldsymbol{\theta} = (\theta_1, \cdots, \theta_r) \in \mathbb{R}^r \;\middle|\; \theta_1 \geq 0, \; \cdots, \theta_r \geq 0, \; \sum_{i=1}^{r} \theta_i \leq 1\right\}. \tag{4.65}$$

Algorithm 4.3 Acceptance–rejection method (ARM).

1: Take simple $G \supset K$, e.g., a box.
2: Pick a random point $\boldsymbol{\eta}$ from uniform distribution in G.
3: If $\boldsymbol{\eta} \in K$, then accept $\boldsymbol{\eta}$; otherwise reject $\boldsymbol{\eta}$.
4: If $\boldsymbol{\eta}$ is accepted, then substitute i with $i+1$ and set $\boldsymbol{\theta}_i = \boldsymbol{\eta}$.
5: Go back to Step 2 until the number of obtained points reaches the desired value n.

First, we set $r = 2$ (2-dimensional space) in (4.65) and number of the desired points $n = 1000$. Scatter graphs obtained by the methods are shown in Fig. 4.13. In theory, marginal cdf of the ith element of $\boldsymbol{\theta}$ can be calculated in a closed form independently of the dimensionality r as follows:

$$\Pr\{\Theta_i \leq t\} = F_{\Theta}(t) = t(2-t) \qquad (i = 1, \cdots, r),$$

where $\boldsymbol{\Theta} = (\Theta_1, \cdots, \Theta_r)$ is a random variable vector uniformly distributed in K. For θ_1, we construct an empirical marginal cdf

$$\hat{F}_{\Theta}(t) = \frac{\text{number of the sample points } \boldsymbol{\theta} = (\theta_1, \cdots \theta_r) \text{ such that } \theta_1 \leq t}{n},$$

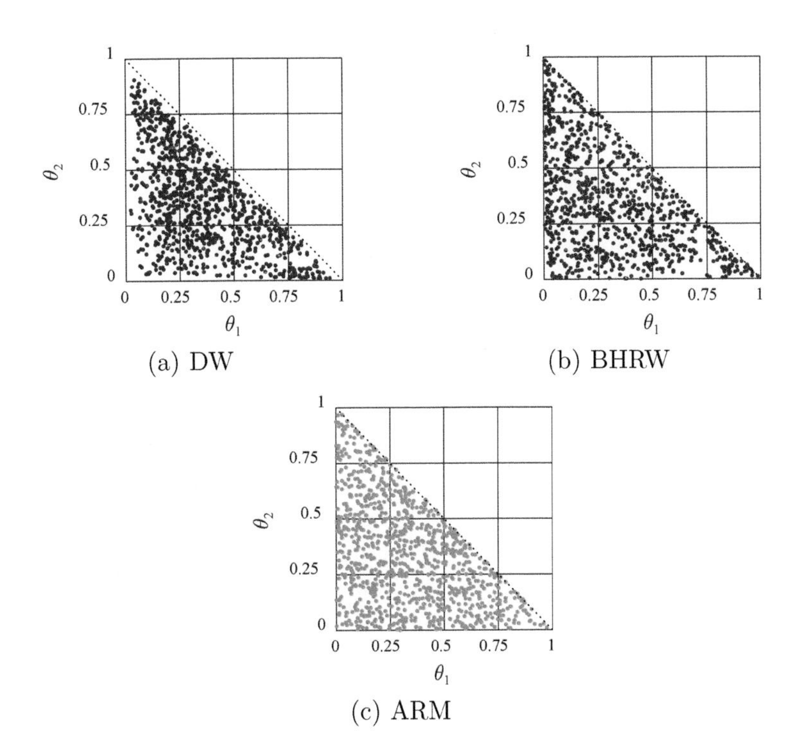

Figure 4.13 Scatter graph on the 2-dimensional polytope.

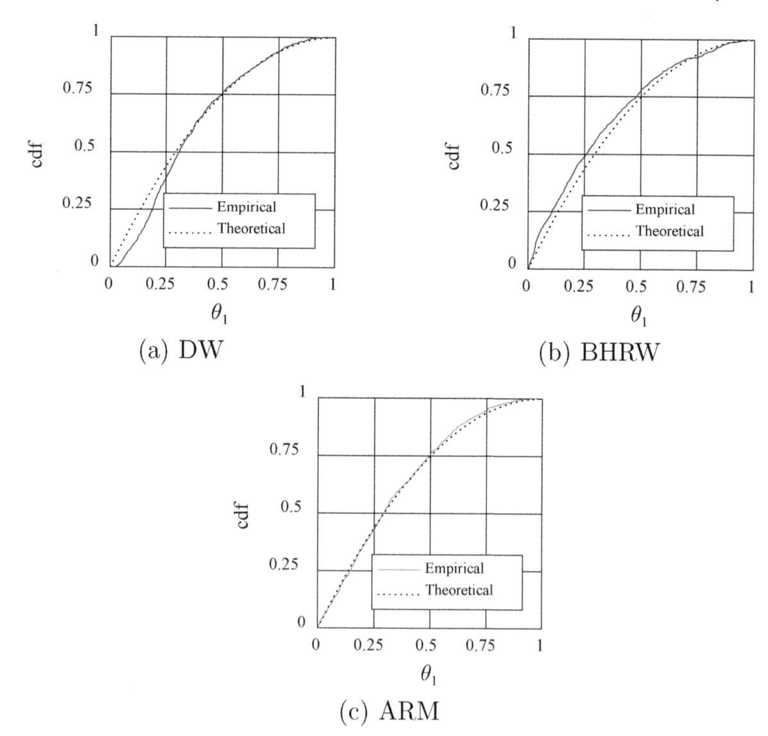

Figure 4.14 Cumulative distribution function on the 2-dimensional polytope.

where $\boldsymbol{\theta}$ is generated by any one of DW, BHRW and ARM. If a method generates random vectors that uniformly distributes on K, then $\hat{F}_\Theta(t)$ almost surely converges to $F_\Theta(t)$ as n approaches infinity. Empirical cdf $\hat{F}_\Theta(t)$ obtained by the methods for $n = 1000$ and corresponding theoretical cdf $F_\Theta(t)$ are compared in Fig. 4.14. From the results in Figs. 4.13 and 4.14, all the methods can generate approximately uniformly distributed points and BHRW shows slightly better performance than DW.

Next, we investigate the dependency on n in each method by evaluating the following error function:

$$e_i = F_\Theta(\theta_1^i) - \hat{F}_\Theta(\theta_1^i) \qquad (i = 1, \cdots, n),$$

where θ_1^i denotes the first element of the ith sample point $\boldsymbol{\theta}_i$. The mean value of the error function is denoted by e^{mean} and the values by each method for $n = 10^2, 10^3, 10^4, 10^5$ are summarized in Table 4.8, from which it is confirmed that the difference between theoretical and empirical cdfs decreases as n increases for all the methods.

Finally, we investigate the relation between the number of accepted points and dimensionality r. In the above example of $r = 2$, ARM shows the best performance; however, it is known that ARM is inefficient in high dimensional

Table 4.8
Summary of the Mean Error on the cdf to Various Values of n.

n	DW	BHRW	ARM
10^2	0.2406	0.0676	0.0179
10^3	0.0401	0.0372	0.0085
10^4	0.0297	0.0049	0.0034
10^5	0.0093	0.0013	0.0004

Table 4.9
Number of Required Iterations for 1000 Feasible Samples ($n = 1000$)

r	2	3	4	5	6	7
DW	3733	4637	5442	5863	6216	6776
BHRW	1377	1649	1809	2039	2284	2370
ARM	1952	6049	23219	124727	738377	4892040

spaces. For the dimensionality $r = 2, \cdots, 7$ of polytope in (4.65), numbers of required iterations for 1000 feasible samples are summarized in Table 4.9, which indicates that ARM is applicable to only a few dimensional problem and not appropriate for our purpose. From all the results, BHRW shows better performance than DW. Hence, we conclude that BHRW is the most appropriate algorithm for sampling in the feasible region with large number of uncertain parameters.

4.4.3 APPLICATION TO SEISMIC DESIGN PROBLEM

Consider a seismic design of 10-story shear frame structure subjected to two input earthquake motions. Configuration of the model is shown in Fig. 4.15(a), where $\boldsymbol{x} = (x_1, \cdots, x_{10})$ represents the story shear coefficients. The relation between the ith interstory drift angle and the ith story shear force is parameterized by x_i as shown in Fig. 4.15(b), where the kinematic hardening is used for the cyclic material property. WCA problem is defined as

$$\underset{\boldsymbol{\theta}}{\text{Maximize}} \quad \delta_{\max}(\boldsymbol{x} + \Delta\boldsymbol{x}; \Delta\boldsymbol{T}) \tag{4.66a}$$

$$\text{subject to} \quad \boldsymbol{\theta} = (\Delta\boldsymbol{x}, \Delta\boldsymbol{T})^\top \in K, \tag{4.66b}$$

where $\delta_{\max}$ is the maximum interstory drift angle among all stories, and $\boldsymbol{\theta} = (\Delta\boldsymbol{x}, \Delta\boldsymbol{T})^\top \in \mathbb{R}^{12}$ is the vector of uncertain parameters which consists of variation of the structural parameters $\Delta\boldsymbol{x} = (\Delta x_1, \cdots, \Delta x_{10}) \in \mathbb{R}^{10}$ and input

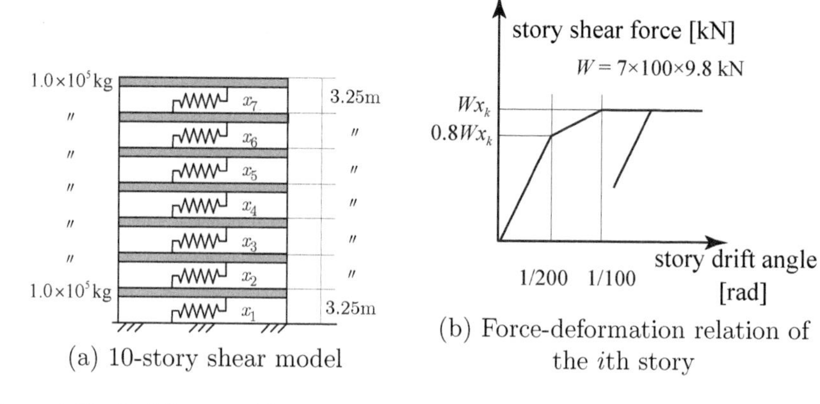

Figure 4.15 Building model.

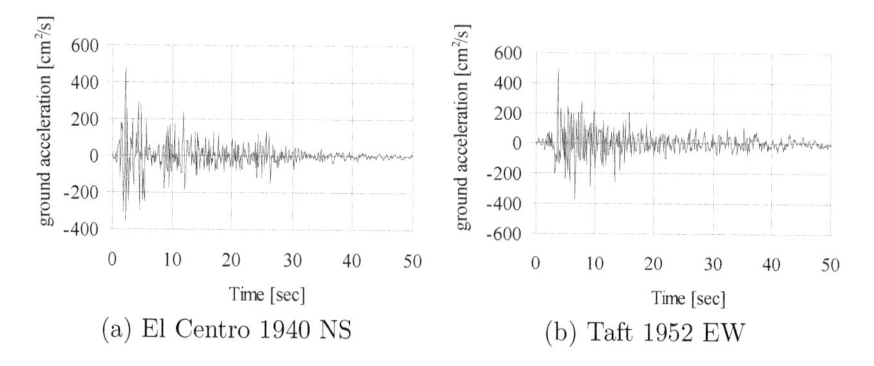

Figure 4.16 Input earthquake motions (ground acceleration).

ground motion parameter $\Delta\boldsymbol{T} = (\Delta T_1, \Delta T_2) \in \mathbb{R}^2$. The uncertain parameter $\Delta\boldsymbol{T}$ represents the variation of the periods in Fourier amplitude spectrum of the input earthquake motions shown in Fig. 4.16. The maximum interstory drift angle is obtained by nonlinear response history analysis, in which ground accelerations of input earthquake motions are generated by assigning random parameters $\Delta\boldsymbol{T}$. The nominal value of design variable vector $\boldsymbol{x}$ is given as

$$\boldsymbol{x} = (0.36, 0.35, 0.34, 0.32, 0.30, 0.29, 0.32, 0.39, 0.40, 0.40),$$

which has been obtained by design optimization without uncertainty, i.e., $\boldsymbol{\theta} = \boldsymbol{0}$. Polytope K in problem (4.66) is given as

$$K = \left\{ (\Delta\boldsymbol{x}, \Delta\boldsymbol{T}) \in \mathbb{R}^{12} \;\middle|\; \begin{array}{ll} -0.05x_i \leq \Delta x_i \leq 0.05x_i & (i = 1, \cdots, 10), \\ -0.1 \leq \Delta T_j \leq 0.1 & (j = 1, 2), \\ \displaystyle\sum_{i=1}^{10} |\Delta x_i| \leq 0.1 & \end{array} \right\}. \tag{4.67}$$

Table 4.10
Observed Sample Values of Order Statistics of 200 Different Sample Sets (Unit: 10^{-3} rad)

Order statistics	$g_{1:44}$	$g_{2:64}$	$g_{3:81}$	$g_{4:97}$	$g_{5:113}$	$g_{20:312}$
Mean value	9.987	9.913	9.877	9.863	9.849	9.767
Std. dev.	0.158	0.120	0.097	0.083	0.075	0.040

Table 4.11
Accuracy of Prediction by Order Statistics of 200 Sets

Order statistics	$g_{1:44}$	$g_{2:64}$	$g_{3:81}$	$g_{4:97}$	$g_{5:113}$	$g_{20:312}$
Mean value of Γ	0.976	0.967	0.962	0.960	0.957	0.935
Std. dev. of Γ	0.022	0.023	0.021	0.020	0.018	0.014
Number of $\Gamma \geq 0.9$	197	196	198	199	198	199
Proportion of $\Gamma \geq 0.9$	98.5%	98.0%	99.0%	99.5%	99.0%	99.5%

Note that the inequality $\sum_{i=1}^{10} |\Delta x_i| \leq 0$ needs $2^{10} = 1024$ linear inequalities and the total number of linear inequalities in (4.67) is calculated as $1024 + 10 \times 2 + 2 \times 2 = 1048$. Therefore, polytope K can be reformulated to linear constraints $\{\boldsymbol{\theta} \mid \boldsymbol{A\theta} \leq \boldsymbol{b}\}$ with a matrix $\boldsymbol{A} \in \mathbb{R}^{1048\times 12}$ and a vector $\boldsymbol{b} \in \mathbb{R}^{1048}$. Thus, problem (4.66) has more than 1000 linear constraints. In this example, we use only BHRW to obtain samples. Furthermore, when n samples are needed, $10n$ samples are generated by BHRW and n samples are then randomly selected from among the $10n$ samples to improve the uniformity. By this approach, we observe the sample values of the kth order statistics of $G = \delta_{\max}$, which are denoted by $G_{k:n}$ and are expected to be the top 10% worst value ($\gamma = 0.9$) with a probability of 99% as shown in Table 4.1(b). For 200 different sample sets, mean values and standard deviations of the observed order statistics denoted by $g_{k:n}$ are summarized in Table 4.10.

To test the validity of prediction by the order statistics, further 10^6 samples $\{g_1, \cdots, g_{1000000}\}$ are obtained on the polytope by the approach stated above. By using this sample set, we compute the following proportion to the total sample size 10^6:

$$\Gamma = \frac{\text{number of samples among } \{g_1, \cdots, g_{1000000}\} \text{ such that } g_i \leq g_{k:m}}{10^6},$$

which represents the empirical proportion of the samples below the order statistics. In theory, the proportion Γ is predicted to be more than 0.9 with a probability of 99%. The mean values and standard deviations of the 200 different sample sets are summarized in Table 4.11. In addition, the number

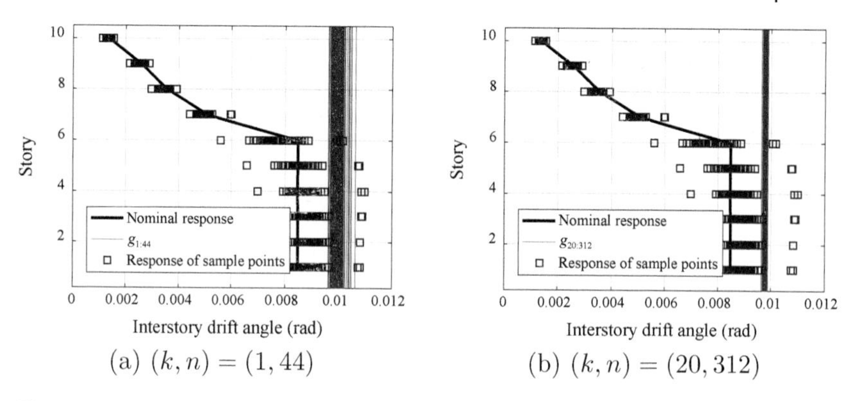

Figure 4.17 Observed maximum interstory drift angles of each story of samples and corresponding order statistics.

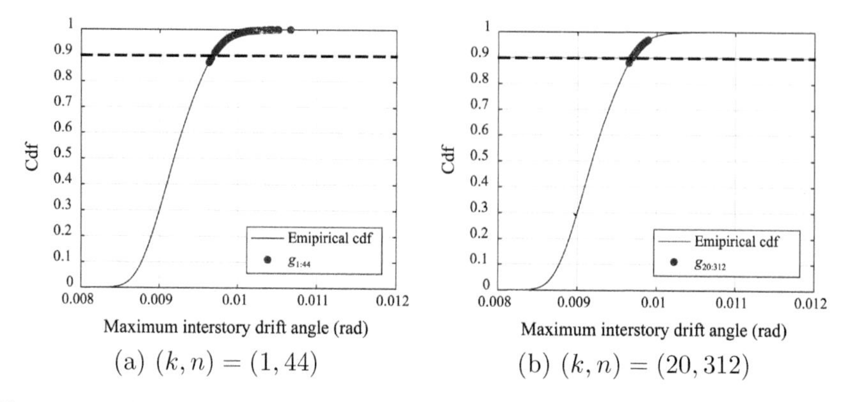

Figure 4.18 Empirical cdf of observed maximum interstory drift angle of samples and corresponding order statistics.

and proportion of sets satisfying $\Gamma \geq 0.9$ among the 200 sets are shown in Table 4.11, which indicates that all the order statistics can predict the 90% worst value with a probability of approximately 99%. If a larger number is used for the order k, the accuracy is further improved although a larger number of sample size n is needed. Interstory drift angles of each story at the sample points and the observed order statistics of the 200 sets are shown in Fig. 4.17. Nominal responses, i.e., interstory drift angles without uncertainties are also shown. As observed in Fig. 4.17, the nominal responses are uniformly distributed in the lower stories because it is obtained by nominal optimization without uncertainty; however, variation of responses in the sample set is not uniform. Thus, uncertainty spoils the optimality of the solution. Empirical cdf of maximum interstory drift angle of 10^6 samples are shown in Fig. 4.18. The observed values of order statistics $g_{1:44}$ and $g_{20:312}$ are also plotted in Figs. 4.17 and 4.18, respectively. In Fig. 4.18, it can be seen that almost all of the values of the order statistics are located over 0.9 in the cdf, and the ratio of which is close to 99% as stated above.

Thus, we can successfully apply the order statistics-based WCA to the problem constrained with more than 1000 linear inequalities, and predict approximately worst values on the polytope with pre-assigned accuracy. This also means that the samples can be uniformly generated on the polytope. Moreover, this approach can be extended to a general type of uncertain variation within a convex set. In general, prediction of the extreme value using small sample set is not easy and we may not need such exact extreme in many practical situations. The presented method would be useful in such situations.

4.5 SUMMARY

Order statistics-based approaches of RDO and WCA have been presented in this chapter. The RDO problem has been first formulated as a two-level optimization problem. Basic definitions and properties of distribution free one-sided tolerance interval have been explained and verified using a mathematical problem. Confidence level of approximation of the worst value can be ensured by selecting appropriate number of samples; thus, leading to an explicit stopping rule of PRS based on the theory of order statistics. It has been shown that the order statistics can be used for practical seismic design problems of building frames considering uncertainty in parameters defining the material properties, seismic motions and ground properties. For application of RS methods utilizing sampling in the feasible region, it is necessary to generate samples uniformly or with the specified distribution in a high dimensional space bounded with some constraints. Therefore, sampling methods based on random work have been introduced in Sec. 4.3, and application to frame design problem with many linear constraints have been presented. Effectiveness of order-statistics based approach will be demonstrated for application to robust geometry and topology optimization in Chap. 5 and multi-objective robust optimization in Chap. 6.

ACKNOWLEDGMENT

We acknowledge the permissions from publishers for use of the licensed contents as follows:

- Figures 4.9, 4.10 and 4.11, and Tables 4.6 and 4.7: Reproduced with permission from John Wiley and Sons [Yamakawa and Ohsaki, 2021].

5 Robust Geometry and Topology Optimization

In geometry and topology optimization of trusses and frames, connectivity of nodes by members and locations of the nodes are simultaneously optimized. Many of such configuration optimization methods have drawbacks; convergence to an unstable structure with possible existence of very long/short and/or overlapping members, and existence of melting or coalescent nodes resulting in a singular stiffness matrix. Accordingly, the optimal solution may not have enough robustness. Order statistics are utilized for approximating the worst response of frames under uncertainty in cross-sectional properties and nodal locations. Quantile-based sequential approach to reliability-based design optimization is introduced. Some results of robust topology optimization are presented for plane frames.

5.1 INTRODUCTION

Geometry and topology optimization, also called configuration optimization, of trusses and frames has been extensively studied since the pioneering work by Maxwell [1890] and Michell [1904]; see Ohsaki [2010] for overview of the research in this field. One of the drawbacks for geometry and topology optimization, where connectivity of nodes by members and locations of the nodes are simultaneously optimized, is that such optimization process often converges to an unstable structure with possible existence of very long/short and/or overlapping members. Accordingly, the optimal solution may not have enough robustness.

Unstable solution can be avoided by incorporating stability constraint in the problem formulation [Rozvany, 1996; Zhou, 1996; Ohsaki and Ikeda, 2007]. Among various definition of stability, the linear buckling is widely considered in structural optimization because it does not incorporate prebuckling deformation, and therefore, does not require higher order geometrical nonlinearity [Tugilimana et al., 2018]. For structures with non-negligible prebuckling deformation, e.g., shallow latticed shells, so called knockdown factor can be used for estimating nonlinear buckling load factor from the linear buckling load factor [Ohsaki and Kanno, 2001].

The truss topology optimization problem with linear buckling constraint may be formulated as a standard nonlinear programming problem, which can be reformulated as a problem of sequentially solving a semidefinite programming problem [Kanno et al., 2001]. It is also well known that the optimal topology exists at an isolated point of the feasible region when stress constraints are considered under multiple loading conditions, and some relaxation

DOI: 10.1201/9781003153160-5

methods should be used for finding the global optimal solution [Cheng and Guo, 1997]. Guo et al. [2001] extended the relaxation method to incorporate local stability constraints corresponding to the existence of members with small cross-sectional areas. Descamps and Filomeno Coelho [2014] used force density, which is defined as the force divided by member length, as an intermediate variable, and avoided the local instability by applying small forces at nodes to produce geometric imperfection. Torii et al. [2015] used beam elements to incorporate global and local instability simultaneously.

Another difficulty in geometry and topology optimization of trusses and frames is the existence of melting or coalescent nodes resulting in a singular stiffness matrix [Achtziger, 2007]. Ohsaki [1998] modeled a regular grid truss with uniform cross-section as a frame, and proposed a method for reducing the cross-sectional areas of colinear members leaving a single member using a Sigmoid function. Wang et al. [2002] proposed a node shift method. Ohsaki and Hayashi [2017] and Hayashi and Ohsaki [2019] utilized force density method (FDM) to avoid solving the stiffness equation for a truss with closely spaced nodes. Shen and Ohsaki [2021] extended this method to frames by introducing an auxiliary truss or cable-net. Note that these methods do not consider uncertainty in structural properties or loading conditions, which inevitably exists in the real world structures.

Among many optimization approaches considering uncertainty, the methods based on order statistics and reliability-based design optimization (RBDO) are utilized in this chapter. The details of order statistics-based approach are explained in Chap. 4. A brief summary of RBDO methods is presented below [Frangopol, 1985; Choi et al., 2007].

According to Valdebenito and Schuëller [2010] and Aoues and Chateauneuf [2010], the methods of RBDO can be classified into the following three categories based on the difference in incorporating probabilistic constraints, and their corresponding schematic procedures are shown in Algorithms 5.1, 5.2 and 5.3.

(1) Double-loop method: This method directly solves the RBDO problem evaluating the probability constraints at each iteration of optimization; therefore, there are two nested cycles: the inner cycle evaluates the probability constraints at each iteration of the outer cycle of optimization procedure.

(2) Single-loop method: This method avoids the inner cycle in double-loop method by converting probability constraints into approximate deterministic constraints at the most probable point (MPP) [Chen et al., 1997]. In this way, it is allowed to avoid the inner cycle and reduces the computational cost. Another construction of single-loop method is to use the Karush-Kuhn-Tucker optimality conditions to convert the RBDO problem to a deterministic optimization problem [Kuschel and Rackwitz, 1997].

Algorithms Strategies for solving RBDO problem

Algorithm 5.1 Double-loop method.

Require: Assign initial values to design variables.
1: **repeat** (outer cycle)
2:     **repeat** (inner cycle)
3:         Perform reliability analysis.
4:         Evaluate probability constraints.
5:     **until** converge
6:     Calculate for updating design variables.
7: **until** converge
8: Output the solution and terminate the process.

Algorithm 5.2 Single-loop method.

Require: Assign initial values to design variables.
1: **repeat**
2:     Convert probability constraint into approximate deterministic. constraint
3:     Calculate for updating design variables.
4: **until** converge
5: Output the solution and terminate the process.

Algorithm 5.3 Decoupling method.

Require: Assign initial values to design variables.
1: **repeat**
2:     Solve deterministic optimization problem.
3:     Output optimal design variables.
4:     Perform reliability analysis and calculate shifted values.
5: **until** converge
6: Output the solution and terminate the process.

(3) Decoupling method: This method asymptotically finds the solution of the RBDO problem by solving a series of sub-deterministic optimization problems, which are formulated based on information from the reliability analysis to avoid implementation of rigorous reliability analysis each time when a new design point is found by the optimization procedure.

Although the double-loop method is easy to implement, the computational cost for design optimization and reliability analysis is very high [Enevoldsen and Sørensen, 1994]. Therefore, great effort has been made to improve computational efficiency of the double-loop method; e.g., reformulation of probability constraints [Tu et al., 1999] and reduction of the number of function evaluations using the dimension reduction method (DRM) [Lee et al., 2008b]. Furthermore, Chen et al. [1997] proposed the single-loop single-vector method for searching MPP in a constant direction. Liang et al. [2008] extended this method by adaptively calculating the search direction. By contrast, the single-loop method is regarded as one of the most efficient approaches for solving RBDO problems [Meng and Keshtegar, 2019]; however, its numerical instability should be resolved when the limit state function is highly nonlinear [Aoues and Chateauneuf, 2010].

The decoupling method decouples the inner reliability analysis from the outer structural optimization, and the probabilistic constraints are to be evaluated at the end of each sub-deterministic optimization problem, e.g., by utilizing the sensitivity information [Weiji and Li, 1994]. Du and Chen [2004] proposed the sequential optimization and reliability assessment (SORA) method utilizing the shifting vector, which is regarded as a sequential deterministic optimization (SDO) including reliability assessment. Since then, SORA has been widely used for solving RBDO problems [Aoues and Chateauneuf, 2010], and has been improved in various ways [Du, 2008; Chen et al., 2013]. However, these SORAs are based on MPP, and may not converge if there are multiple MPPs. To overcome this difficulty, a quantile-based approach is proposed by Li et al. [2019, 2020], which is extended using fractional moment-based maximum entropy method (MEM) [He et al., 2021], Laplace transform and inverse saddlepoint approximation [Do et al., 2021b].

Another difficulty in practical application of RBDO is that the exact probability information of uncertainty is usually unknown, and the reliability need to be calculated from a given set of random samples [Kanno, 2019; Yamakawa and Ohsaki, 2020]. However, accurate estimation of the sample central moments is very difficult [Hosking, 1990; Vapnik, 2000]. To overcome such difficulties, Pandey [2000] estimated the quantile function using MEM for specified sample probability weighted moments (PWMs) that are obtained from random samples, and showed better results than the method using sample central moments when sample size is small. Deng and Pandey [2008] extended this method utilizing fractional PWM. On the other hand, Pandey [2001a,b] presented an approach using minimum cross-entropy principle subject to PWM constraints. Hosking [2007] proposed a method for finding the distribution with maximum entropy for specified linear moments (L-moments), and showed that such distribution has a polynomial density-quantile function which can be obtained by numerical integration; see Sec. 1.4.5. It has also been shown by Hosking [2007] that specifying the first r L-moments is equivalent to specifying the expected values of the r order statistics.

In this chapter, some results of robust topology optimization are presented for plane frames. Order statistics are utilized for approximating the worst response of frames under uncertainty in cross-sectional properties and nodal locations. Quantile-based sequential deterministic approach to RBDO is introduced.

5.2 ORDER STATISTICS-BASED ROBUST TOPOLOGY OPTIMIZATION FOR PLANE FRAMES

5.2.1 ROBUST GEOMETRY AND TOPOLOGY OPTIMIZATION PROBLEM

In this section, we summarize the results in Shen et al. [2021] on robust geometry and topology optimization of plane frames to minimize the maximum stress under volume and linear buckling constraints. The order statistics in Sec. 1.4.2 are used for obtaining approximate worst values of the objective and constraint functions. The method in Shen and Ohsaki [2021] is used for determination of geometry utilizing the force densities of the auxiliary truss.

Uncertainty in member stiffness

Uncertainty is considered in the nodal locations and the cross-sectional areas. Figure 5.1(a) illustrates the ith frame member with the length ℓ_i in the global coordinates (x, y). Uncertainty is incorporated into the x- and y-coordinates of the end nodes, which are simply denoted by nodes 1 and 2, as

$$(X_j, Y_j) = (x_j, y_j) + (\Delta x_j, \Delta y_j) \quad (j = 1, 2), \tag{5.1}$$

where (X_j, Y_j) is the uncertain location of node j which is indicated by the filled circle in Fig. 5.1(a), and (x_j, y_j) is the corresponding value without uncertainty. The increment $(\Delta x_j, \Delta y_j)$ of the nodal location is taken as random variable.

The ith member with length L_i, after incorporating uncertainty, is divided into four beam elements with the same length by the intermediate nodes 3, 4 and 5, as shown in Fig. 5.1(b). The locations of intermediate nodes (X_j^0, Y_j^0) $(j = 2, 3, 4)$ are first obtained without considering eccentricity (geometrical

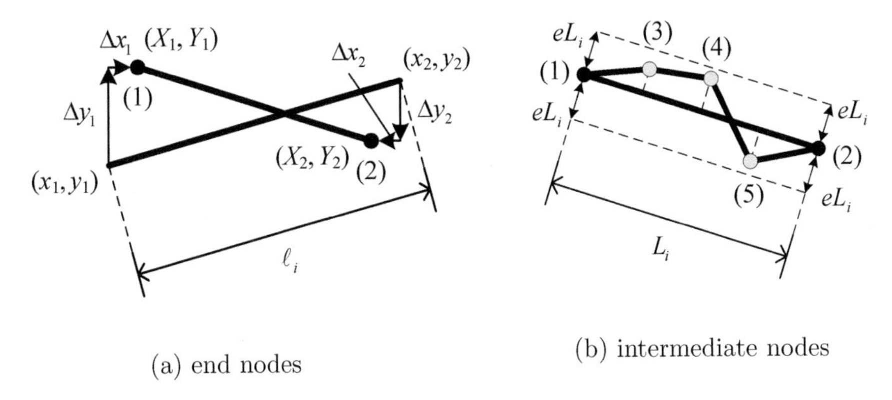

Figure 5.1 Uncertainty in nodal locations of the ith member.

imperfection, out-of-straightness) of members as

$$(X_j^0, Y_j^0) = (X_1, X_1) + (j-2)\left[\frac{(X_2, Y_2) - (X_1, Y_1)}{4}\right] \quad (j = 3, 4, 5), \quad (5.2)$$

where (X_1, Y_1) and (X_2, Y_2) are obtained from (5.1). For the specified upper-bound eccentricity e of a member, the uncertain locations of intermediate nodes, denoted by (X_j, Y_j) $(j = 3, 4, 5)$ and indicated by the blank circles in Fig. 5.1(b), are located randomly in the perpendicular direction of the member within the range of eL_i as

$$(X_j, Y_j) = (X_j^0, Y_j^0) + (\Delta x_j, \Delta y_j), \quad (5.3a)$$

$$\sqrt{\Delta x_j^2 + \Delta y_j^2} \leq eL_i, \quad (5.3b)$$

$$(\Delta x_j, \Delta y_j) \perp (X_2 - X_1, Y_2 - Y_1) \quad (j = 3, 4, 5). \quad (5.3c)$$

Accordingly, uncertainty in the locations of intermediate nodes 3, 4 and 5 consists of uncertainty in the locations of nodes 1 and 2 due to construction error and the randomness in eccentricity of the member due to manufacturing error. Therefore, correlation among uncertainties due to both errors should be appropriately incorporated; see Shen et al. [2021] for details.

A random value Δa_i is also added to the cross-sectional area a_i of the ith element, as follows, in a similar manner as (5.1), to obtain the uncertain cross-sectional area A_i:

$$A_i = a_i + \Delta a_i \quad (i = 1, \cdots, m_{\mathrm{e}}), \quad (5.4)$$

where m_{e} is the number of elements. Note that the four elements in each member has the same uncertain cross-sectional area.

Problem formulation

Consider a problem of minimizing the maximum stress under constraints on the structural volume and linear buckling load factor. Let $\sigma_{i,j}$ denote the von Mises stress evaluated at point $j \in \{1, \cdots, p\}$ of the ith element. The selection of stress evaluation points in each element will be explained in Sec. 5.2.2.

The nodes of members are classified into free and fixed nodes. Note that the intermediate nodes along the member are not included, and *fixed node* means that the location of the node is fixed in the optimization process although its displacements may not be constrained. Let $\boldsymbol{x}_{\mathrm{f}}$, $\boldsymbol{y}_{\mathrm{f}}$ and $\boldsymbol{a}$ denote the vectors consisting of x- and y-coordinates of free nodes and the vector of cross-sectional areas of beam elements, respectively, which are the design variables. The elastic stiffness matrix and the geometrical stiffness matrix corresponding to the unit load factor are denoted by $\boldsymbol{K}$ and $\boldsymbol{K}_{\mathrm{G}}$, respectively [McGuire et al., 2000]. The linear buckling load factor λ^{cr} is calculated as the smallest positive eigenvalue of the following eigenvalue problem:

$$(\boldsymbol{K}(\boldsymbol{x}_{\mathrm{f}}, \boldsymbol{y}_{\mathrm{f}}, \boldsymbol{a}) + \lambda \boldsymbol{K}_{\mathrm{G}}(\boldsymbol{x}_{\mathrm{f}}, \boldsymbol{y}_{\mathrm{f}}, \boldsymbol{a}))\boldsymbol{\Phi} = \boldsymbol{0}. \quad (5.5)$$

Let λ^{L} denote the lower bound for λ^{cr}. The linear buckling constraint can be assigned with respect to the reciprocal of λ^{cr} denoted by $\gamma = 1/\lambda^{\mathrm{cr}}$ and its upper bound $\gamma^{\mathrm{U}} = 1/\lambda^{\mathrm{L}}$ to ensure that λ^{cr} is either larger than λ^{L} or negative [Kanno et al., 2001]. Then, the deterministic worst-case design problem of geometry and topology optimization of a frame is formulated as follows:

$$\underset{\boldsymbol{x}_{\mathrm{f}},\boldsymbol{y}_{\mathrm{f}},\boldsymbol{a}}{\text{Minimize}} \quad \sigma = \max_{\substack{i\in\{1,\cdots m_{\mathrm{e}}\}\\ j\in\{1,\cdots p\}}} \sigma_{i,j}\left(\boldsymbol{x}_{\mathrm{f}},\boldsymbol{y}_{\mathrm{f}},\boldsymbol{a}\right) \tag{5.6a}$$

$$\text{subject to} \quad \gamma\left(\boldsymbol{x}_{\mathrm{f}},\boldsymbol{y}_{\mathrm{f}},\boldsymbol{a}\right) \leq \gamma^{\mathrm{U}}, \tag{5.6b}$$

$$v\left(\boldsymbol{x}_{\mathrm{f}},\boldsymbol{y}_{\mathrm{f}},\boldsymbol{a}\right) \leq v^{\mathrm{U}}, \tag{5.6c}$$

$$\left(\boldsymbol{x}_{\mathrm{f}},\boldsymbol{y}_{\mathrm{f}},\boldsymbol{a}\right) \in \mathcal{X}_0, \tag{5.6d}$$

where v^{U} is the upper bound for the structural volume $v\left(\boldsymbol{x}_{\mathrm{f}},\boldsymbol{y}_{\mathrm{f}},\boldsymbol{a}\right)$, and $\mathcal{X}_0$ is the box domain defined by the upper and lower bounds for $\boldsymbol{x}_{\mathrm{f}}$, $\boldsymbol{y}_{\mathrm{f}}$ and $\boldsymbol{a}$.

By utilizing the FDM, the locations of free nodes can be obtained from the force density vector $\boldsymbol{t}$ of a truss. However, it is not possible to define nodal location of a frame using FDM because a frame has shear forces and bending moments. Here we utilize an auxiliary truss for determining nodal locations of a frame [Shen and Ohsaki, 2021]. See Appendix A.4 for details of FDM. Since the nodal coordinates are functions of $\boldsymbol{t}$, the functions $\sigma_{i,j}\left(\boldsymbol{x}_{\mathrm{f}}(\boldsymbol{t}),\boldsymbol{y}_{\mathrm{f}}(\boldsymbol{t}),\boldsymbol{a}\right)$, $\gamma\left(\boldsymbol{x}_{\mathrm{f}}(\boldsymbol{t}),\boldsymbol{y}_{\mathrm{f}}(\boldsymbol{t}),\boldsymbol{a}\right)$ and $v\left(\boldsymbol{x}_{\mathrm{f}}(\boldsymbol{t}),\boldsymbol{y}_{\mathrm{f}}(\boldsymbol{t}),\boldsymbol{a}\right)$ can be simply written as $\sigma_{i,j}\left(\boldsymbol{t},\boldsymbol{a}\right)$, $\gamma\left(\boldsymbol{t},\boldsymbol{a}\right)$ and $v\left(\boldsymbol{t},\boldsymbol{a}\right)$, respectively. Hence, optimization problem (5.6) is restated as

$$\underset{\boldsymbol{t},\boldsymbol{a}}{\text{Minimize}} \quad \sigma = \max_{\substack{i\in\{1,\cdots m_{\mathrm{e}}\}\\ j\in\{1,\cdots p\}}} \sigma_{i,j}\left(\boldsymbol{t},\boldsymbol{a}\right) \tag{5.7a}$$

$$\text{subject to} \quad \gamma\left(\boldsymbol{t},\boldsymbol{a}\right) \leq \gamma^{\mathrm{U}}, \tag{5.7b}$$

$$v\left(\boldsymbol{t},\boldsymbol{a}\right) \leq v^{\mathrm{U}}, \tag{5.7c}$$

$$\left(\boldsymbol{t},\boldsymbol{a}\right) \in \mathcal{X}, \tag{5.7d}$$

where $\mathcal{X}$ is a box domain defined by the upper and lower bounds for $\boldsymbol{t}$ and $\boldsymbol{a}$. As stated in Shen and Ohsaki [2021], problems (5.6) and (5.7) are basically the same and will have the same optimal solution if there exists a vector $\boldsymbol{t}$ that can define the optimal nodal locations of problem (5.6), i.e., optimal solution of problem (5.6) can be found by solving problem (5.7) if it exists in the feasible domain of problem (5.7).

When uncertainty discussed in Sec. 5.2.1 is introduced to problem (5.7), the optimization problem is stated as

$$\underset{\boldsymbol{t},\boldsymbol{a}}{\text{Minimize}} \quad \sigma^{\max} = \max_{\boldsymbol{\theta}\in\Omega} \sigma\left(\boldsymbol{t},\boldsymbol{a};\boldsymbol{\theta}\right) \tag{5.8a}$$

$$\text{subject to} \quad \gamma^{\max} = \max_{\boldsymbol{\theta}\in\Omega} \gamma\left(\boldsymbol{t},\boldsymbol{a};\boldsymbol{\theta}\right) \leq \gamma^{\mathrm{U}}, \tag{5.8b}$$

$$v\left(\boldsymbol{t},\boldsymbol{a}\right) \leq v^{\mathrm{U}}, \tag{5.8c}$$

$$\left(\boldsymbol{t},\boldsymbol{a}\right) \in \mathcal{X}, \tag{5.8d}$$

where $\boldsymbol{\theta} = (\Delta \boldsymbol{x}, \Delta \boldsymbol{y}, \Delta \boldsymbol{a})$ is the vector representing the uncertainty in nodal locations and the cross-sectional areas; Ω is the uncertainty set of $\boldsymbol{\theta}$; $\sigma^{\max}$ and $\gamma^{\max}$ are the maximum values of von Mises stress and reciprocal of the linear buckling load factor, respectively. Note that problem (5.8) can be regarded as a worst-case design problem with semi-infinite constraints with respect to design variables and uncertain parameters, since it can be equivalently transformed into

$$\begin{aligned}
\underset{\boldsymbol{t},\boldsymbol{a}}{\text{Minimize}} \quad & \sigma^{\max} = \max_{\boldsymbol{\theta}\in\Omega} \sigma(\boldsymbol{t},\boldsymbol{a};\boldsymbol{\theta}) && \text{(5.9a)}\\
\text{subject to} \quad & \gamma(\boldsymbol{t},\boldsymbol{a};\boldsymbol{\theta}) \leq \gamma^{\mathrm{U}} \quad \text{for } \forall\boldsymbol{\theta}\in\Omega, && \text{(5.9b)}\\
& v(\boldsymbol{t},\boldsymbol{a}) \leq v^{\mathrm{U}}, && \text{(5.9c)}\\
& (\boldsymbol{t},\boldsymbol{a}) \in \mathcal{X}. && \text{(5.9d)}
\end{aligned}$$

Using the slack variable s, problem (5.9) can be further reformulated as follows:

$$\begin{aligned}
\underset{s,\boldsymbol{t},\boldsymbol{a}}{\text{Minimize}} \quad & s && \text{(5.10a)}\\
\text{subject to} \quad & \sigma(\boldsymbol{t},\boldsymbol{a};\boldsymbol{\theta}) \leq s \quad \text{for } \forall\boldsymbol{\theta}\in\Omega, && \text{(5.10b)}\\
& \gamma(\boldsymbol{t},\boldsymbol{a};\boldsymbol{\theta}) \leq \gamma^{\mathrm{U}} \quad \text{for } \forall\boldsymbol{\theta}\in\Omega, && \text{(5.10c)}\\
& v(\boldsymbol{t},\boldsymbol{a}) \leq v^{\mathrm{U}}, && \text{(5.10d)}\\
& (\boldsymbol{t},\boldsymbol{a}) \in \mathcal{X}, && \text{(5.10e)}
\end{aligned}$$

where the constraints are to be satisfied for all parameter values in the uncertainty set, which means that the constraints should be satisfied for the worst values of constraint functions. However, it is difficult to directly solve problem (5.10) because the worst values of constraint functions to be obtained by structural analysis and linear buckling analysis are difficult to find. Therefore, the exact worst values are relaxed to the βth quantile values, which are approximated by order statistics, for a β value slightly less than 1.

Worst-value approximation using order statistics

As explained in Chap. 4, an order statistics-based approach has been developed for estimating the worst value with certain confidence level [Shen et al., 2020; Ohsaki et al., 2019; Yamakawa and Ohsaki, 2016], where the worst value is approximated by the kth order statistics and the order k is regarded as representing the robustness level according to the theory of distribution-free tolerance interval.

Suppose $\boldsymbol{\theta}_1, \cdots, \boldsymbol{\theta}_n$ are the n realizations of independent and identically distributed random variables (iidrvs) of uncertain nodal locations and cross-sectional areas with unknown distributions. The n structural responses of design $(\boldsymbol{t}, \boldsymbol{a})$ in (5.10) corresponding to $\boldsymbol{\theta}_1, \cdots, \boldsymbol{\theta}_n$ are denoted by $\sigma_1 = \sigma(\boldsymbol{t}, \boldsymbol{a}; \boldsymbol{\theta}_1), \cdots, \sigma_n = \sigma(\boldsymbol{t}, \boldsymbol{a}; \boldsymbol{\theta}_n)$ and $\gamma_1 = \gamma(\boldsymbol{t}, \boldsymbol{a}; \boldsymbol{\theta}_1), \cdots, \gamma_n = \gamma(\boldsymbol{t}, \boldsymbol{a}; \boldsymbol{\theta}_n)$, respectively. Furthermore, we define $\sigma_{1:n}, \cdots, \sigma_{n:n}$ as a permutation of $\sigma_1, \cdots, \sigma_n$ in a descending order, i.e., $\sigma_{1:n} \geq \cdots \geq \sigma_{n:n}$. The permutation $\gamma_{1:n}, \cdots, \gamma_{n:n}$ is defined in the same manner. Note that the descending order is used for convenience for approximating the worst values.

Since $\sigma_{1:n}, \cdots, \sigma_{n:n}$ are obtained from the same function with different vectors of uncertain parameters, it is reasonable to assume that they are n realizations of the random variable because the function of random variables is also a random variable that is denoted by Y_σ, and the cumulative distribution function (cdf) of Y_σ is denoted by $F_\sigma(y) = \Pr\{Y_\sigma \leq y\}$. As explained in Sec. 4.2, the probability α_k of $100\beta\%$ of the stress σ less than the kth order statistic of stress $\sigma_{k:n}$ is formulated as

$$\Pr\{F_\sigma(\sigma_{k:n}) \geq \beta\} = \alpha_k, \tag{5.11}$$

where the minimum sample size requirements are summarized in Table 4.1. Similarly, $\gamma_{1:n}, \cdots, \gamma_{n:n}$ denote n realizations of the random variables Y_γ with cdf $F_\gamma(y) = \Pr\{Y_\gamma \leq y\}$, and we have

$$\Pr\{F_\gamma(\gamma_{k:n}) \geq \beta\} = \alpha_k. \tag{5.12}$$

The probability of the structural response falling into the one-side interval $(-\infty, \sigma_{k:n})$ or $(-\infty, \gamma_{k:n})$ is free of the definition of F_σ or F_γ, respectively, which is called distribution-free one-side tolerance interval. The kth order statistics $\sigma_{k:n}$ and $\gamma_{k:n}$ can be seen as the βth quantile response in probabilistic sense and α_k is regarded as the confidence level. Obviously, if α_k and β are large enough, e.g., 0.95 or 0.99, $\sigma_{k:n}$ and $\gamma_{k:n}$ can provide accurate quantile responses as relaxation of the worst values required in problems (5.8) and (5.10).

Moreover, as demonstrated by Shen et al. [2020] and Ohsaki et al. [2019], β is a decreasing function of k for given sample size n and confidence level α_k, which will be discussed in detail in Sec. 6.2. This property indicates that a higher order k corresponds to a less βth quantile response under uncertainty. Therefore, the order k and its corresponding order statistic approximating the βth quantile response can represent the robustness level of the structure. Suppose the worst responses in problem (5.10) are approximated by $\sigma_{k:n}$ and $\gamma_{k:n}$. Then problem (5.10) can be approximately rewritten as

$$\begin{aligned}
&\underset{s,\boldsymbol{t},\boldsymbol{a}}{\text{Minimize}} && s && (5.13\text{a})\\
&\text{subject to} && \sigma_{k:n}(\boldsymbol{t}, \boldsymbol{a}; \boldsymbol{\theta}_1, \cdots, \boldsymbol{\theta}_n) \leq s, && (5.13\text{b})\\
& && \gamma_{k:n}(\boldsymbol{t}, \boldsymbol{a}; \boldsymbol{\theta}_1, \cdots, \boldsymbol{\theta}_n) \leq \gamma^{\mathrm{U}}, && (5.13\text{c})\\
& && v(\boldsymbol{t}, \boldsymbol{a}) \leq v^{\mathrm{U}}, && (5.13\text{d})\\
& && (\boldsymbol{t}, \boldsymbol{a}) \in \mathcal{X}. && (5.13\text{e})
\end{aligned}$$

The difference between the approximate worst values in problem (5.13) and the exact worst values in problem (5.10) will become smaller if k is closer to

1, and the robustness levels of the solution with respect to both stress and linear buckling also increase as k decreases. Note that the constraints in the worst-case design problem (5.10) are relaxed, and accordingly, the robustness for satisfying the constraints is relaxed, by using the order statistics in problem (5.13) at specified confidence level α_k, and such robustness is referred to as *statistical feasibility robustness* which handles the semi-infinite constraints stochastically [Beyer and Sendhoff, 2007]; see also *feasibility uncertainty* in Sec. 1.3.1. Although similar formulation to incorporate uncertainty in the constraints can be found in RBDO problem and referred to as risk or chance constraint [Ito et al., 2018; Moon et al., 2018; Kanno, 2019], there is no general consensus that RBDO should not be considered as part of the RDO methodology, and vice versa [Beyer and Sendhoff, 2007], and the connection between robustness and the stochastic or probability theory has also been exploited recently [Gabrel et al., 2014]. The equivalence between the probabilistic optimization problem and the RDO with uncertain-but-bounded variables are also discussed by Elishakoff and Ohsaki [2010]. Hence, we can consider problem (5.13) as RDO problem due to the fact that it is a relaxed version of the worst-case design problem (5.10).

As pointed out by Shen and Ohsaki [2021], closely spaced nodes may still exist in the solution of problem (5.13), although the member length is indirectly constrained by the bounds of force densities. Such nodes in the solution of problem (5.13) are merged into one node to obtain a simplified structural layout. Moreover, if the cross-sectional area of a short member is large, a large von Mises stress may appear due to its large bending stiffness. Therefore, the stress of a short member is ignored in the process of solving problem (5.13). Furthermore, a short member is modeled by only one beam element to avoid singularity due to its large stiffness unlike other members divided into four beam elements. The process of solving problem (5.13) is summarized in Algorithm. 5.4, where the penalization methods for stress and geometrical stiffness matrix are explained below.

5.2.2 PENALIZATION OF STRESS AND GEOMETRICAL STIFFNESS

It is well known that topology optimization under stress constraint is difficult because the stress constraints need not be satisfied by the vanishing elements/members; i.e., the constraint suddenly dissapears when a cross-sectional variable is used as a continuous design variable and it is decreased to 0. Therefore, the stress constraint is categorized as a design dependent constraint, and the optimal solution is located at a singular point in the feasible domain. To alleviate this difficulty, some relaxation and penalization methods have been proposed [Guo et al., 2001; Le et al., 2010]. Since singularity phenomena exist also in the member buckling for linear buckling constraint, a similar penalization or relaxation approach may be used for the geometrical stiffness of frame member [Shen et al., 2021].

Algorithm 5.4 Robust geometry and topology optimization of plane frame.

Require: Give the initial values of cross-sectional areas $\boldsymbol{a}$ and force densities $\boldsymbol{t}$ based on the initial structure. Select the order k.
1: **repeat**
2: Determine the structural geometry by utilizing the FDM for the auxiliary truss.
3: Model the short member by one beam element and the others by four beam elements.
4: Generate n vectors of uncertain parameters $\boldsymbol{\theta} = (\Delta\boldsymbol{x}, \Delta\boldsymbol{y}, \Delta\boldsymbol{a})$ to obtain the sample set for order statistics.
5: Generate n finite element models based on the structural geometry, topology and the n vectors of $\boldsymbol{\theta}$ for structural analysis.
6: **for** $i \leftarrow 1$ to n **do**
7: Implement stress evaluation and linear buckling analysis, while the thin elements are penalized using (5.14) and (5.15).
8: **end for**
9: Sort the n results in descending order and output the kth order statistics $\sigma_{k:n}$ and $\gamma_{k:n}$.
10: Compute the total structural volume v.
11: Update design variables $\boldsymbol{a}$ and $\boldsymbol{t}$ according to the optimization algorithm.
12: **until** converge
13: Output the solution and terminate the process.

Penalization of stress

Based on the method for stress-based optimization in Le et al. [2010], the stress of a thin member is underestimated, as follows, using a penalization parameter $\eta \in [0, 1]$:

$$\hat{\sigma}_i = \left(a_i/a_i^{\mathrm{U}}\right)^{\eta} \max_{j=1,\cdots p} \sigma_{i,j}, \tag{5.14}$$

where i, j, p and $\sigma_{i,j}$ have the same meaning as (5.6a) and (5.7a); and a_i^{U} is the upper bound of a_i.

To verify the effectiveness of using (5.14), the optimal topology is found for minimizing the maximum stress of a simple frame as shown in Fig. 5.2(a) under structural volume constraint. Only the cross-sectional areas are considered as design variables and uncertainty is not incorporated; therefore, the second moment of area is a function of the cross-sectional area. In Fig. 5.2(a), the node number and the member number are indicated by those with and without parentheses, respectively, and the crossing diagonal members are not connected at their centers. The frame is pin-supported at nodes 1 and 2, and the frame has $m_{\mathrm{e}} = 5 \times 4 = 20$ beam elements. Young's modulus is 3×10^{11} Pa, and the von Mises stress for the load $F = 2000$ N is calculated at the six points indicated in Fig. 5.2(b) for each member, i.e., $p = 6$. The elements have solid

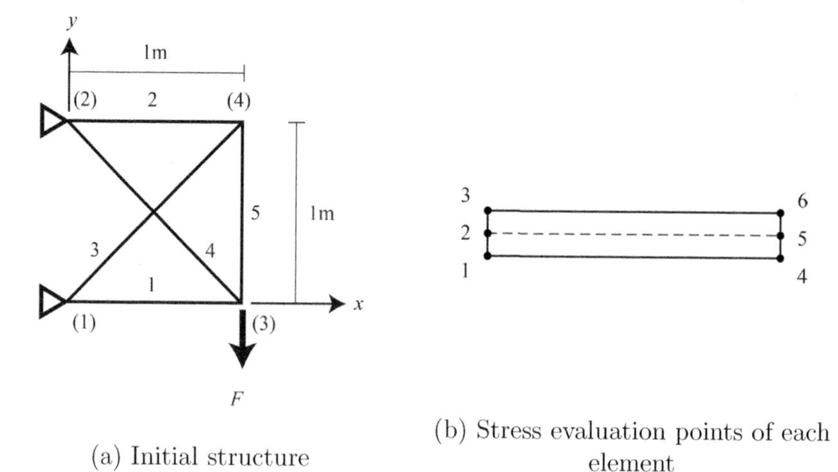

(a) Initial structure

(b) Stress evaluation points of each element

Figure 5.2 A simple frame model.

circular section, and the upper and lower bounds of cross-sectional areas are $0.05\,\mathrm{m}^2$ and $1 \times 10^{-7}\,\mathrm{m}^2$, respectively, for all elements. The elements with $a_i < 0.01a_i^{\mathrm{U}}$ are regarded as thin and the stress constraints are neglected. The generalized reduced gradient method (GRG) [Holmström et al., 2007] available in MATLAB [Mathworks, 2018c] is used for optimization.

The optimal solution is shown in Fig. 5.3(a), where the width of each member is proportional to its cross-sectional area. The optimal cross-sectional areas of members 1–5 are 0.03478, 1×10^{-7}, 1×10^{-7}, 0.04611 and 1×10^{-7}, respectively, which are shown in Fig. 5.3(b) after removing the thin elements. The maximum von Mises stresses before and after removing thin elements are 68199 Pa and 68200 Pa, respectively, which are almost the same. This result

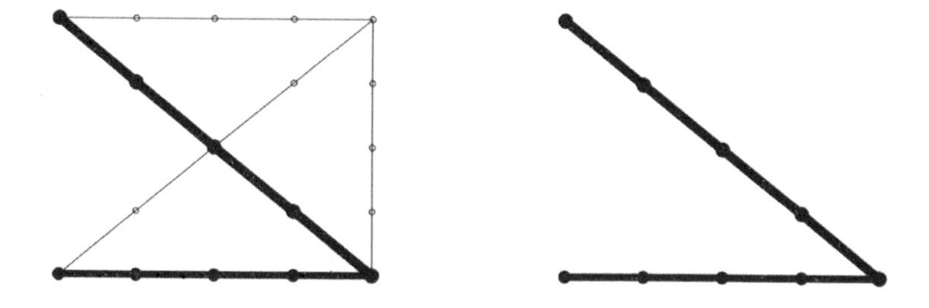

(a) Before removing thin elements

(b) After removing thin elements

Figure 5.3 Optimal solution of simple frame model.

confirms that the stress singularity phenomenon has been successfully avoided using the penalization approach.

Penalization of geometrical stiffness

The penalization approach to alleviate stress singularity can also be used for avoiding local instability for a problem under linear buckling constraint. Since existence of a slender member in compression will lead to a violation of linear buckling constraint due to the negative contribution in stiffness matrix [Tugilimana et al., 2018], this superfluous linear buckling should be eliminated for accurate estimation of linear buckling load factor after removing the thin elements. Therefore, the geometrical stiffness of the thin element i is penalized as follows:

$$\hat{\boldsymbol{K}}_{G,i} = \left(a_i/a_i^{\mathrm{U}}\right)^{\rho} \boldsymbol{K}_{G,i} \tag{5.15}$$

where $\boldsymbol{K}_{G,i}$ is the original geometrical stiffness matrix of the ith element and ρ is the penalization parameter. This way, the negative geometrical stiffness of a slender member is underestimated, and superfluous member buckling is avoided without removing any thin element.

To illustrate the existence of superfluous member buckling, we compute the linear buckling load factor of the solution in Fig. 5.3(a) that was obtained without considering linear buckling constraint. The values of λ^{cr} before and after removing thin elements are 0.797 and 238441, respectively. Figure 5.4 shows the relation between λ^{cr} before removing thin elements and the penalization parameter ρ in (5.15), where the red line shows the value of λ^{cr} after removing the thin elements. The figure shows that the value of λ^{cr} can be accurately estimated without removing the thin elements when ρ is greater than 1.

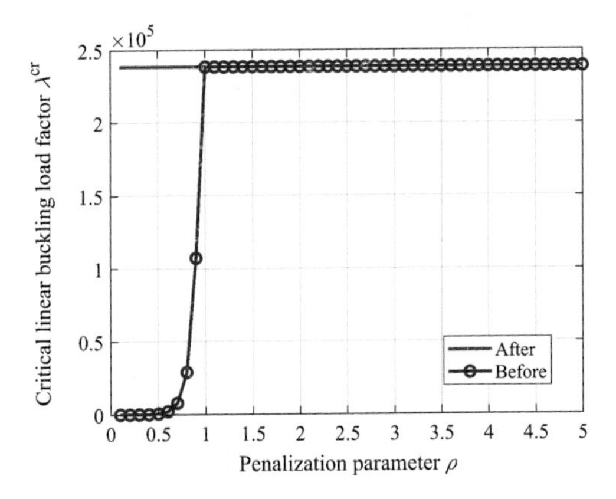

Figure 5.4 Variation of linear buckling load factor with respect to penalization parameter.

5.2.3 NUMERICAL EXAMPLE

A numerical example is presented in this section to demonstrate effectiveness of the method utilizing FDM and the penalization approaches. The nonlinear optimization problem (5.13) is solved using GRG. The uncertain parameter vector $\boldsymbol{\theta} = (\Delta \boldsymbol{x}, \Delta \boldsymbol{y}, \Delta \boldsymbol{a})$ is defined as increment from the solution without considering uncertainty, and uncertainty is assumed to be uniformly distributed between the lower and upper bounds. Note that only the free nodes at the member ends have the uniform distributions, and the locations of intermediate nodes 3, 4 and 5 in Fig. 5.1(b) are defined from the locations of free nodes and the eccentricity e. The following correlation coefficient using exponential decay function is used in the examples below [Jalalpour et al., 2013]:

$$c_{ij} = \exp\left(-\frac{\|(X_i, Y_i) - (X_j, Y_j)\|}{l_{\text{exp}}}\right) \quad (i, j = 1, \cdots, 5), \tag{5.16}$$

where c_{ij} is the correlation coefficient between locations of nodes i and j in Fig. 5.1(b), and l_{exp} is the correlation length, where a larger l_{exp} indicates a stronger correlation. Because the correlation among nodal locations of frames is relatively weak than that of continua [Jalalpour et al., 2013], a small value is used for l_{exp}. The correlated uncertain nodal locations are generated by using copulas [Park et al., 2015; Noh et al., 2009] in MATLAB [Mathworks, 2018c] using the correlation coefficients in (5.16). The frame member is not divided into four beam elements if its length is less than 0.1 m. A member is regarded as thin if its cross-sectional area is less than 1% of the maximum value of the frame. The stress is evaluated at $p = 6$ points in Fig. 5.2(b). The parameter values are listed in Table 5.1, where $\mathbf{1}$ is the vector with all entries equal to 1, and the superscripts L and U denote the lower and upper bounds, respectively, of the corresponding parameter. Note that $\boldsymbol{a}^{\text{L}}$ in Table 5.1 is the lower bound for $\boldsymbol{a}$ without considering uncertainty.

Since the robustness level β is a decreasing function of order k with given sample size n, as discussed in Sec. 5.2.1, and higher values of β and confidence level α_k lead to a better approximation of the worst structural response, we assume $k = 1$, $n = 150$ and $\alpha_k = 0.9995$ for problem (5.13), which lead to $\beta = 0.95$ for both stress and linear buckling load factor. Therefore, the order statistics in problem (5.13) are written as $\sigma_{1:150}$ and $\gamma_{1:150}$, respectively. The eccentricity e for each member in Sec. 5.2.1 is 0.01 in accordance with Pedersen [2003].

A plane frame as shown in Fig. 5.5 is optimized. The frame is pin-supported at nodes 1 and 2, a downward vertical load $F = 200\,\text{kN}$ is applied at node 5, and the crossing members are not connected at their centers. The fixed nodes are the supports and the loaded node 5, and the others are free nodes. The members have solid circular sections. The penalization parameter for stress in (5.14) is $\eta = 0.5$ as suggested by Le et al. [2010], and that for geometrical stiffness matrix in (5.15) is $\rho = 2$.

Table 5.1
Parameter Settings of the Example

Parameters	Values	Parameters	Values
Lower bound $\Delta \boldsymbol{x}_{\mathrm{f}}^{\mathrm{L}}$	$-0.02 \cdot \mathbf{1}\,\mathrm{m}$	Lower bound $\boldsymbol{a}^{\mathrm{L}}$	$1 \times 10^{-7} \cdot \mathbf{1}\,\mathrm{m}^2$
Upper bound $\Delta \boldsymbol{x}_{\mathrm{f}}^{\mathrm{U}}$	$0.02 \cdot \mathbf{1}\,\mathrm{m}$	Upper bound $\boldsymbol{a}^{\mathrm{U}}$	$0.05 \cdot \mathbf{1}\,\mathrm{m}^2$
Lower bound $\Delta \boldsymbol{y}_{\mathrm{f}}^{\mathrm{L}}$	$-0.02 \cdot \mathbf{1}\,\mathrm{m}$	Lower bound $\boldsymbol{t}^{\mathrm{L}}$	$-1000 \cdot \mathbf{1}\,\mathrm{N/m}$
Upper bound $\Delta \boldsymbol{y}_{\mathrm{f}}^{\mathrm{U}}$	$0.02 \cdot \mathbf{1}\,\mathrm{m}$	Upper bound $\boldsymbol{t}^{\mathrm{U}}$	$1000 \cdot \mathbf{1}\,\mathrm{N/m}$
Lower bound $\Delta \boldsymbol{a}^{\mathrm{L}}$	$-0.02 \cdot \boldsymbol{a}\,\mathrm{m}^2$	Upper bound v^{U}	$0.02\,\mathrm{m}^3$
Upper bound $\Delta \boldsymbol{a}^{\mathrm{U}}$	$0.02 \cdot \boldsymbol{a}\,\mathrm{m}^2$	Lower bound λ^{L}	3.4
Young's modulus $\boldsymbol{E}$	$2 \times 10^{11} \cdot \mathbf{1}\,\mathrm{Pa}$	Upper bound γ^{U}	0.29
Sample size n	150	Correlation length l_{exp}	$0.1\,\mathrm{m}$
Order k	1	Number of members n_{m}	10
Confidence level α_k	0.9995	Number of nodes n_{n}	6
Robustness level β	0.95		
Eccentricity e	0.01		

The solution of RDO problem (5.13) and deterministic optimization problem (5.7) are denoted as solutions R and D, respectively, and shown in Fig. 5.6, where the contour represents the value of von Mises stress. The force densities, cross-sectional areas and member lengths of solutions R and D are listed in Table 5.2. To give a more intuitive comparison, the worst stress distributions of solutions R and D are shown in Fig. 5.7, and the values of σ, $\sigma_{1:150}$, γ, λ^{cr}, $\gamma_{1:150}$, $\lambda^{\mathrm{cr}}_{1:150}$ and v are listed in Table 5.3.

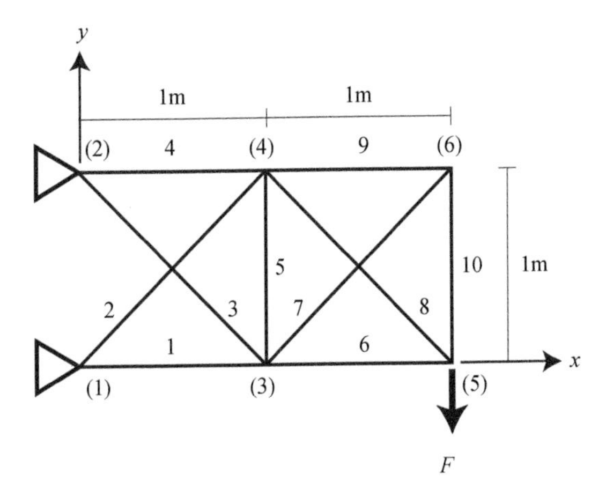

Figure 5.5 A plane frame with two square units.

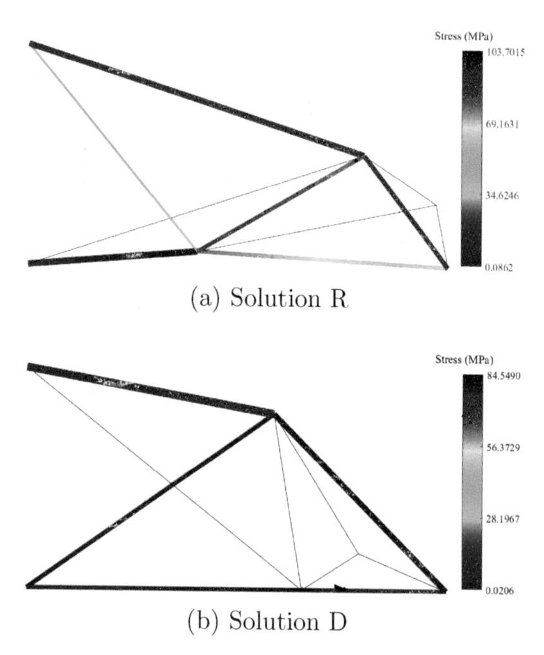

(a) Solution R

(b) Solution D

Figure 5.6 Solutions and stress distribution without uncertainty.

It can be seen from Fig. 5.6 and Table 5.3 that solution D has a smaller maximum stress σ than solution R, and $\lambda^{\rm cr}$ of solution D is close to $\lambda^{\rm L} = 3.4$. Moreover, for solution D, the stresses of members with moderate cross-sectional areas have values close to σ, which means that solution D is a kind of fully stressed design. However, as seen from Table 5.2, members 1 and 6 which sequentially connect nodes 1, 3 and 5 are almost colinear, and only slender members are connected to node 3. Therefore, members 1 and 6 can be regarded as a single long member connecting nodes 1 and 5, resulting in a much larger approximate worst stress $\sigma_{1:150}$ than that of solution R. Accordingly, the linear buckling constraint is violated when uncertainty is incorporated as shown in Table 5.3.

It is seen from Fig. 5.6 that node 3 is connected by more members with moderately large cross-sectional areas in solution R than in solution D. Therefore, the long member in solution D no longer exists in solution R, increasing the redundancy of the frame to reduce the effect of uncertainty on structural performance. Furthermore, for the worst stress values in Fig. 5.7, solution R has more members with large stress level close to $\sigma_{1:150}$ than solution D, in which a significantly large stress is observed in the members 1 and 6. This demonstrates that the solution R is a kind of fully stressed design for the worst parameter values. Table 5.3 also shows that $\sigma_{1:150}$ of solution R is smaller than that of solution D, and the linear buckling load of solution R is far above its

Table 5.2
Cross-sectional Area, Force Density and Member Length of Solutions

(a) Solution R

Member	Cross-sectional area (m^2)	Force density (N/m^2)	Length (m)
1	0.004302	0.2617	0.8059
2	1×10^{-7}	0.0933	1.6767
3	0.001560	0.1924	1.2332
4	0.004069	0.1263	1.6729
5	0.002682	−0.2225	0.9096
6	0.002719	0.5726	1.1983
7	1×10^{-7}	−0.1260	1.1603
8	0.003228	0.0078	0.6462
9	1×10^{-7}	0.4978	0.4081
10	1×10^{-7}	0.4798	0.2921

(b) Solution D

Member	Cross-sectional area (m^2)	Force density (N/m^2)	Length (m)
1	0.002678	0.4489	1.3101
2	0.002933	−0.4132	1.4202
3	1×10^{-7}	−0.1647	1.6476
4	0.005220	0.7074	1.1920
5	1×10^{-7}	0.0980	0.8085
6	0.002675	0.3539	0.6898
7	1×10^{-7}	0.5198	0.3196
8	0.003710	0.2497	1.1480
9	1×10^{-7}	0.3099	0.7499
10	1×10^{-7}	0.6410	0.4504

lower bound. Specifically, $\sigma_{1:150}$ of solution R is 126% larger than the nominal value σ, and $\lambda^{cr}_{1:150}$ decreases 5.02% from the nominal value λ^{cr}. By contrast, $\sigma_{1:150}$ of solution D is 320% larger than σ, and $\lambda^{cr}_{1:150}$ is 9.02% less than λ^{cr}. This result indicates that solution R is more robust than solution D in view of reducing the influence of uncertainty in nodal locations and cross-sectional areas on structural performance. Although the details are not shown here, a solution similar to solution D is obtained if linear buckling constraint is not considered. This fact emphasizes importance of an appropriate incorporation of linear buckling constraint to obtain the optimal solution that is not influenced by the superficial local buckling of the slender members.

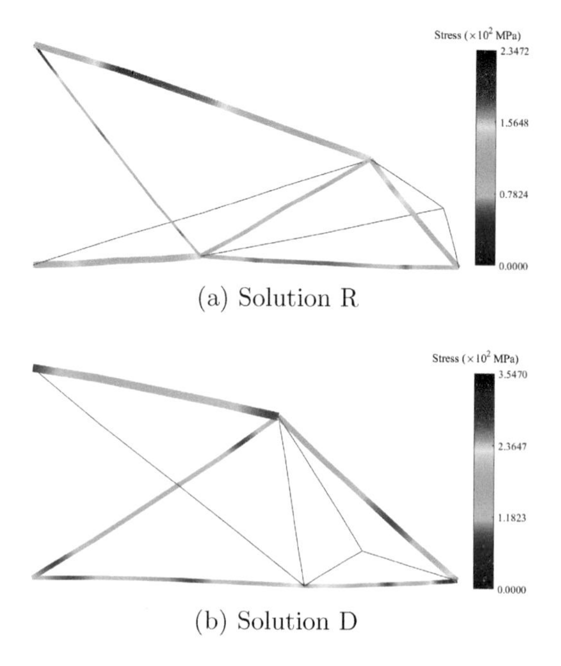

(a) Solution R

(b) Solution D

Figure 5.7 Solutions and worst stress distribution under uncertainty.

Table 5.3
Response Properties of Solutions R and D

Solution	σ (MPa)	$\sigma_{1:150}$ (MPa)	γ	λ^{cr}	$\gamma_{1:150}$	$\lambda^{\mathrm{cr}}_{1:150}$	v (m^3)
R	103.7	234.7	0.0878	11.5	0.0914	10.9	0.02
D	84.5	354.7	0.290	3.45	0.319	3.13	0.02

5.3 QUANTILE-BASED ROBUST FRAME TOPOLOGY OPTIMIZATION

In this section, a quantile-based approach is presented for simultaneous shape and topology optimization of plane frames under reliability constraints [Shen et al., 2022].

5.3.1 QUANTILE-BASED SORA AND ESTIMATION OF QUANTILE

Let $\boldsymbol{d} = (d_1, \cdots, d_m)$ and $\boldsymbol{\Theta} = (\Theta_1, \cdots, \Theta_r)$ denote the vectors of m design variables and r random variables, respectively. The RBDO problem for minimizing the objective function $w(\boldsymbol{d})$ can be generally given as

[Choi et al., 2007]

$$
\begin{aligned}
&\text{Minimize} && w(\boldsymbol{d}) && (5.17\text{a})\\
&\text{subject to} && \Pr\left\{g_i(\boldsymbol{d};\boldsymbol{\Theta}) \leq g_i^{\mathrm{U}}\right\} \geq \bar{r}_i \quad (i=1,\cdots,l), && (5.17\text{b})\\
& && \boldsymbol{d} \in \mathcal{D}, && (5.17\text{c})
\end{aligned}
$$

where $g_i(\boldsymbol{d};\boldsymbol{\Theta})$ is the ith function among l performance functions under uncertainty; $\bar{r}_i$ is the target probability of $g_i(\boldsymbol{d};\boldsymbol{\Theta})$ not to exceed the specified upper bound g_i^{U}; and $\mathcal{D}$ is the box domain defined by the upper and lower bounds for $\boldsymbol{d}$. Note that $\boldsymbol{d}$ in problem (5.17) can include deterministic design variables as well as the nominal values of random design variables. Problem (5.17) is solved using the quantile-based SORA, and its basic concept is briefly introduced as follows [Li et al., 2020; He et al., 2021].

Unlike the traditional SORA, the quantile-based SORA computes the shifting values utilizing quantile, instead of the MPP, based on the equivalence between the constraint on reliability and the quantile response [Moustapha et al., 2016]. The quantile $Q_i(\boldsymbol{d})$ corresponding to the ith performance function $g_i(\boldsymbol{d};\boldsymbol{\Theta})$ and the target probability $\bar{r}_i$ is defined as

$$
Q_i(\boldsymbol{d}) = \inf\left\{Q : \Pr\left\{g_i(\boldsymbol{d};\boldsymbol{\Theta}) \leq Q\right\} \geq \bar{r}_i\right\} \quad (i=1,2,\cdots,l). \tag{5.18}
$$

Then, we can rewrite an RBDO problem (5.17) in terms of quantile as follows:

$$
\begin{aligned}
&\text{Minimize} && w(\boldsymbol{d}) && (5.19\text{a})\\
&\text{subject to} && Q_i(\boldsymbol{d}) \leq \bar{g}_i \quad (i=1,\cdots,l), && (5.19\text{b})\\
& && \boldsymbol{d} \in \mathcal{D}, && (5.19\text{c})
\end{aligned}
$$

$\bar{g}_i = g_i^{\mathrm{U}}$ is the target quantile value for the ith response function g_i. It can be observed that the target values of probability $\bar{r}_i$ $(i=1,\cdots,l)$ are now implicitly incorporated into problem (5.19) by the corresponding quantiles defined in (5.18), bridging the equivalence between problems (5.17) and (5.19) [Moustapha et al., 2016]. Although the random variables in problem (5.19) are assumed to be mutually independent, correlated random variables can be transformed into independent variables using, e.g., Rosenblatt or Nataf transformations [Der Kiureghian and Liu, 1986; Melchers and Beck, 2017] or other transformation methods without utilizing the marginal probability density function (pdf) [Tong et al., 2021].

Based on the quantile-based SORA [Li et al., 2020; He et al., 2021], the optimal solution of problem (5.19) is obtained by successively solving deterministic optimization problems shifting the upper bounds of performance functions so that the quantile $Q_i(\boldsymbol{d})$ is on the boundary of the reliability constraint. Let $\bar{c}_i^{(j)}$ denote the shifting value of $\bar{g}_i$ at the jth iteration. The jth deterministic optimization problem is formulated as

$$\begin{aligned} &\text{Minimize} && w(\boldsymbol{d}) && \text{(5.20a)}\\ &\text{subject to} && Q_i(\boldsymbol{d}) \leq \bar{g}_i - \bar{c}_i^{(j)} \quad (i = 1, \cdots, l), && \text{(5.20b)}\\ & && \boldsymbol{d} \in \mathcal{D}. && \text{(5.20c)} \end{aligned}$$

Let $\boldsymbol{d}^{(j)}$ denote the solution of problem (5.20) at the jth iteration. The shifting value $\bar{c}_i^{(j)}$ is calculated by

$$\bar{c}_i^{(j)} = Q_i(\boldsymbol{d}^{(j-1)}) - g_i(\boldsymbol{d}^{(j-1)}) \quad (i = 1, \cdots, l;\ j = 1, \cdots), \tag{5.21}$$

where $g_i(\boldsymbol{d}^{(j-1)})$ is the nominal value of the ith response function. Note that $\bar{c}_i^{(0)} = 0$ at the initial iteration because the solution of problem (5.20) has not been found yet and no information is available for the quantile value [Du and Chen, 2004]. Problem (5.20) with (5.21) is solved at each iteration of quantile-based SORA, and to prevent obtaining a too conservative result, the process is terminated if all the reliability constraints are satisfied with inequality except one constraint satisfied with equality as follows:

Condition 5.1 *For all $i = 1, \cdots, l$, one has $Q_i(\boldsymbol{d}^{(j)}) \leq \bar{g}_i$ and there exists e in $\{1, \cdots, l\}$ such that $Q_e(\boldsymbol{d}^{(j)}) = \bar{g}_e$.*

Estimation of the quantile $Q_i^{(j)}(\boldsymbol{d}^{(j)})$ in (5.21) after solving problem (5.20) at each iteration is one of the most important process in the quantile-based SORA. Shen et al. [2022] proposed a method for obtaining the desired quantile $Q_i^{(j)}(\boldsymbol{d})$ utilizing MEM subject to constraints on the sample L-moments [Zhao et al., 2020]. More accurately, $\tilde{Q}_i^{(j)}(\boldsymbol{d}; \boldsymbol{\theta}_1, \cdots, \boldsymbol{\theta}_n)$ is estimated by the MEM and is used in (5.21) instead of $Q_i^{(j)}(\boldsymbol{d})$, where $\boldsymbol{\theta}_1, \cdots, \boldsymbol{\theta}_n$ are n realizations of $\boldsymbol{\Theta}$. The shifting value is calculated by

$$\bar{c}_i^{(j)} = \tilde{Q}_i(\boldsymbol{d}^{(j-1)}; \boldsymbol{\theta}_1, \cdots, \boldsymbol{\theta}_n) - g_i(\boldsymbol{d}^{(j-1)}) \quad (i = 1, \cdots, l;\ j = 1, \cdots). \tag{5.22}$$

Thus, Condition 5.1 is replaced with the following condition:

Condition 5.2 *For all $i = 1, \cdots, l$, one has $\tilde{Q}_i(\boldsymbol{d}^{(j)}; \boldsymbol{\theta}_1, \cdots, \boldsymbol{\theta}_n) \leq \bar{g}_i$ and there exists e in $\{1, \cdots, l\}$ such that $|\tilde{Q}_e(\boldsymbol{d}^{(j)}) - \bar{g}_e| \leq \varepsilon$ for pre-specified ε $(0 < \varepsilon \ll 1)$.*

Algorithm of the quantile-based SORA using MEM is given in Algorithm 5.5. See Sec. 1.4.5 for details of MEM.

Suppose after the jth iteration the ith performance function under uncertainty is a continuous random variable $Z_i^{(j)} = g_i(\boldsymbol{d}^{(j)}; \boldsymbol{\Theta})$. For the sake of simplicity, the indices i and j will be omitted hereafter, i.e., $Z_i^{(j)} = g_i(\boldsymbol{d}^{(j)}; \boldsymbol{\Theta})$ will be simply denoted by $Z = g(\boldsymbol{d}; \boldsymbol{\Theta})$. Let $Q(u)$ and $q(u) = dQ/du$ denote the quantile function of Z and its derivative for $0 < u < 1$, respectively.

Algorithm 5.5 Algorithm of quantile-based SORA using MEM.

Require: Set $j = 0$ and $\bar{c}_i^{(0)} = 0$ $(i = 1, \cdots, l)$.
1: **loop**
2: **if** $j \geq 1$ **then**
3: Calculate $\bar{c}_i^{(j)} = \tilde{Q}_i(\boldsymbol{d}^{(j-1)}; \boldsymbol{\theta}_1, \cdots, \boldsymbol{\theta}_n) - g_i(\boldsymbol{d}^{(j-1)})$ for $i = 1, \cdots, l$.
4: **end if**
5: Solve problem (5.20) and obtain the optimal solution $\boldsymbol{d}^{(j)}$.
6: Approximate the quantile function of $g_i(\boldsymbol{d}^{(j)}; \boldsymbol{\theta})$ using MEM.
7: Calculate the target quantile $\tilde{Q}_i(\boldsymbol{d}^{(j)}; \boldsymbol{\theta}_1, \cdots, \boldsymbol{\theta}_n)$ for $i = 1, \cdots, l$.
8: **if** stopping criterion of Condition 5.2 is satisfied **then**
9: **Break.**
10: **end if**
11: Substitute $j \leftarrow j + 1$.
12: **end loop**
13: Output the solution and terminate the process.

According to Hosking [2007], the MEM estimates the quantile by maximizing the entropy defined in (1.59) for specified first S L-moments in the form of (1.60), i.e., PWM constraints, and the problem reads

$$\underset{q}{\text{Maximize}} \quad \int_0^1 \log q(u) du \tag{5.23a}$$

$$\text{subject to} \quad \int_0^1 K_s(u) q(u) du = h_s \quad (s = 1, \cdots, S). \tag{5.23b}$$

as stated in Sec. 1.4.5. Based on the Lagrangian multiplier method, the solution to optimization problem (5.23) is obtained by maximizing the following Lagrangian functional of the entropy:

$$\bar{H}(q) = \int_0^1 \log q(u) du - \sum_{s=1}^{S} a_s \left(\int_0^1 K_s(u) q(u) du - h_s \right),$$

where a_s $(r = 1, \cdots, S)$ are the unknown Lagrangian multipliers. The stationary condition (Euler-Lagrangian equation) of $\bar{H}(q)$ yields the estimation

$$q(u) = \frac{1}{\sum_{s=1}^{S} a_s K_s(u)}, \tag{5.24}$$

as mentioned in (1.65). The Lagrangian multipliers $\boldsymbol{a} = (a_1, \cdots, a_S)$ are determined by finding the stationary point of

$$D(\boldsymbol{a}) = -\int_0^1 \log \left(\sum_{s=1}^{S} a_s K_s(u) \right) du + \sum_{s=1}^{S} a_s h_s. \tag{5.25}$$

as presented in (1.68). Since $K_s(1) = 0$ for any arbitrary order s; as seen in (1.71), it is expected that the estimated $q(u)$ in (5.24) will diverge to infinity as u approaches 1. Therefore, in the following numerical examples, (5.25) is only integrated to $u = 0.9999$ to avoid numerical difficulty as follows

$$D(\boldsymbol{a}) \approx -\int_0^{0.9999} \log\left(\sum_{s=1}^{S} a_s K_s(u)\right) du + \sum_{s=1}^{S} a_s h_s. \tag{5.26}$$

The Lagrange multipliers can be found efficiently using numerical algorithms because $D(\boldsymbol{a})$ is a convex function. Once the values of $\boldsymbol{a}$ and $q(u)$ are determined, the quantile function $Q(u)$ can be obtained by integrating $q(u)$ as

$$Q(u) = Q(0) + \int_0^u q(v)dv. \tag{5.27}$$

as derived in (1.72). Because in general the exact value of $Q(0)$ is unknown beforehand, one can estimate $Q(u)$ by the sample value of the corresponding smallest order statistic $z_{1:n}$, and $Q(u)$ in (5.27) is obtained by

$$Q(u) \approx z_{1:n} + \int_0^u q(v)dv. \tag{5.28}$$

Thus, the quantile Q in (5.22) can be estimated for the target probability as

$$Q(\bar{r}) \approx z_{1:n} + \int_0^{\bar{r}} q(v)dv. \tag{5.29}$$

Note that the maximization problem of $\bar{H}(q)$, or alternatively $D(\boldsymbol{a})$, is an unconstrained convex optimization problem which is similar to the traditional MEM for pdf estimation [Xi et al., 2012]. However, unlike the method in He et al. [2021], the method in this section does not depend on the initial guess of the Lagrangian multipliers because a convex optimization problem is solved instead of solving the stationary condition.

5.3.2 SHAPE AND TOPOLOGY OPTIMIZATION OF PLANE FRAMES UNDER UNCERTAINTY

Once the quantile in (5.21) is estimated and the shifting value is computed, problem (5.20) can be iteratively solved until Condition 5.2 is satisfied. The design variables are the vectors of x- and y-coordinates of nodes, denoted by $\boldsymbol{x}$ and $\boldsymbol{y}$, and the cross-sectional areas $\boldsymbol{a}$. Since nodal locations and cross-sectional areas of members in the frame may deviate from their nominal values due to manufacturing and/or construction error, uncertainty is assumed in the design variables. Uncertainty is also considered in the vector $\boldsymbol{E}$ consisting of Young's modulli of members. Thus, uncertainty in the vectors $\boldsymbol{x}$, $\boldsymbol{y}$, $\boldsymbol{a}$ and $\boldsymbol{E}$ are denoted by $\Delta\boldsymbol{x}$, $\Delta\boldsymbol{y}$, $\Delta\boldsymbol{a}$ and $\Delta\boldsymbol{E}$, respectively, which are combined to a random vector $\boldsymbol{\Theta} = (\Delta\boldsymbol{x}, \Delta\boldsymbol{y}, \Delta\boldsymbol{a}, \Delta\boldsymbol{E})$. Based on problem (5.20), the $(j+1)$th deterministic problem of quantile-based SORA for shape and topology optimization of plane frames can be formulated as

$$\begin{aligned} \text{Minimize} \quad & w(\boldsymbol{x}, \boldsymbol{y}, \boldsymbol{a}) && (5.30\text{a}) \\ \text{subject to} \quad & g_i(\boldsymbol{x}, \boldsymbol{y}, \boldsymbol{a}) \leq \bar{g}_i - \bar{c}_i^{(j+1)} \quad (i = 1, \cdots, l), && (5.30\text{b}) \\ & (\boldsymbol{x}, \boldsymbol{y}, \boldsymbol{a}) \in \mathcal{X}_0, && (5.30\text{c}) \end{aligned}$$

where $\mathcal{X}_0$ is a box domain defined by the upper and lower bounds for $\boldsymbol{x}$, $\boldsymbol{y}$ and $\boldsymbol{a}$.

In the same manner as Sec. 5.2, FDM is applied to an auxiliary truss to prevent numerical difficulty due to existence of extremely short members [Shen and Ohsaki, 2021]. Let $\boldsymbol{x}_{\mathrm{f}}$ and $\boldsymbol{y}_{\mathrm{f}}$ denote the x- and y-coordinates of free nodes, respectively, of the auxiliary truss, which are regarded as functions of the force density vector $\boldsymbol{t}$. Problem (5.30) is rewritten as

$$\begin{aligned} \text{Minimize} \quad & w(\boldsymbol{x}_{\mathrm{f}}(\boldsymbol{t}), \boldsymbol{y}_{\mathrm{f}}(\boldsymbol{t}), \boldsymbol{a}) && (5.31\text{a}) \\ \text{subject to} \quad & g_i(\boldsymbol{x}_{\mathrm{f}}(\boldsymbol{t}), \boldsymbol{y}_{\mathrm{f}}(\boldsymbol{t}), \boldsymbol{a}) \leq \bar{g}_i - \bar{c}_i^{(j+1)} \quad (i = 1, \cdots, l), && (5.31\text{b}) \\ & (\boldsymbol{t}, \boldsymbol{a}) \in \mathcal{X}, && (5.31\text{c}) \end{aligned}$$

where $\mathcal{X}$ is a box domain defined by the upper and lower bounds for $\boldsymbol{t}$ and $\boldsymbol{a}$. Note that the design variables are the force density vector $\boldsymbol{t}$ and the cross-sectional areas $\boldsymbol{a}$, and $\boldsymbol{x}_{\mathrm{f}}$ and $\boldsymbol{y}_{\mathrm{f}}$ can be determined by solving the equilibrium equations with respect to $\boldsymbol{t}$; see Appendix A.4 for details. The shifting value is calculated by

$$\bar{c}_i^{(j+1)} = \tilde{Q}_i\left(\boldsymbol{x}_{\mathrm{f}}(\boldsymbol{t}^{(j)}), \boldsymbol{y}_{\mathrm{f}}(\boldsymbol{t}^{(j)}), \boldsymbol{a}^{(j)}; \boldsymbol{\theta}_1, \cdots, \boldsymbol{\theta}_n\right) - g_i\left(\boldsymbol{x}_{\mathrm{f}}(\boldsymbol{t}^{(j)}), \boldsymbol{y}_{\mathrm{f}}(\boldsymbol{t}^{(j)}), \boldsymbol{a}^{(j)}\right) \quad (i = 1, \cdots, l).$$

NUMERICAL EXAMPLE

A numerical example is presented to investigate the effectiveness of the quantile-based SORA using MEM, and the results in Pandey [2000] and Deng and Pandey [2008] are also given for comparison. The deterministic optimization problem (5.20) of quantile-based SORA is solved by sequential quadratic programming (SQP) using *fmincon* with default settings in Optimization Toolbox of MATLAB [Mathworks, 2018c]. The first four sample L-moments (i.e., $S = 4$) are calculated using the equations in Sec. 1.4.4, and are used in MEM to estimate the derivative of quantile function with sample size $n = 50$. The Monte Carlo simulation (MCS) with sample size $n_{\mathrm{MCS}} = 1 \times 10^5$ is implemented to investigate the accuracy of the quantile function using MEM with sample L-moments, and the solution at the initial iteration is also presented to compare it with the final solution.

Table 5.4
Parameter Settings of a 3×2 Cantilever Frame

Parameters	Values	Parameters	Values
Lower bound $\Delta \boldsymbol{x}_{\mathrm{f}}^{\mathrm{L}}$	$-0.02 \cdot \mathbf{1}\,\mathrm{m}$	Lower bound $\boldsymbol{a}^{\mathrm{L}}$	$1 \times 10^{-7} \cdot \mathbf{1}\,\mathrm{m}^2$
Upper bound $\Delta \boldsymbol{x}_{\mathrm{f}}^{\mathrm{U}}$	$0.02 \cdot \mathbf{1}\,\mathrm{m}$	Upper bound $\boldsymbol{a}^{\mathrm{U}}$	$0.02 \cdot \mathbf{1}\,\mathrm{m}^2$
Lower bound $\Delta \boldsymbol{y}_{\mathrm{f}}^{\mathrm{L}}$	$-0.02 \cdot \mathbf{1}\,\mathrm{m}$	Lower bound $\boldsymbol{t}^{\mathrm{L}}$	$-1000 \cdot \mathbf{1}\,\mathrm{N/m}$
Upper bound $\Delta \boldsymbol{y}_{\mathrm{f}}^{\mathrm{U}}$	$0.02 \cdot \mathbf{1}\,\mathrm{m}$	Upper bound $\boldsymbol{t}^{\mathrm{U}}$	$1000 \cdot \mathbf{1}\,\mathrm{N/m}$
Lower bound $\Delta \boldsymbol{a}^{\mathrm{L}}$	$-0.02 \cdot \boldsymbol{a}\,\mathrm{m}^2$	Sample size n	50
Upper bound $\Delta \boldsymbol{a}^{\mathrm{U}}$	$0.02 \cdot \boldsymbol{a}\,\mathrm{m}^2$	Sample size n_{MCS}	1×10^5
Lower bound $\Delta \boldsymbol{E}^{\mathrm{L}}$	$-0.05 \cdot \boldsymbol{E}\,\mathrm{Pa}$		
Upper bound $\Delta \boldsymbol{E}^{\mathrm{U}}$	$0.05 \cdot \boldsymbol{E}\,\mathrm{Pa}$		
Nominal value $\boldsymbol{E}$	$3 \times 10^{11} \cdot \mathbf{1}\,\mathrm{Pa}$		

Assuming that the fixed nodes are precisely located for simplicity, the uncertainties in x- and y-coordinates are considered in the free nodes. The uncertain parameters are uniformly distributed between their lower and upper bounds. Each member has a solid circular section. The parameter values are listed in Table 5.4. The initial configuration of a 3×2 grid cantilever frame with 12 nodes and 27 members is shown in Fig. 5.8, where the crossing diagonal members are not connected at their intersections. The frame is pin-supported at nodes 1, 2 and 3, and a downward vertical load $F = 1000$ kN is applied at node 11. Therefore, the fixed nodes of the auxiliary truss for FDM are nodes 1, 2, 3 and 11. The total structural volume $w\left(\boldsymbol{x}_{\mathrm{f}}(\boldsymbol{t}), \boldsymbol{y}_{\mathrm{f}}(\boldsymbol{t}), \boldsymbol{a}\right)$ is minimized under reliability constraint on the downward vertical displacement of node 11, for which the upper bound is 3.0×10^{-3} m. The target probability $\bar{r}$ is 0.99.

Optimal solution of problem (5.31) has been found with the three steps of sequential deterministic optimization. The results at the initial and final

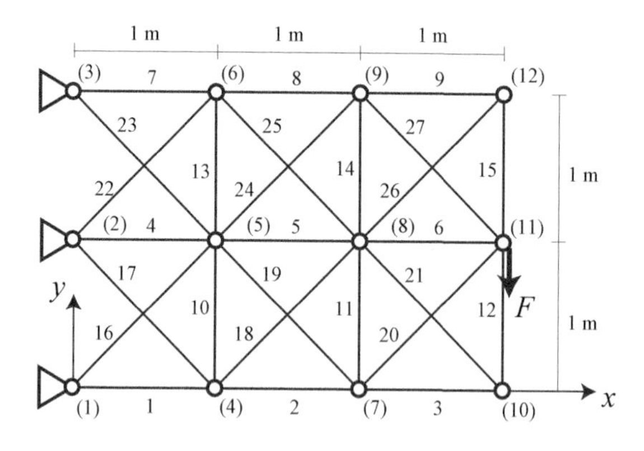

Figure 5.8 A 3×2 cantilever frame.

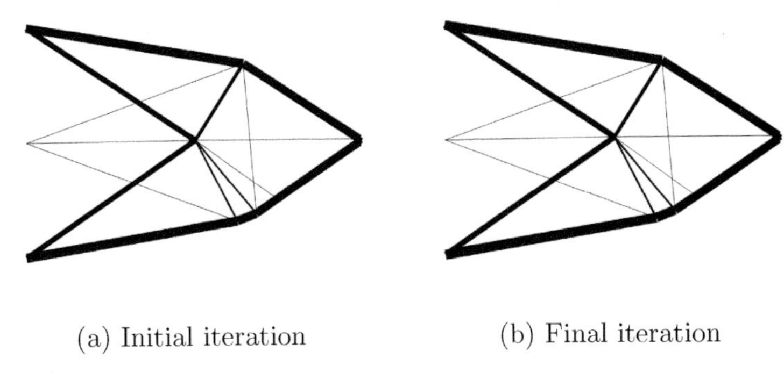

(a) Initial iteration (b) Final iteration

Figure 5.9 Optimal solutions.

iterations are shown in Fig. 5.9. The structural volume w and quantile $Q(\bar{r})$ at the initial and final iterations are listed in Table 5.5, where the quantiles obtained by MCS with sample size $n_{\mathrm{MCS}} = 1 \times 10^5$ are also listed in the parentheses. The nodal locations of the solutions at the initial and final iterations are shown in Tables 5.6. As can be seen from Tables 5.5, 5.6 and Fig. 5.9, the two solutions have similar shapes; however, the cross-sectional areas slightly increase from the initial to final solution to satisfy the reliability constraint on the displacement of node 11, leading to a larger structural volume and a smaller quantile.

The quantile functions of vertical displacement of node 11 obtained by MCS and MEM with sample L-moments are given in Fig. 5.10 for the solutions at the initial and final iterations. It can be observed from Fig. 5.10(a) and Table 5.5 that when uncertainty is taken into consideration at the initial iteration, probability of the vertical displacement of node 11 exceeding the upper bound 3×10^{-3} m is about 0.8, which means the structure has low reliability. By contrast, at the final iteration, the reliability constraint is satisfied, and the quantile estimated by MEM has good agreement with the value obtained by MCS with the relative error about 2% as shown in Fig. 5.10(b), with the increase of structural volume about 7% from the initial iteration. This

Table 5.5
Structural Volume w and Quantile $Q(\bar{r})$ at the Initial and Final Iterations

	Initial iteration	Final iteration
w (m^3)	9.2167×10^{-2}	9.8017×10^{-2}
$Q(\bar{r})$ (m)	3.199×10^{-3}	3.0×10^{-3}
	(3.175×10^{-3})	(2.973×10^{-3})

Table 5.6
Locations of Nodes at the Initial and Final Iteration (m)

Node	Initial iteration		Final iteration	
	x-coordinate	y-coordinate	x-coordinate	y-coordinate
1	0	0	0	0
2	0	1	0	1
3	0	2	0	2
4	1.8857	0.3067	1.8852	0.3068
5	1.5154	1.0144	1.5149	1.0144
6	1.9414	1.6754	1.9409	1.6753
7	2.2633	0.4784	2.2625	0.4787
8	2.0518	0.3718	2.0511	0.3718
9	2.9785	1.0144	2.9776	1.0148
10	2.4708	0.6290	2.4701	0.6291
11	3	1	3	1
12	2.9785	0.9886	2.9777	0.9882

result indicates that the method based on SORA and MEM can estimate the quantile function with good accuracy.

To refine the optimal solution and obtain a frame with small numbers of nodes and members, the three nodes 9, 11 and 12 are combined to a single node, and members 2, 3, 4, 6, 11, 12, 14, 17, 19, 20, 21, 22, 24 and 25 with very small cross-sectional areas are removed as shown in Fig. 5.11(a). The quantile functions of vertical displacement of the loaded node (node 7 after refinement) obtained by MCS and MEM for the refined frame are also given in Fig. 5.11(b). The locations of nodes, cross-sectional areas and member lengths after refinement are listed in Tables 5.7 and 5.8, and the structural volume, nominal value and quantile of the displacement constraint function before and after refinement are given in Table 5.9 in which the quantiles obtained by MCS are listed in the parentheses.

It is seen from Table 5.9 that due to the regularity of the stiffness matrix before refinement, the nominal value of the displacement at the loaded node is only slightly smaller than that before refinement. However, the quantile becomes slightly smaller due to removal of members with small cross-sectional areas, resulting in a frame with a little higher reliability than required, because only 7 nodes and 10 members are left in the refined frame as shown in Fig. 5.11(a) and Table 5.6. Accordingly, the quantile of the displacement decreases slightly due to less uncertainty involved in the frame. This way, an optimal frame with small numbers of nodes and members satisfying reliability constraint can be obtained by combining FDM, SORA and MEM.

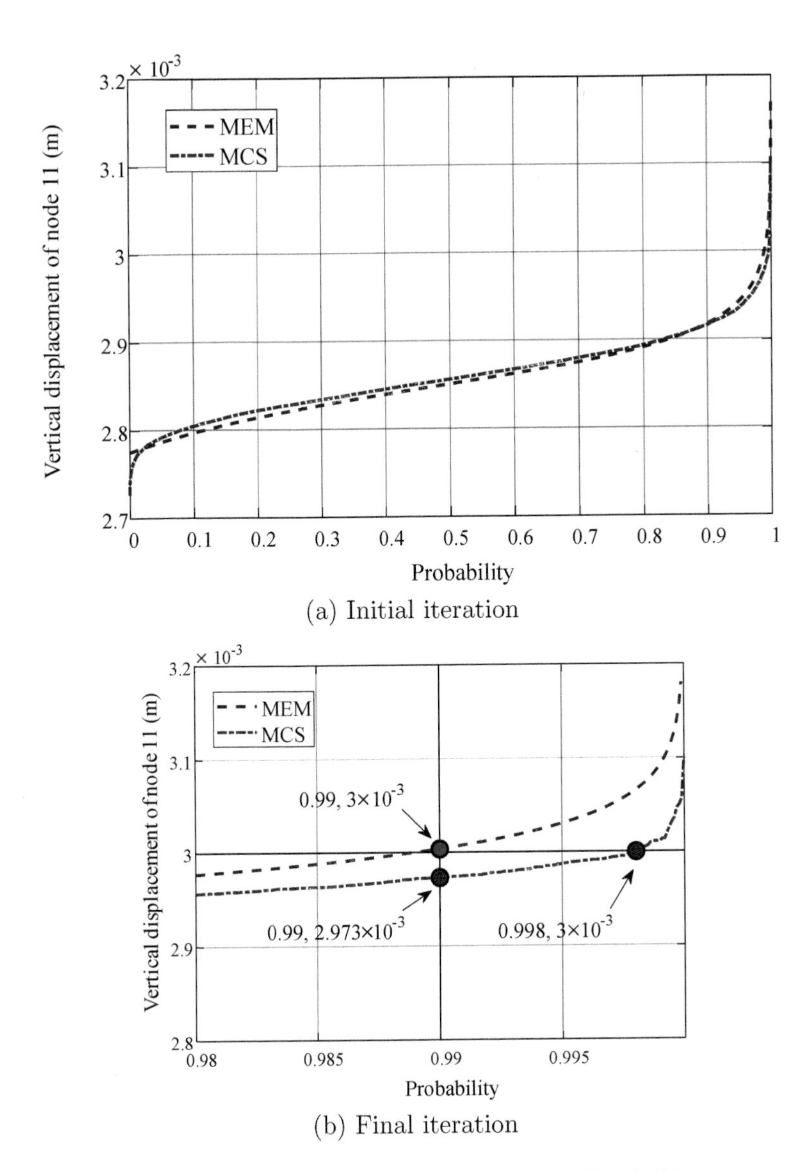

Figure 5.10 Quantile functions of vertical displacement of node 11.

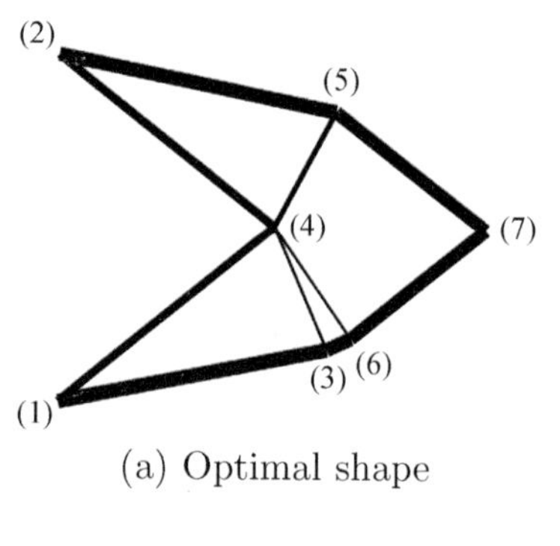

(a) Optimal shape

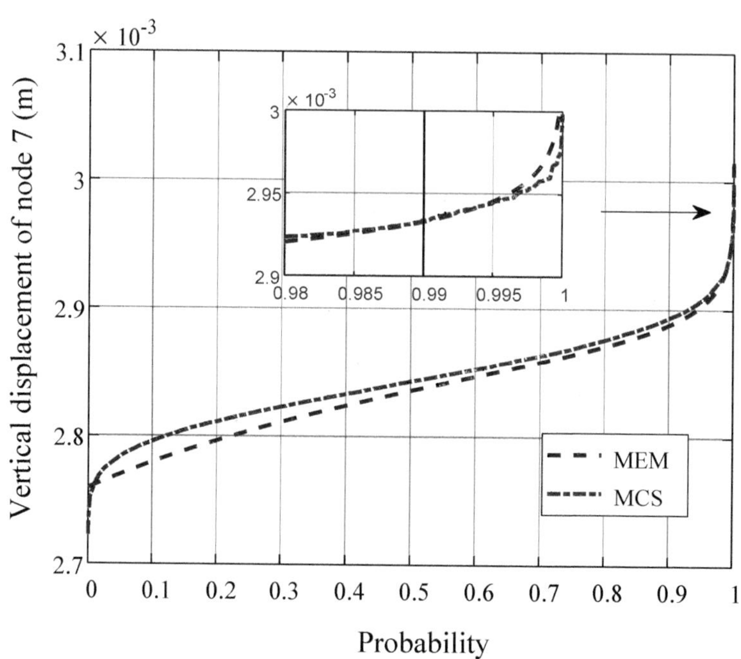

(b) Quantile functions of vertical displacement of the loaded node

Figure 5.11 Result after refinement.

Table 5.7
Location of Nodes of Optimal Result After Refinement

Node	1	2	3	4	5	6	7
x-coordinate (m)	0	0	1.8852	1.5149	1.9409	2.0511	3
y-coordinate (m)	0	2	0.3068	1.0144	1.6753	0.3718	1

Table 5.8
Cross-sectional Areas and Member Lengths of Optimal Result After Refinement

Element			
Node 1	Node 2	Cross-sectional area (m^2)	Member length (m)
1	3	0.0109	1.9100
4	6	0.00218	0.8368
2	5	0.0107	1.9678
5	7	0.00982	1.2293
3	4	0.00230	0.7986
4	5	0.00417	0.7863
1	4	0.00645	1.8232
3	6	0.0105	0.1782
2	4	0.00667	1.8073
6	7	0.00996	1.1128

Table 5.9
Structural Volume, Quantile and Nominal Value of Constraint Before and After Refinement

	Before modification	After modification
Structural volume (m^3)	9.8017×10^{-2}	9.8016×10^{-2}
Nominal value of constraint (m)	2.820×10^{-3}	2.828×10^{-3}
Quantile of constraint (m)	3.0×10^{-3} (2.973×10^{-3})	2.932×10^{-3} (2.933×10^{-3})

5.4 STOCHASTIC FAIL-SAFE TOPOLOGY OPTIMIZATION

5.4.1 FAIL-SAFE TOPOLOGY OPTIMIZATION

It is important to take redundancy into consideration in structural design, which helps us to design a structure against the force beyond our estimation and/or in accidental events. Generally, it is difficult to ensure the safety against unknown accidental events. In order to prevent such danger, we should consider redundancy of structural design. One of approaches linked with topology optimization is referred to as fail-safe topology optimization (FSTO). Recent studies deal with FSTO against local failure of continuum structure and truss structure [Jansen et al., 2014; Kanno, 2017; Kohta et al., 2014]. Finite scenario-based worst-case approach is most frequently used in this field. Generally, a worst-case design problem includes non-smooth and nondifferentiable functions, and requires high computational cost. Finding the exact worst value even from a small number of scenarios is not a so easy task. To make matters worse, number of possible scenarios may be infinite and preparation for extremely rare events may not be practically admissible. To reduce the difficulties in FSTO, the order statistics-based approach is potentially applicable. This approach, i.e., FSTO linked with order statistics, has stochastic properties and will be referred to as stochastic fail-safe topology optimization (SFSTO). Generally, SFSTO tends to have several thousands or more design variables. Therefore, we need a scalable stochastic method for handling SFSTO. Even in SFSTO, we can perform sensitivity analysis of the functions with respect to the design variables. The sensitivity inevitably inherits stochastic properties of the functions. Recently, the use of stochastic gradient descent (SGD) and its variants has become popular for solving large-scale problems of finding the minimizer of an objective function. Effectiveness of the methods has been studied intensively in the field of machine learning [Kasai, 2018]. In this section, we apply SGD to SFSTO.

Deterministic nominal topology optimization

The nominal topology optimization of continuum structure is briefly introduced [Bendsøe and Sigmund, 2003; Sigmund and Maute, 2013]. The density-based approach is a popular approach to describe the material distribution in topology optimization. We adopt the solid isotropic material with penalization (SIMP) method [Bendsøe, 1989]. The design domain is assumed to be rectangular and discretized into unit size square finite elements. The number of elements in the horizontal and vertical directions are denoted by m_{X} and m_{Y}, respectively, and the total number of elements is $m = m_{\mathrm{X}} \times m_{\mathrm{Y}}$. An example of the design domain is shown in Fig. 5.12. Each element e is assigned a density where 0 and 1 indicate the absence and presence of material, respectively. Young's modulus of element e is given by

$$E_e = E_{\min} + (\rho_e)^p (E_0 - E_{\min}) \quad (e = 1, \cdots, m), \tag{5.32}$$

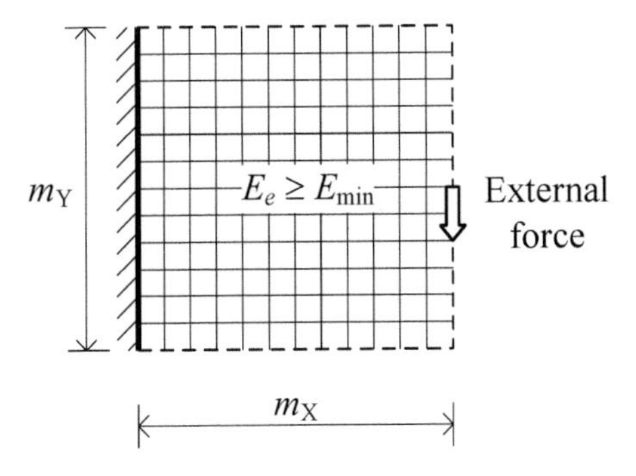

Figure 5.12 Design domain of topology optimization.

where E_0 is the Young's modulus of the material, $E_{\min} = 10^{-9}$ is the small value corresponding to void regions, and $p = 3$ is a penalization factor. The design variable vector is denoted by $\boldsymbol{x} = (x_1, \cdots, x_m)$, whose eth component is denoted by x_e and assigned with the density, i.e., $\rho_e = x_e$.

Without smoothing or filtering, topology optimization typically shows poor performance and mesh-dependent solutions. To prevent this, we adopt combination of the following two filtering operators.

Density filter: For element e, the density variable is evaluated as a weighted average of the densities over the neighbors within a radius $r_{\min}$:

$$\tilde{x}_e = \frac{\sum_{i=1}^{m} H_{ei} x_i}{\sum_{i=1}^{m} H_{ei}}, \tag{5.33a}$$

$$H_{ei} = \max\left\{0, r_{\min} - \sqrt{(\xi_i - \xi_e)^2 + (\eta_i - \eta_e)^2}\right\}, \tag{5.33b}$$

where $\tilde{x}_e$ is the transformed density variable of element e, H_{ei} is a weight factor, (ξ_e, η_e) and (ξ_i, η_i) are the center coordinates of elements e and i, respectively [Bruns and Tortorelli, 2001; Bourdin, 2001].

Threshold projection: In order to remove the gray transition zones, we also use the parameterized projection function:

$$\bar{\tilde{x}}_e = \frac{\tanh(\beta\eta) + \tanh(\beta(\tilde{x}_e - \eta))}{\tanh(\beta\eta) + \tanh(\beta(1 - \eta))}, \tag{5.34}$$

where β is a steepness parameter and η $(0 \leq \eta \leq 1)$ is the threshold value of the projection [Guest, 2009; Wang et al., 2011].

We assign the filtered variable $\bar{\tilde{x}}_e$ to the density ρ_e, i.e., $\rho_e = \bar{\tilde{x}}_e$. From (5.32) to (5.34), Young's modulus of the element e is regarded as a function of design variable $\boldsymbol{x}$ as $E_e(\boldsymbol{x})$. The global stiffness matrix $\boldsymbol{K}(\boldsymbol{x})$ is assembled from the local stiffness matrix $E_e(\boldsymbol{x})\boldsymbol{k}_e^0$, where $\boldsymbol{k}_e^0$ is the constant element stiffness matrix for an element with unit Young's modulus. By using $\boldsymbol{K}$, we write the global equilibrium as

$$\boldsymbol{F} = \boldsymbol{K}\boldsymbol{U}$$

where $\boldsymbol{F}$ and $\boldsymbol{U}$ are the global force and displacement vectors, respectively. The nominal topology optimization problem is mathematically formulated as

$$\begin{aligned} &\text{Minimize} && w(\boldsymbol{x}) = \boldsymbol{F}^\top \boldsymbol{U} = \sum_{e=1}^{m} E_e(\boldsymbol{x}) \boldsymbol{u}_e^\top \boldsymbol{k}_e^0 \boldsymbol{u}_e && (5.35\text{a}) \\ &\text{subject to} && v(\boldsymbol{x}) \leq \mu v_0, && (5.35\text{b}) \\ & && \boldsymbol{x} \in \mathcal{X}, && (5.35\text{c}) \end{aligned}$$

where $w(\boldsymbol{x})$ is the structural compliance, $\boldsymbol{u}_e$ is the element displacement vector, $v(\boldsymbol{x})$ is the material volume, v_0 is the design domain volume, μ is the volume fraction, and $\mathcal{X} = \{\boldsymbol{x} \in \mathbb{R}^m \mid 0 \leq x_e \leq 1,\ e = 1, \cdots, m\}$ is the feasible set.

Fail-safe topology optimization

In the context of topology optimization of discretized continuum-type structure, fail-safe design was proposed by Jansen et al. [2014]. A simplified model of material removal and worst-case formulation to a scenario-based problem has been adopted. To avoid the difficulties in robust design optimization, the order statistics can be used instead of the exact worst value [Ohsaki et al., 2019; Shen et al., 2020]. The formulation of FSTO is briefly introduced here with a slight change of definition of the local failure from Jansen et al. [2014]. Let us consider a set of local failure elements as all elements whose centroid is located within a circle with the center $(\xi_{\mathrm{d}}, \eta_{\mathrm{d}}) \in [0, m_{\mathrm{X}}] \times [0, m_{\mathrm{Y}}]$ and radius r, which is referred to as local failure zone as illustrated in Fig. 5.13.

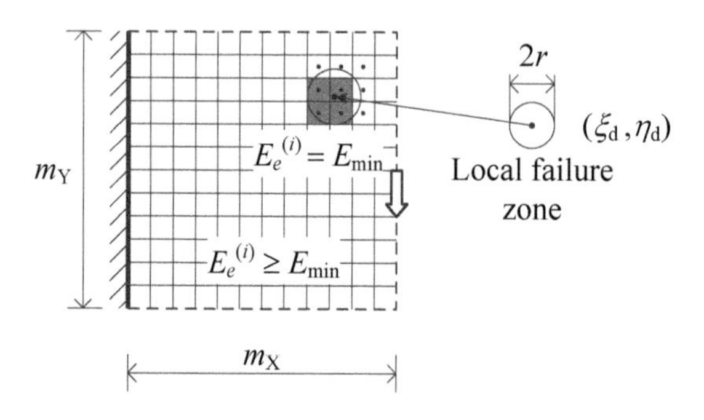

Figure 5.13 Local failure elements.

Thus, the set of local failure elements is defined as

$$\mathcal{D} = \{e \mid \sqrt{(\xi_e - \xi_{\mathrm{d}})^2 + (\eta_e - \eta_{\mathrm{d}})^2} \leq r,\ e = 1, \cdots, m\}, \tag{5.36}$$

where (ξ_e, η_e) is the centroid coordinate of element e as shown in Fig. 5.13. Local failure is modeled by changing Young's modulus of the material to its minimum value in the SIMP interpolation as

$$E_e^{\mathrm{d}} = \begin{cases} E_{\min} & \text{if } e \in \mathcal{D}, \\ E_{\min} + (\rho_e)^p (E_0 - E_{\min}) & \text{otherwise.} \end{cases} \tag{5.37}$$

By using (5.37) instead of (5.32), we can evaluate compliance of the structure. Uncertainty related to the occurrence of local failure can be represented by an appropriate set of prescribed n scenarios of local failure. For $i = 1, \cdots, n$, center of the local failure zone and the corresponding compliance are denoted by $(\xi_{\mathrm{d}}^{(i)}, \eta_{\mathrm{d}}^{(i)})$ and $w_{\mathrm{d}}^{(i)}(\boldsymbol{x})$, respectively. The robust optimization problem can be formulated as a scenario-based problem as

$$\text{Minimize} \quad w_{\mathrm{d}}^{\mathrm{U}}(\boldsymbol{x}) = \max_{i=1,\cdots,n} w_{\mathrm{d}}^{(i)}(\boldsymbol{x}) \tag{5.38a}$$

$$\text{subject to} \quad v(\boldsymbol{x}) \leq \mu v_0, \tag{5.38b}$$

$$\boldsymbol{x} \in \mathcal{X}. \tag{5.38c}$$

This type of FSTO approach is known as worst-case formulation [Ben-Tal et al., 2009], which has some drawbacks. First, the number of elements m tends to be large. The number of local failure scenarios must be the same as the number of elements, which means m may be thousands or more and this approach needs computational cost of m times as large as that of the usual nominal topology optimization method if all the possible scenarios are considered. Second, the solution must be stable and resist to any external forces in prescribed scenarios. This leads to high dependence of the results on the prescribed scenarios. For example, we must carefully exclude scenarios such that the structure inevitably becomes unstable. Otherwise, FSTO fails to find a meaningful solution.

5.4.2 STOCHASTIC GRADIENT DESCENT

Formulation of stochastic fail-safe topology optimization (SFSTO)

To avoid the difficulties in FSTO, a probabilistic approach is available. Let us consider random variables X_{d} and Y_{d} that follow uniform distributions:

$$X_{\mathrm{d}} \sim U(0, m_{\mathrm{X}}), \quad Y_{\mathrm{d}} \sim U(0, m_{\mathrm{Y}})$$

where $U(a, b)$ denotes a uniform distribution in (a, b). In (5.36), we regard $(X_{\mathrm{d}}, Y_{\mathrm{d}})$ as the center of local failure zone, i.e., $(\xi_{\mathrm{d}}, \eta_{\mathrm{d}}) = (X_{\mathrm{d}}, Y_{\mathrm{d}})$, and thus the corresponding structural compliance also becomes a random variable and is denoted by

$$W_{\mathrm{d}} = W_{\mathrm{d}}(\boldsymbol{x}) = w_{\mathrm{d}}(\boldsymbol{x} \mid X_{\mathrm{d}}, Y_{\mathrm{d}}).$$

This formulation enables us to consider the set $\{(X_{\mathrm{d}}^{(i)}, Y_{\mathrm{d}}^{(i)})\}_{i=1}^{n}$ as a sample of size n taken from $(X_{\mathrm{d}}, Y_{\mathrm{d}})$, and the corresponding structural compliance is denoted by

$$W_{\mathrm{d}}^{(i)} = W_{\mathrm{d}}^{(i)}(\boldsymbol{x}) = w_{\mathrm{d}}(\boldsymbol{x} \mid X_{\mathrm{d}}^{(i)}, Y_{\mathrm{d}}^{(i)}) \qquad (i = 1, \cdots, n).$$

We assume $W_{\mathrm{d}}^{(i)}$ are independent and identically distributed random variables, which are arranged in decreasing order of magnitude as

$$W_{1:n} \geq W_{2:n} \geq \cdots \geq W_{k:n} \geq \cdots \geq W_{n:n} \qquad (1 \leq k \leq n),$$

where $W_{k:n} = W_{k:n}(\boldsymbol{x})$ is the kth order statistics [Arnold et al., 2008]; see Secs. 1.4.2 and 4.2.3 for details. Thus, the stochastic optimization problem is formulated as

$$\begin{aligned} &\text{Minimize} && w_{k:n}(\boldsymbol{x}) && (5.39a)\\ &\text{subject to} && v(\boldsymbol{x}) \leq \mu v_0, && (5.39b)\\ &&& \boldsymbol{x} \in \mathcal{X}, && (5.39c) \end{aligned}$$

where $w_{k:n}(\boldsymbol{x})$ is a sample value of $W_{k:n}(\boldsymbol{x})$. Note that $W_{k:n}$ is a random variable, and hence takes a different value at each observation. This clearly involves a stochastic property in FSTO, and solving problem (5.39) is referred to as SFSTO. Due to the stochastic nature of $W_{k:n}$, we can decrease sample size n and have flexibility in assuming scenarios.

Stochastic optimization methods

Recently, the use of SGD and its variants has become popular for solving a large-scale optimization problem, and their effectiveness has been studied extensively. An application of SGD to robust topology optimization has been studied by De et al. [2019]. A method without estimation of the expected value and/or variance is described below. Reddi et al. [2018] provided a generic framework of stochastic optimization that covers popular variants of SGD, which is shown in Algorithm 5.6. For a given vector $\boldsymbol{g} \in \mathbb{R}^m$, we denote its ith component by g_i, and the value at the tth iteration as $\boldsymbol{g}_t$. For given two vectors $\boldsymbol{a}, \boldsymbol{b} \in \mathbb{R}^m$, $\boldsymbol{a} \odot \boldsymbol{b}$ denotes element-wise product that is also called Hadamard product, i.e., $(\boldsymbol{a} \odot \boldsymbol{b})_i = a_i b_i$. For a symmetric matrix $\boldsymbol{M} \in \mathbb{R}^{m \times m}$ and a vector $\boldsymbol{y} \in \mathbb{R}^m$, the projection operation $\Pi_{\mathcal{X},\boldsymbol{M}}$ is defined as $\mathrm{argmin}_{\boldsymbol{x} \in \mathcal{X}} \|\boldsymbol{M}^{1/2}(\boldsymbol{x} - \boldsymbol{y})\|$. Algorithm 5.6 is still abstract because $\boldsymbol{m}_t$ and $\boldsymbol{V}_t$ for functions $\phi_t : \mathbb{R}^{m \times t} \to \mathbb{R}^m$ and $\psi_t : \mathbb{R}^{m \times t} \to \mathbb{R}^{m \times m}$ have not been explicitly specified. We refer to α_t as step size and $\alpha_t \boldsymbol{V}_t^{-1/2}$ as learning rate of the algorithm. The

Algorithm 5.6 Generic framework of stochastic gradient descent (SGD).

Require: initial guess $\boldsymbol{x}_1 \in \mathcal{X}$, sequence of functions $\{\phi_t, \psi_t\}_{t=1}^T$, step sizes $\{\alpha_t\}_{t=1}^T$, maximum number of iterations T, objective function $f : \mathbb{R}^m \to \mathbb{R}$.

1: **for** $t \leftarrow 1$ to T **do**
2: $\quad \boldsymbol{g}_t = \nabla f(\boldsymbol{x}_t)$.
3: $\quad \boldsymbol{m}_t = \phi_t(\boldsymbol{g}_1, \cdots, \boldsymbol{g}_t)$ and $\boldsymbol{V}_t = \psi_t(\boldsymbol{g}_1, \cdots, \boldsymbol{g}_t)$.
4: $\quad \hat{\boldsymbol{x}}_{t+1} = \boldsymbol{x}_t - \alpha_t \boldsymbol{V}_t^{-1/2} \boldsymbol{m}_t$.
5: $\quad \boldsymbol{x}_{t+1} = \Pi_{\mathcal{X}, \boldsymbol{V}_t^{1/2}}(\hat{\boldsymbol{x}}_{t+1})$.
6: **end for**

standard SGD algorithm falls into the framework of Algorithm 5.6 by giving the averaging functions as

$$\phi_t(\boldsymbol{g}_1, \cdots, \boldsymbol{g}_t) = \boldsymbol{g}_t, \qquad \psi_t(\boldsymbol{g}_1, \cdots, \boldsymbol{g}_t) = \boldsymbol{I},$$

where $\boldsymbol{I}$ is the identity matrix of size m. Momentum is used for ϕ_t and ψ_t to help accelerate SGD in the relevant direction and prevent oscillations [Qian, 1999]. Adaptive methods have been proposed as averaging approaches of the squared values of the gradients in the previous step. Adaptive moment estimation (Adam) is a method that computes adaptive learning rates for each parameter [Kingma and Ba, 2015]. Adam uses an exponential moving average to both ϕ_t and ψ_t as

$$\phi_t(\boldsymbol{g}_1, \cdots, \boldsymbol{g}_t) = (1 - \beta_1) \sum_{i=1}^{t} \beta_1^{t-i} \boldsymbol{g}_i, \tag{5.40a}$$

$$\psi_t(\boldsymbol{g}_1, \cdots, \boldsymbol{g}_t) = (1 - \beta_2) \mathrm{diag} \left(\sum_{i=1}^{t} \beta_2^{t-i} \boldsymbol{g}_i \odot \boldsymbol{g}_i \right) + \epsilon \boldsymbol{I}, \tag{5.40b}$$

for some $\beta_1, \beta_2 \in [0, 1)$, and ϵ is a smoothing term that avoids division by zero (usually in the order of 10^{-8}). Kingma and Ba [2015] proposed default values of 0.9 for β_1 and 0.999 for β_2.

Warmup schedule for learning rate

We focus on Adam as a stochastic method. It has been reported that the stability of Adam is often improved with a *warmup schedule* [Ma and Yarats, 2019]. We adopt a trapezoidal schedule [Xing et al., 2018; Schmidt et al., 2020]. The step size α_t in the tth iteration is given by

$$\alpha_t = \alpha_0 \omega_t, \tag{5.41a}$$

$$\omega_t = \begin{cases} \frac{t}{\rho_1 T} & (0 \le t \le \rho_1 T), \\ 1 & (\rho_1 T \le t \le \rho_2 T), \\ \frac{t-T}{(\rho_2 - 1)T} & (\rho_2 T \le t \le T), \end{cases} \tag{5.41b}$$

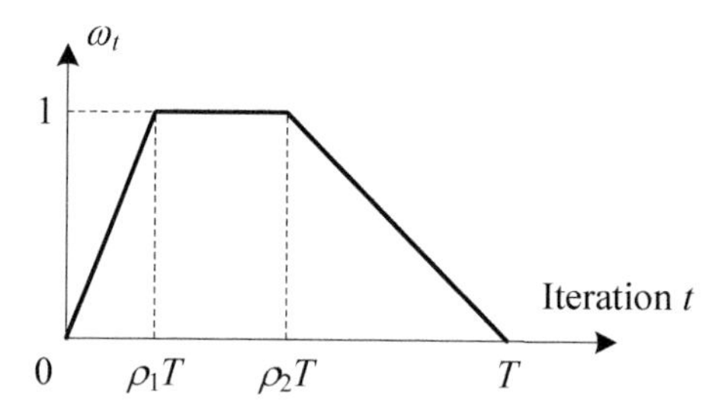

Figure 5.14 Trapezoidal learning rate warmup schedule.

where α_0 is the upper bound of step size, $\omega_t \in [0, 1]$ is the warmup factor, T is the maximum number of iterations, and $\rho_1, \rho_2 \in (0, 1)$ $(\rho_1 < \rho_2)$ are the warmup period factors. The schedule in (5.41b) has three stages as illustrated in Fig. 5.14:

Stage 1: Warmup stage, start from a low learning rate and linearly increase to maximum $(0 \le t \le \rho_1 T)$.

Stage 2: Highway training stage, stay at a maximum learning rate $(\rho_1 T \le t \le \rho_2 T)$.

Stage 3: Converging stage, learning rate linearly decreases to its minimum $(\rho_2 T \le t \le T)$.

We set $\alpha_0 = 1$, $\rho_1 = 0.2$ and $\rho_2 = 0.5$ in the example from some trials.

Quadratic penalty method

To apply Adam to problem (5.39), i.e., SFSTO, problem (5.39) is reformulated using the quadratic penalty method as follows:

$$\text{Minimize} \quad f(x) = w_{k:n}(\boldsymbol{x}) + \lambda \max\left[0, v(\boldsymbol{x})/v_0 - \mu\right]^2 \tag{5.42a}$$

$$\text{subject to} \quad \boldsymbol{x} \in \mathcal{X}. \tag{5.42b}$$

Instead of problem (5.39), the SFSTO problem (5.42) is solved by Adam using parameter $\lambda = 5 \times 10^4$ after some trials.

5.4.3 NUMERICAL EXAMPLE

Let us consider a design domain where the length of each element is unit length and the number of elements in two directions are given by $m_X = m_Y = 90$, i.e., the total number of elements is $m = 90 \times 90 = 8100$. One third of the left side of the domain is supported, and a unit force is applied at the center of the right side as shown in Fig. 5.15(a). Young's modulus of the material and the volume fraction are given as $E_0 = 1$ and $\mu = 0.2$, respectively. We set $T = 200$

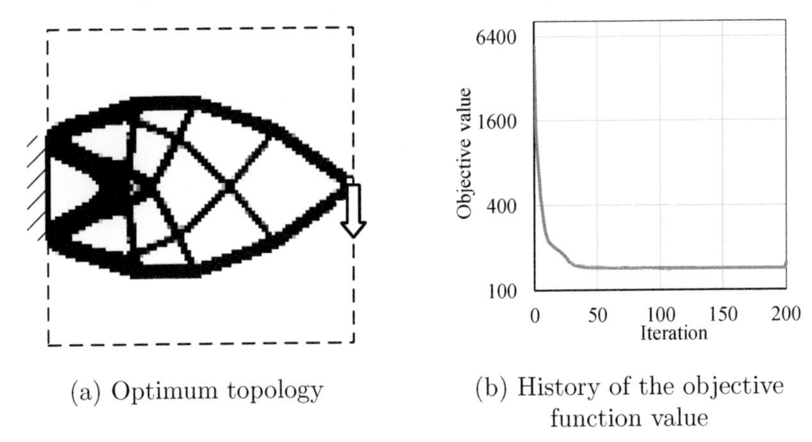

(a) Optimum topology

(b) History of the objective function value

Figure 5.15 Result of nominal topology optimization.

as the maximum number of iterations. Without considering local failure, a solution of nominal topology optimization is obtained by SIMP method with sensitivity filter using optimality criteria method [Andreassen et al., 2011]. The optimum topology is shown in Fig. 5.15(a), and history of the objective value is shown in Fig. 5.15(b).

We set $r = 3.6$ as the radius of the local failure zone; $k = 10$ and $n = 100$ are given in problem (5.42). The order statistics $W_{10:100}$ corresponds to estimator of the 90th percentile response, which is to be minimized as the objective function. This means that we expectedly minimize the top 10% of the worst value. Problem (5.42) is solved with Adam solver, and the density filter in (5.33a) and the threshold projection in (5.34) are applied. Since Adam is a stochastic method, we obtain a different solution at each run. Hence, problem (5.42) is solved independently five times with different random seeds.

Histories of the values of penalized objective function $f(\boldsymbol{x})$ for five cases and their moving average of the 10 data points in the process of SFSTO are shown in Fig. 5.16. It is observed from Fig. 5.16(b) that the differences among five cases are not large; hence, it has been confirmed that Adam shows good convergence even for SFSTO. The four topologies obtained for SFSTO are shown in Fig. 5.17, where circles represent the local failure zone corresponding to each scenario of order statistic $w_{10:100}$. It is seen from Fig. 5.17 that different frame structures are generated for four cases by SFSTO unlike similarity of the curves in Fig. 5.16(b). This implies that there are many different fail-safe designs that have almost the same value of objective function $w_{10:100}$. The design problem is regarded as multimodal and/or close to indeterminate. Compared with the nominal optimal topology in Fig. 5.15(a), more members and reinforcement of the joints are seen in Fig. 5.17. This suggests that the fail-safe design improves the robustness by generating multiple load transmission paths to prepare for the accidental local failure.

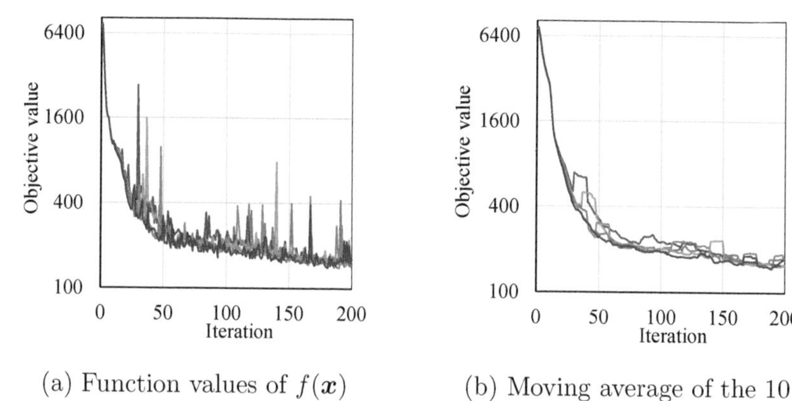

(a) Function values of $f(\boldsymbol{x})$

(b) Moving average of the 10 data points

Figure 5.16 History of the objective function values of five cases in SFSTO.

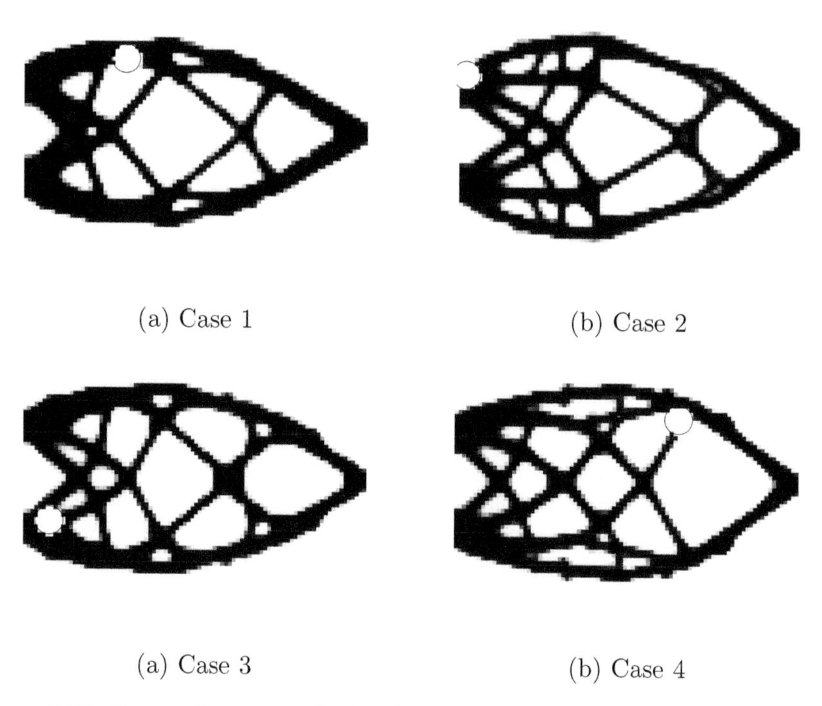

(a) Case 1

(b) Case 2

(a) Case 3

(b) Case 4

Figure 5.17 Four optimum topologies of SFSTO problem.

The nominal objective function values of the solutions, i.e., those without local failure, are summarized in Table 5.10. Obviously, the solution of the nominal topology optimization has the smallest nominal value; however, the differences from the others are not large. To investigate the effect of uncertainty,

Table 5.10
Summary of the Observed Objective Function Values After Trimming the Worst 10% of the Data Points

	Case	Nominal	Mean	Std. dev.	Max.
Nominal topology optimization	**-**	**111.7**	**118.0**	**14.71**	**283.9**
SFSTO	1	112.2	115.1	**5.51**	**135.6**
	2	117.3	122.1	8.53	150.7
	3	115.8	119.1	6.21	142.3
	4	115.4	119.9	8.08	143.4
	5	114.0	116.9	6.03	140.8

MCS is carried out with 100000 randomly generated local failure scenarios. The exact worst scenarios are nearly unstable and the corresponding compliance values are very large. Since the outliers of the random responses have a strong influence on statistics, we exclude the worst 10% values and compute the maximum value, mean value and standard deviation from the remaining 90% of the data, and summarize them in Table 5.10. In Case 1, the smallest mean, standard deviation and maximum values are observed. Compared with those of the nominal topology optimization, the standard deviation and the maximum value decrease by 63% and 52%, respectively. Note again that the maximum value corresponds to the 90th percentile response. Therefore, the results indicate minimizing the order statistics is effective to reduce the 90th percentile response. We can say that the approach based on order statistics enhances the redundancy and robustness of the structure.

5.5 SUMMARY

Some results utilizing order statistics and quantile-based approach have been presented for robust geometry and topology optimization of plane frames. Large computational cost for estimating accurate worst response can be successfully alleviated by relaxing the worst value to a worse value using order statistics. Singularity of the optimal solution in the feasible region under stress constraints can be avoided by underestimating the stress in thin elements. Local instability due to existence of slender members can also be alleviated by penalizing the geometrical stiffness of slender members. A quantile-based SORA has been presented for RBDO of plane frames to avoid computational cost for accurate estimation of limit state functions and to show that MEM is effective for estimating quantile values for specified sample L-moments. Order statistics-based approach is also applicable to SFSTO of a plate subjected to static loads. Through the example of SFSTO, it has been confirmed that Adam, which is one of the popular stochastic gradient-based methods, can find optimal topologies accurately, and robustness against local failure can be improved by generating multiple load transmission paths, which can reduce the risk for the accidental local failure.

ACKNOWLEDGMENT

We acknowledge the permissions from publishers for use of the licensed contents as follows:

- Figures 5.3, 5.4, 5.6 and 5.7, and Tables 5.2 and 5.3: Reproduced with permission from John Wiley and Sons [Shen et al., 2021].
- Figures 5.9, 5.10 and 5.11, and Tables 5.5, 5.6, 5.7, 5.8 and 5.9: Reprinted from Shen et al. [2022] with permission from Elsevier.

6 Multi-objective Robust Optimization Approach

It is very difficult in practical design process to specify an appropriate bound for the structural responses, and solutions to various bounds are preferred to be obtained. Therefore, the design problem can be naturally formulated as a multi-objective optimization (MOO). The weighted sum of the mean and the standard deviation is minimized in the standard formulation of robust design. Two objective functions representing exploration and exploitation may be maximized in the optimization process. In this chapter, it is demonstrated that robustness level can be represented by the order of the solution in the framework of order statistics, and an MOO problem can be solved to obtain Pareto optimal solutions for various levels of robustness.

6.1 INTRODUCTION

In the previous chapters, single-objective formulations have been presented for the design methods considering uncertainty such as robust design, reliability-based design and randomization approach to structural design. In the single-objective formulations, robustness level, confidence level and reliability bound are specified by the designers. However, it is very difficult in practical design process to specify an appropriate bound, and solutions to various bounds are preferred to be obtained. Therefore, the design problem can be naturally formulated as a multi-objective optimization (MOO) problem to find several candidate solutions as Pareto optimal solutions. The designers can find the most preferred solution from the Pareto optimal solutions based on trade-off relations of the objective functions and other performance measures that have not been explicitly incorporated in the MOO problem [Cohon, 1978; Marler and Arora, 2004].

MOO approaches appear at many aspects of structural design under uncertainty. An MOO problem can be simply formulated to find a trade-off solution between the structural cost and structural response [Ohsaki, 2010]. In Sec. 7.4, some results will be presented for minimizing the mean and standard deviation of the uncertain function. Two objective functions representing exploration and exploitation may be maximized in the process of Bayesian optimization [Do et al., 2021a]. For the robust design optimization, it is desired to find solutions corresponding to various levels of robustness because it is difficult to assign the desired robustness level *a priori*. Therefore, the appropriate robustness level can be found in view of trade-off relation among the structural cost, manufacturability, complexity of the design and robustness

DOI: 10.1201/9781003153160-6

against parameter uncertainty. In this chapter, an MOO problem is formulated for minimizing the upper-bounds of the worst values of the representative response with various levels of robustness based on the distribution-free tolerance intervals of order statistics described in Chap. 4.

6.2 ROBUSTNESS LEVEL USING ORDER STATISTICS

Consider a problem for finding approximate worst responses of structures under various sources of uncertainty including those of material property, geometry and external load. As discussed in Chap. 4, order statistics can be effectively utilized for obtaining approximate upper bounds of the worst responses such as displacements and stresses of structures against various external loads including static and seismic loads. The upper bound of the γth quantile of the representative response for specified confidence level can be obtained using the theory of order statistics. Since the upper bound is obviously an increasing function of the parameter γ, the representative response is most likely to be less than the specified value, if the approximate worst representative response is less than the specified value with γ close to 1. Therefore, γ may be regarded as a parameter of robustness as shown in Fig. 6.1, where $\gamma = F(r)$ is the cumulative distribution function (cdf) of the response r. The high-robustness design A satisfies $\gamma_A = F(\bar{r})$ for the specified response $\bar{r}$ and the parameter value γ_A close to 1, while the low-robustness design B satisfies $\gamma_B = F(\bar{r})$ for the parameter value γ_B that is sufficiently smaller than 1.

Let $\boldsymbol{\theta} = (\theta_1, \cdots, \theta_t) \in \Omega$ denote a vector of t uncertain parameters representing, e.g., yield stress, Young's modulus, cross-sectional area and external load. The uncertain parameters are assumed to be continuous random values distributed in the bounded t-dimensional domain Ω. The representative response of a structure such as maximum stress under specified loading con-

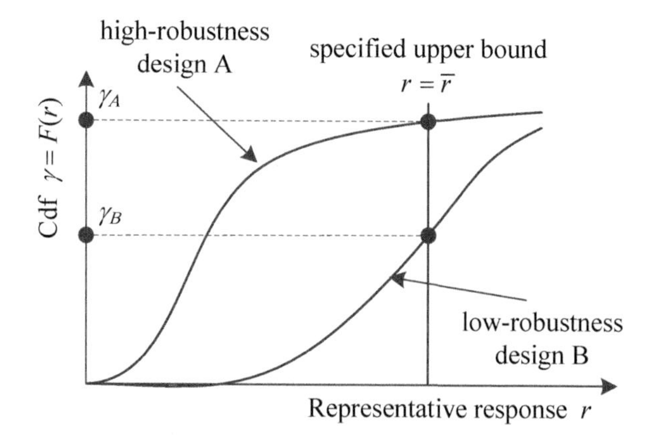

Figure 6.1 Relation between parameter and robustness for specified $r = \bar{r}$.

ditions is denoted by $g(\boldsymbol{\theta})$. For solving a robust design problem, the worst (largest) value of $g(\boldsymbol{\theta})$ for $\boldsymbol{\theta} \in \Omega$ need to be obtained. However, finding the worst value accurately demands a large computational cost even for the simple case where Ω is defined by the interval or box regions. Therefore, we can relax the requirement of *worst response* to the γth ($0 \leq \gamma \leq 1$) *quantile response*, where γ is slightly less than 1, e.g., 0.9, 0.95 or 0.99. For this purpose, we can use order statistics effectively, and the final equations of order statistics for distribution-free tolerance intervals are shown below. See Secs. 1.4.2 and 4.2 for details.

Let $\boldsymbol{\theta}_1, \cdots, \boldsymbol{\theta}_n$ denote a set of t-dimensional vectors of uncertain parameters generated corresponding to the same probability distribution in domain Ω, where n is the number of samples. Accordingly, the representative responses $G_1 = g(\boldsymbol{\theta}_1), \cdots, G_n = g(\boldsymbol{\theta}_n)$ are random variables that have the same probability distribution, for which the cdf is denoted by $F(g(\boldsymbol{\theta})) = \Pr\{G \leq g(\boldsymbol{\theta})\}$. These responses are renumbered in decreasing order as $G_{1:n} \geq G_{2:n} \geq \cdots \geq G_{n:n}$. The kth response $G_{k:n}$ among n responses is a probabilistic value called kth order statistics.

We can select k and n ($1 \leq k \leq n$) so that there exist real values α and γ to state that *"The probability of $g(\boldsymbol{\theta})$ so that its $100\gamma\%$ is less than $G_{k:n}$ is at least $100\alpha\%$"*, i.e.,

$$\Pr\{F(G_{k:n}) \geq \gamma\} \geq \alpha. \tag{6.1}$$

If k is close to 1 without changing n, then the response close to the worst (extreme) value can be obtained. When the values of k, α and γ are specified, the number of required samples n can be found as the smallest integer satisfying the following inequality:

$$1 - I_\gamma(n-k+1, k) \geq \alpha, \tag{6.2}$$

where $I_\gamma(\cdot, \cdot)$ is the incomplete beta function defined in (1.40) and (4.13). Alternatively, if n is specified as the minimum required value satisfying the inequality (6.2) and k is also given, one of the remaining two parameters α and γ are obtained from the following equation by assigning the other parameter:

$$1 - I_\gamma(n-k+1, k) = \alpha, \tag{6.3}$$

because α and γ are real numbers.

Relations between α and γ are plotted in Figs. 6.2(a)–(c) for $n = 50$, 100 and 150, respectively. The curves in each figure correspond to $k = 1, \cdots, 20$ from top-right to bottom-left. The following properties are observed from the figure [Ohsaki et al., 2019]:

(1) For specified values of n and k, α is a decreasing function of γ, and has a larger value for a smaller value of k.

(2) For specified values of n and α, k is a discretely decreasing function of γ.

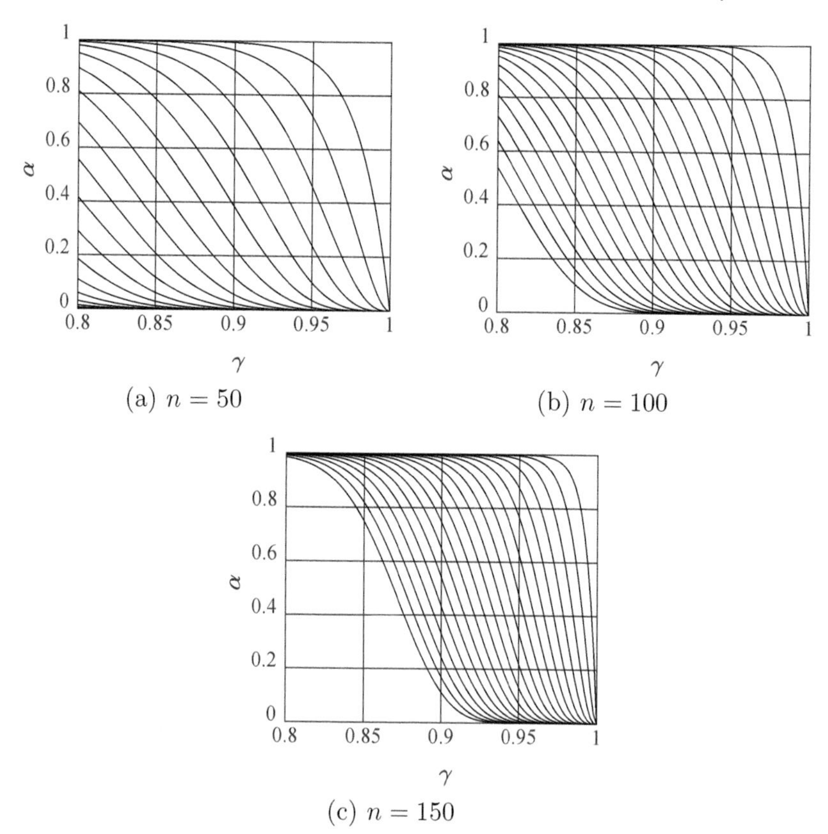

Figure 6.2 Relation between α and γ for some fixed values of n and various values of k.

(3) For specified values of n and γ, k is a discretely decreasing function of α.

(4) The curves move to top-right as n is increased.

Therefore, the confidence level α can be increased by increasing n for a fixed k, or by assigning smaller value for k for a fixed n. The values of k and γ for specified n and α can be listed in a tables or plotted as graphs. The values of γ obtained from (6.3) for $n = 100$ and 200 are listed, respectively, in Tables 6.1(a) and (b), where $\alpha = 0.9$, and k is varied in each table. We can confirm from these tables that γ decreases as k increases while n and α are fixed. Therefore, k can be regarded as a parameter of robustness level; i.e., a smaller value of k leads to a design of a larger robustness as design A in Fig. 6.1.

Table 6.1
Relation Between k and γ for Order Statistics of $\alpha = 0.9$

(a) $n = 100$

k	1	2	3	4	5	6	7	8	9	10
γ	0.977	0.962	0.948	0.934	0.922	0.909	0.897	0.885	0.873	0.862
k	11	12	13	14	15	16	17	18	19	20
γ	0.850	0.839	0.827	0.816	0.805	0.794	0.783	0.772	0.761	0.750

(b) $n = 200$

k	1	2	3	4	5	6	7	8	9	10
γ	0.989	0.981	0.974	0.967	0.960	0.954	0.948	0.942	0.936	0.930
k	11	12	13	14	15	16	17	18	19	20
γ	0.924	0.918	0.912	0.907	0.901	0.895	0.890	0.884	0.878	0.873

6.3 FORMULATION OF MULTI-OBJECTIVE OPTIMIZATION PROBLEM

Consider a structural optimization problem for minimizing the total structural volume under upper-bound constraint on the representative response. Let $\boldsymbol{x}$ denote the vector of design variables such as stiffness of members, nodal coordinates and damping coefficients of viscous dampers. The representative response $g(\boldsymbol{x}; \boldsymbol{\theta})$ is a function of $\boldsymbol{x}$ and uncertainty exists in the parameter vector $\boldsymbol{\theta}$. The kth largest value among n samples $g(\boldsymbol{x}; \boldsymbol{\theta}_i)$ $(i = 1, \cdots, n)$ is denoted by $G_{k:n}(\boldsymbol{x})$. If deviation of response is not a major factor for the design, the median value $G_{(n+1)/2:n}$ (if n is an odd number) may be minimized. By contrast, in the context of robust design optimization, an upper bound is given for the worst value $G_{1:n}$ of the representative response. However, it is natural to assume that the *worst* value may be relaxed to some extent to have a design with slightly smaller robustness with smaller total structural volume. Accordingly, the problem for obtaining a design with an appropriate robustness level $k = k^*$ is formulated as follows:

$$\text{Minimize} \quad f(\boldsymbol{x}) \tag{6.4a}$$

$$\text{subject to} \quad G_{k^*:n}(\boldsymbol{x}) \leq g^{\mathrm{U}}, \tag{6.4b}$$

$$\boldsymbol{x} \in \mathcal{X}, \tag{6.4c}$$

where $f(\boldsymbol{x})$ is the total structural volume; $\mathcal{X}$ is the feasible region of $\boldsymbol{x}$; and g^{U} is the upper bound for $G_{k^*:n}(\boldsymbol{x})$.

As discussed in Sec. 6.2, the design becomes more robust if k becomes smaller. However, larger robustness leads to larger estimate of the response, and the design obtained by solving problem (6.4) becomes more conservative.

For example, the nominal response of the solution for $k^* = 1$ is smaller than that for $k^* = 10$. Therefore, it is desired to generate and compare the solutions with various robustness levels to decide the appropriate robustness level. Another important point is that it is very difficult to select an appropriate value of k reflecting the practical situation, because the possibility of occurrence of the exact extreme value is extremely small. For some cases it may be important to minimize the median and/or quartiles of the representative response. Therefore, it is practically useful to obtain the solutions minimizing the structural response with different values of robustness level k.

For this purpose, we first exchange the objective and constraint functions for a problem with single response constraint as

$$\text{Minimize} \quad G_{k^*:n}(\boldsymbol{x}) \tag{6.5a}$$
$$\text{subject to} \quad f(\boldsymbol{x}) \leq f^{\mathrm{U}}, \tag{6.5b}$$
$$\boldsymbol{x} \in \mathcal{X}, \tag{6.5c}$$

where f^{U} is the upper bound for $f(\boldsymbol{x})$. Solutions for various levels of robustness can be obtained by varying k^* in the objective function. For different values k_1 and k_2, let $\boldsymbol{x}^{(k_1)}$ and $\boldsymbol{x}^{(k_2)}$ denote the optimal solutions of problem (6.5) corresponding to $k^* = k_1$ and k_2, respectively. Then, obviously the inequalities $G_{k_1:n}(\boldsymbol{x}^{(k_1)}) \leq G_{k_1:n}(\boldsymbol{x}^{(k_2)})$ and $G_{k_2:n}(\boldsymbol{x}^{(k_2)}) \leq G_{k_2:n}(\boldsymbol{x}^{(k_1)})$ hold. Therefore, for example, if we compare $\boldsymbol{x}^{(k_1)}$ and $\boldsymbol{x}^{(k_2)}$ for $k_1 = 1$ and $k_2 = \lfloor (n+1)/2 \rfloor$, where $\lfloor \cdot \rfloor$ is the floor function that represents the largest integer less than the variable, $\boldsymbol{x}^{(k_2)}$ is likely to have larger worst value than $\boldsymbol{x}^{(k_1)}$; i.e., $\boldsymbol{x}^{(k_1)}$ is more robust than $\boldsymbol{x}^{(k_2)}$. However, $\boldsymbol{x}^{(k_1)}$ is likely to have larger nominal value than $\boldsymbol{x}^{(k_2)}$.

To obtain the solutions corresponding to various levels of robustness, we formulate an MOO problem to minimize the order statistics $G_{k:n}(\boldsymbol{x})$ $(k \in \mathcal{K})$ as

$$\text{Minimize} \quad G_{k:n}(\boldsymbol{x}) \ (k \in \mathcal{K}) \tag{6.6a}$$
$$\text{subject to} \quad f(\boldsymbol{x}) \leq f^{\mathrm{U}}, \tag{6.6b}$$
$$\boldsymbol{x} \in \mathcal{X}, \tag{6.6c}$$

where $\mathcal{K}$ is a subset of $\{1, \cdots, n\}$. When we are interested in approximate worst values, for example, $\mathcal{K} = \{1, \cdots, \lfloor n/10 \rfloor\}$ may be given, where $k = \lfloor n/10 \rfloor$ approximately corresponds to the 90 percentile of the response.

6.4 APPLICATION TO ENGINEERING PROBLEMS

6.4.1 PLANE TRUSS SUBJECTED TO STATIC LOAD

Before presenting solutions of MOO problems for robust design, properties of distribution of order statistics of responses are investigated using a simple example. Consider a plane grid truss, as shown in Fig. 6.3, subjected to a static vertical load. Note that the crossing diagonal members are not connected at

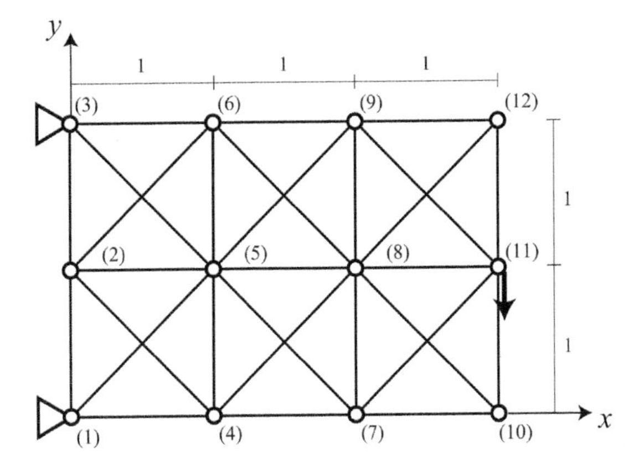

Figure 6.3 A 3 × 2 plane grid truss.

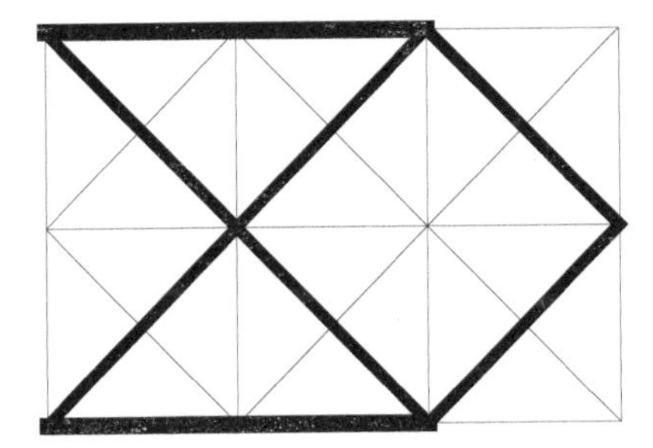

Figure 6.4 Optimal solution for Case 1; no visible difference exists among the plots of Cases 1–5.

their centers; i.e., the truss has 29 members. The units are omitted because they are not important. Young's modulus of the material is 1.0, a downward load of the magnitude 1.0 is applied at node 11, and the unit size is 1.0. The compliance defined as (1.6) in Sec. 1.2.2 is minimized under constraint on the total structural volume.

The design variables are the cross-sectional areas of members that have continuous values. Five optimal solutions, denoted by Cases 1–5, are found without considering uncertainty corresponding to the lower-bound cross-sectional area $a^{\mathrm{L}} = 0.1$, 0.2, 0.3, 0.4 and 0.5, respectively, where the upper bound 100 is assigned for the total structural volume. The nonlinear programming library *SLSQP* in SciPy package [Virtanen et al., 2020] is used for optimization. The optimal solution for Case 1 is shown in Fig. 6.4, where the width of each member is proportional to its cross-sectional area. We can see from the figure that eight members have cross-sectional areas larger than a^{L}. The cross-sectional areas of those members decrease as a^{L} is increased to have the same total structural volume. The horizontal members connected to the supports in Fig. 6.4 have the largest cross-sectional area 9.9104, which is quite larger than the cross-sectional areas of thin members which are equal to the

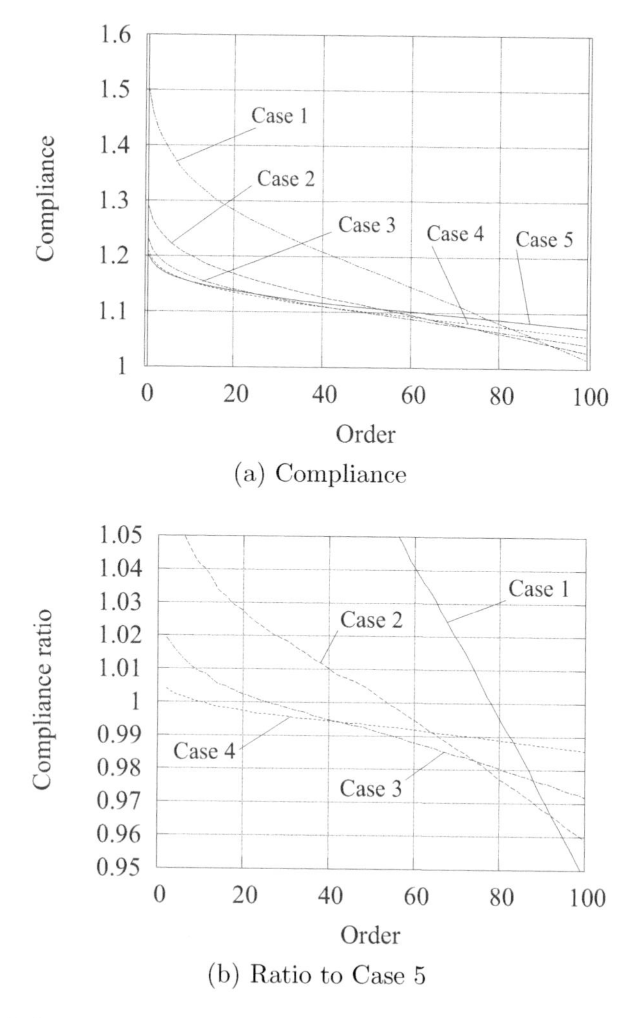

Figure 6.5 Relation between k and compliance for solutions of Cases 1–5.

lower-bound value. Therefore, no visible difference exists among the plots of Cases 1–5.

Uncertainty is considered in the vertical coordinates y_4 and y_6 of nodes 4 and 6, respectively. The coordinates are uniformly distributed in the ranges $[y_i^0 - 0.1, y_i^0 + 0.1]$ $(i = 4, 6)$, where y_i^0 is the coordinate in Fig. 6.3. Let $\boldsymbol{a}^{(i)}$ denote the vector of optimal cross-sectional areas for Case i. The order statistics of compliance of the optimal solutions among n samples are denoted by $G_{k:n}(\boldsymbol{x}^{(i)})$. Suppose $n = 100$, and we generate 100 sets of 100 samples. The mean values of $G_{k:100}(\boldsymbol{x}^{(i)})$, denoted by $\bar{G}_{k:100}(\boldsymbol{x}^{(i)})$, for Cases 1–5 are plotted with respect to k in Fig. 6.5(a). Note that the optimal objective values without considering uncertainty are 1.0141, 1.0286, 1.0435, 1.0590 and 1.0748

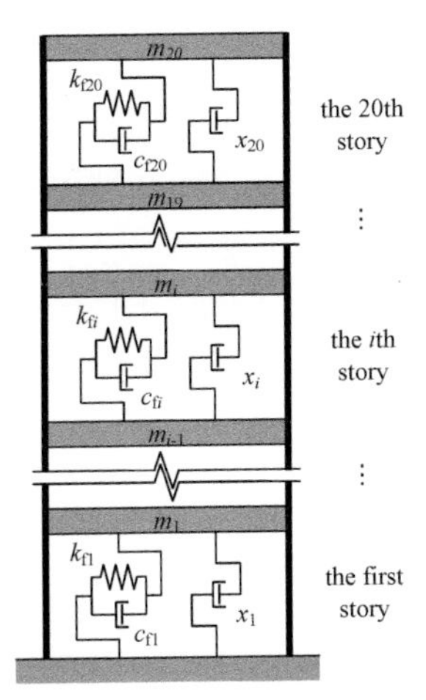

Figure 6.6 A 20-story shear frame model.

for Cases 1–5, respectively. It is seen from the figure that Case 1 with small lower-bound cross-sectional area $a^{\mathrm{L}} = 0.1$ has very large extreme value of compliance for $k = 1$; i.e., the compliance is very sensitive to imperfection of nodal locations, because the optimal solution is unstable if the members with lower-bound cross-sectional area are removed. The compliance approaches the optimal value without uncertainty as k is increased to 100. To clearly observe the cases with the smallest value of $\bar{G}_{k:100}(\boldsymbol{x}^{(i)})$, the ratios of $\bar{G}_{k:100}(\boldsymbol{x}^{(i)})$ to $\bar{G}_{k:100}(\boldsymbol{x}^{(5)})$ are plotted in Fig. 6.5(b). It is seen from the figure that the case with the smallest compliance changes with the variation of k: Case 5 for $k = 1, \cdots, 10$, Case 4 for $k = 11, \cdots, 42$, Case 3 for $k = 43, \cdots, 73$, Case 2 for $k = 74, \cdots, 92$, and Case 1 for $k = 93, \cdots, 100$. Therefore, if we consider a problem of selecting the best solution among the five designs, the optimal solution depends on the order k, which is regarded as the robustness level.

6.4.2 SHEAR FRAME SUBJECTED TO SEISMIC EXCITATION

Problem (6.6) is solved for a 20-story shear frame with viscous dampers [Ohsaki et al., 2019]; see also Yamakawa and Ohsaki [2021]. The first, top and one of the other intermediate stories of the shear frame model are illustrated in Fig. 6.6, and the intermediate stories can be modeled similarly.

The additional damping coefficients due to viscous dampers [Tsuji and Nakamura, 1996] in the stories are taken as the design variables that are denoted by the vector $\boldsymbol{x} = (x_1, \cdots, x_{20})$ in problem (6.6). Uncertainty exists in story mass and stiffness as well as damping coefficient that is also a design variable. The nominal values of mass m_i at the ith story ($i = 1, \cdots, 20$) are 4.0×10^5 kg, and the nominal values of story stiffness $k_{\mathrm{f}i}$ of the ith story ($i = 1, \cdots, 20$) are proportional to the seismic shear coefficient defined by

Table 6.2
Nominal Values of Story Stiffness of Shear Frame Model ($\times 10^9$ N/m)

i	1	2	3	4	5	6	7	8	9	10
$k_{\mathrm{f}i}$	1.244	1.234	1.220	1.202	1.180	1.154	1.123	1.088	1.048	1.004
i	11	12	13	14	15	16	17	18	19	20
$k_{\mathrm{f}i}$	0.955	0.901	0.841	0.777	0.706	0.629	0.545	0.452	0.347	0.223

Japanese building code, which are listed in Table 6.2. The 1st natural period T_{f} is 1.6 sec. for the frame with nominal mass and stiffness. Structural damping of the frame, which is proportional to the stiffness, is also considered as

$$c_{\mathrm{f}i} = \frac{0.02T_{\mathrm{f}}}{\pi} k_{\mathrm{f}i} \quad (i = 1, \cdots, 20).$$

Thus, the nominal value of the damping coefficient of the ith story denoted by c_i is given by

$$c_i = c_{\mathrm{f}i} + x_i \quad (i = 1, \cdots, 20).$$

Seismic responses are evaluated using the design displacement response spectrum $S_{\mathrm{d}}(T, h)$ and acceleration response spectrum $S_{\mathrm{a}}(T, h)$, which are functions of the natural period T and the damping factor h as

$$S_{\mathrm{d}}(T, h) = \left(\frac{2\pi}{T}\right)^2 S_{\mathrm{a}}(T, h),$$

$$S_{\mathrm{a}}(T, h) = \frac{1.5}{1 + 10h} S_{\mathrm{a}0},$$

$$S_{\mathrm{a}0} = \begin{cases} 0.96 + 9T & (T < 0.16), \\ 2.4 & (0.16 \leq T < 0.864) \ , \\ 2.074/T & (0.864 \leq T), \end{cases}$$

which are defined by Japanese building code. Figure 6.7 shows $S_{\mathrm{a}}(T, h)$ for $h = 0.02$, 0.05 and 0.10.

The structural damping matrix is non-proportional due to existence of viscous damper in each story. Therefore, the extended complete quadratic combination (CQC) method [Iyenger and Manohar, 1987; Yang et al., 1990] is used for response evaluation. The maximum value among all stories of the maximum interstory drift is chosen as the representative response $g(\boldsymbol{x})$. All 20 eigenmodes are used for evaluation by the extended CQC method.

Since uncertainty is considered for the vectors of mass, stiffness and damping coefficient, which have 20 components, respectively, the vector $\boldsymbol{\theta}$ of uncertain parameters have 60 components in total. Uniform distribution is assumed

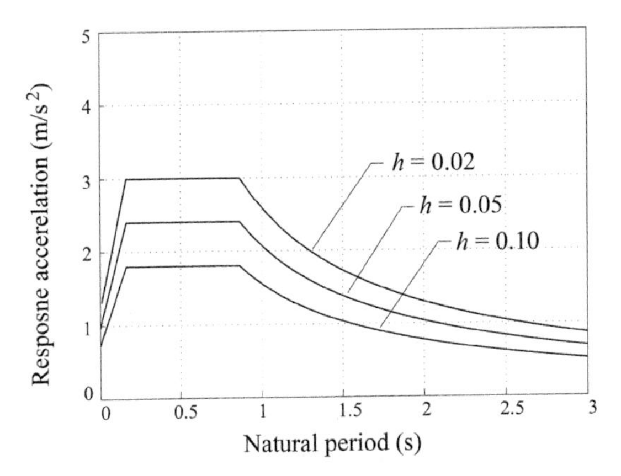

Figure 6.7 Design acceleration response spectra for damping factors 0.02, 0.05 and 0.10.

for each parameter in the range of $\pm 10\%$ to the mean values in 20 stories. Hence, $\boldsymbol{\theta}$ is defined as

$$\boldsymbol{\theta} = (\theta_1, \cdots, \theta_{60}) \in \Omega, \tag{6.8a}$$

$$\Omega = \{(\theta_1, \cdots, \theta_{60}) \mid -0.1 \leq \theta_i \leq 0.1;\ i = 1, \cdots, 60\}. \tag{6.8b}$$

The uncertain parameters θ_i are multiplied to the mean values of story mass, stiffness and damping coefficient, respectively, and added to the nominal values to obtain random variables as

$$M_i = m_i + \theta_i \hat{m}, \quad K_{\mathrm{f}i} = k_{\mathrm{f}i} + \theta_{20+i} \hat{k}_{\mathrm{f}}, \quad C_i = c_i + \theta_{40+i} \hat{c} \quad (i = 1, \cdots, 20),$$

where

$$\hat{m} = \frac{1}{20} \sum_{i=1}^{20} m_i, \quad \hat{k}_{\mathrm{f}} = \frac{1}{20} \sum_{i=1}^{20} k_{\mathrm{f}i}, \quad \hat{c} = \frac{1}{20} \sum_{i=1}^{20} c_i.$$

The 60 parameters are generated from uniform distribution in the intervals, and eigenvalue analysis is carried out to obtain the natural periods and eigenmodes for each solution. Then the representative response is computed using the extended CQC method, and the representative responses among 150 samples are sorted in descending order for generating order statistics $G_{1:150}, \cdots, G_{150:150}$. The MOO problem for minimizing 20 order statistics is formulated as

$$\text{Minimize} \quad G_{1:150}(\boldsymbol{x}), \ \cdots, \ G_{20:150}(\boldsymbol{x}) \tag{6.9a}$$

$$\text{subject to} \quad \sum_{i=1}^{20} x_i \leq c^{\mathrm{U}}, \tag{6.9b}$$

$$x_i \geq 0 \quad (i = 1, \cdots, 20), \tag{6.9c}$$

where c^{U} is the upper bound for the total value of damping coefficients.

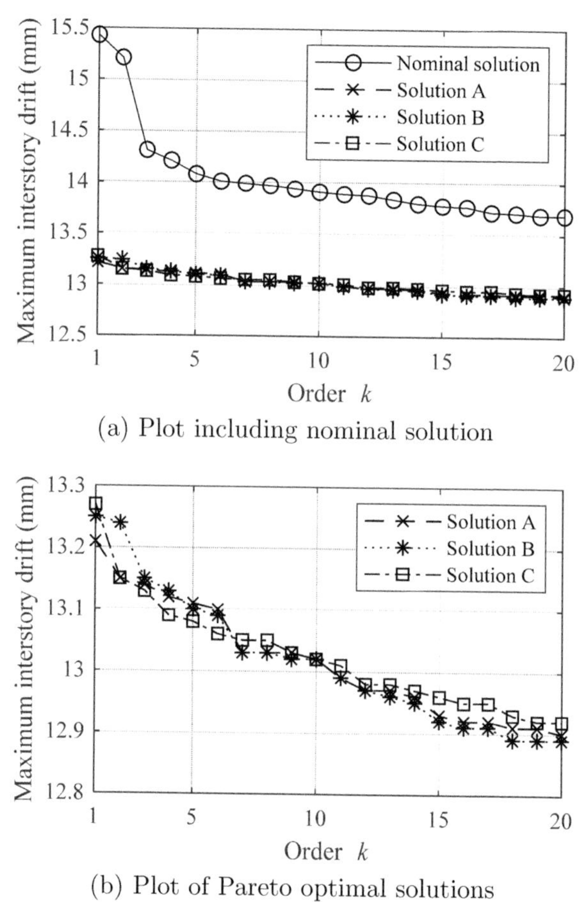

Figure 6.8 Relation between maximum interstory drift and order k for nominal solution and three Pareto optimal solutions.

Pareto optimal solutions are found using NSGA-II [Deb, 2001] available in Global Optimization Toolbox of Matlab [Mathworks, 2018a]. The population size is 200, the number of generations is 117, and the elitist strategy is used. The value of α is 0.9; therefore, as seen from Table 6.1(b), the robustness level γ decreases from 0.985 to 0.831 as k is increased from 1 to 20. Response spectrum analysis is performed 150 times for each solution by generating 150 different sets of parameter values with uniform distribution. Note that $\boldsymbol{\theta}$ is shared by all individuals in a generation, and updated at each generation.

As a result of optimization, the 200 solutions converged to a set of 70 different Pareto optimal solutions. Among them, the values of $G_{1:150}(\boldsymbol{x})$, $\cdots$, $G_{20:150}(\boldsymbol{x})$ of three Pareto optimal solutions A, B and C are plotted in Figs. 6.8(a) and (b). Note that there are nine different Pareto optimal solutions that minimize $G_{k:150}(\boldsymbol{x})$ for different values of $k \in \{1, \cdots, 20\}$.

The nominal solution is also obtained by assigning nominal values to all uncertain parameters. Random parameter values are then assigned for the nominal solution to compute $G_{1:150}(\boldsymbol{x}), \cdots, G_{20:150}(\boldsymbol{x})$ as plotted in Fig. 6.8(a). Figure 6.8(b) shows the detailed view of the values of $G_{1:150}(\boldsymbol{x}), \cdots, G_{20:150}(\boldsymbol{x})$ of the three Pareto optimal solutions. The following properties are observed from these figures:

1. The nominal solution is far from optimal because the responses are larger than those of other Pareto optimal solutions for all orders corresponding to $k = 1, \cdots, 20$.
2. Solution A has the smallest value for $k = 1$ among all 70 Pareto optimal solutions.
3. Solution B has the smallest value for $k = 16, 18, 19$.
4. Solution C has the smallest value for $k = 2, \cdots, 6$.
5. Decrease of $G_{k:150}(\boldsymbol{x})$ for larger k is very small for solutions A and B.
6. For solution C, $G_{k:150}(\boldsymbol{x})$ has a large value if k is small; however, it rapidly decreases as k is increased.

The values of additional damping coefficients in the stories, which are the design variables, are plotted in Fig. 6.9(a) for solutions A, B and C. As seen in the figure, the three solutions have similar distributions with small difference in the middle stories. Figure 6.9(b) compares the story damping coefficients of solution A and those of the nominal solution, which shows the nominal solution has smaller/larger damping coefficients in the upper/lower stories than the Pareto optimal solution A.

6.5 SHAPE AND TOPOLOGY OPTIMIZATION OF PLANE FRAME

6.5.1 MULTI-OBJECTIVE OPTIMIZATION PROBLEM

The multi-objective formulation of robust optimization is applied to shape and topology optimization of plane frames. This section summarizes the results by Shen et al. [2020]. The worst structural response under uncertainty is to be relaxed to a quantile response, and the robustness level is represented by the order statistics in the same manner as Sec. 6.4. Let $g(\boldsymbol{x})$ denote the representative response that is a function of the vector of design variables $\boldsymbol{x}$. The structural response with uncertain parameter vector $\boldsymbol{\theta}$ is denoted by $g(\boldsymbol{x}; \boldsymbol{\theta})$, which can be regarded as a realization of random variable $G(\boldsymbol{x}) = g(\boldsymbol{x}; \boldsymbol{\Theta})$, where $\boldsymbol{\Theta}$ is a random variable vector within a feasible region Ω and $\boldsymbol{\theta}$ is also regarded as a realization of $\boldsymbol{\Theta}$. As stated in Sec. 6.3, the worst structural response $\max_{\boldsymbol{\theta} \in \Omega} g(\boldsymbol{x}; \boldsymbol{\theta})$ is approximated by the kth order statistic among n samples denoted by $G_{k:n}(\boldsymbol{x}) = g_{k:n}(\boldsymbol{x}; \boldsymbol{\Theta}_1, \cdots, \boldsymbol{\Theta}_n)$ with independent and

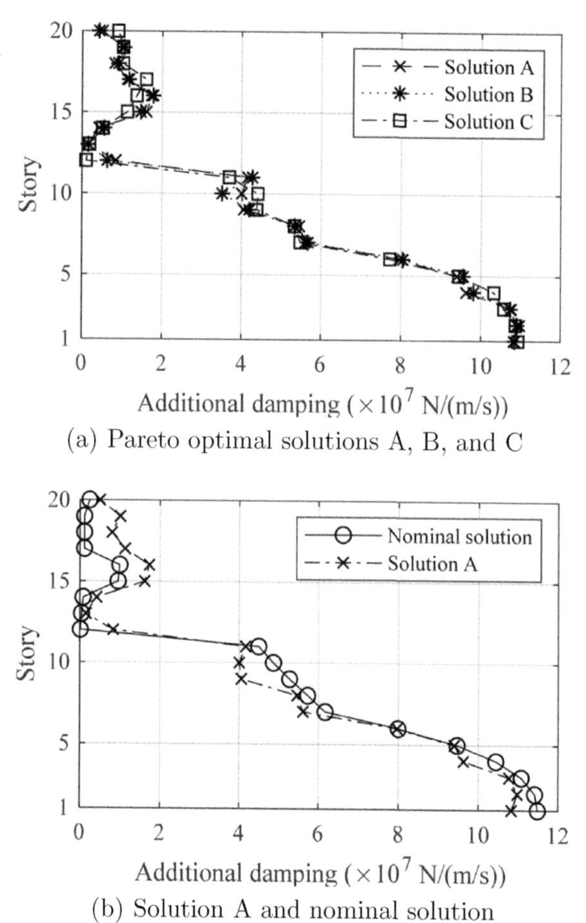

(a) Pareto optimal solutions A, B, and C

(b) Solution A and nominal solution

Figure 6.9 Distributions of additional damping coefficients of viscous dampers.

identically distributed random variables (iidrvs) $\boldsymbol{\Theta}_i$ $(i = 1, \cdots, n)$. Note that $G_{k:n}$ also becomes random variable because $\boldsymbol{\Theta}_i$ is iidrv. According to the dependence properties of order statistics [Boland et al., 1996; Avérous et al., 2005], two order statistics are non-negatively correlated if they are iidrvs. Since the probability distribution of response is unknown, only distribution-free measurements of dependence such as Spearman's correlation and Kendall's tau can be applied; see Navarro and Balakrishnan [2010] for details.

For formulating an MOO problem to obtain solutions with various robustness levels, the response without considering uncertainty denoted by $g(\boldsymbol{x}) = g(\boldsymbol{x}; \bar{\boldsymbol{\theta}})$ for the nominal parameter values $\bar{\boldsymbol{\theta}}$ is also included in the objective function. Hence, an MOO problem is formulated as

$$\underset{\boldsymbol{x}}{\text{Minimize}} \quad g(\boldsymbol{x}),\ g_{k:n}(\boldsymbol{x}; \boldsymbol{\Theta}_1, \cdots, \boldsymbol{\Theta}_n) \quad (k \in \mathcal{K}) \tag{6.10a}$$

$$\text{subject to} \quad h_i(\boldsymbol{x}) \leq 0 \quad (i = 1, \cdots, n_h), \tag{6.10b}$$

$$\boldsymbol{x} \in \mathcal{X}, \tag{6.10c}$$

where $\mathcal{K}$ is a given subset of $\{1, \cdots, n\}$ and $h_i(\boldsymbol{x}) \leq 0$ is the ith deterministic constraint.

6.5.2 ROBUST SHAPE AND TOPOLOGY OPTIMIZATION OF PLANE FRAME

Generally, in shape and topology optimization of plane frame, the design variables are the vector of nodal coordinates including x- and y-coordinates, and cross-sectional areas, which are denoted by

$$\boldsymbol{x} = \left(x_1, \cdots, x_{n_x}, y_1, \cdots, y_{n_y}\right), \quad \boldsymbol{a} = \left(a_1, \cdots, a_{n_e}\right),$$

respectively, where n_e, n_x and n_y are the number of variables of members and nodal locations in x- and y-coordinates, respectively. We consider a robust optimization problem with uncertainty in Young's modulus of each member. To alleviate difficulties due to complexity of shape and topology optimization considering uncertainty in parameters, the following problem without uncertainty is first solved:

$$\underset{\boldsymbol{a},\boldsymbol{x}}{\text{Minimize}} \quad g(\boldsymbol{a}, \boldsymbol{x}; \bar{\boldsymbol{e}}) \tag{6.11a}$$

$$\text{subject to} \quad h_i(\boldsymbol{a}, \boldsymbol{x}) \leq 0 \quad (i = 1, \cdots, n_h), \tag{6.11b}$$

$$(\boldsymbol{a}, \boldsymbol{x}) \in \mathcal{X}, \tag{6.11c}$$

where $\bar{\boldsymbol{e}}$ is the vector that consists of the nominal values of Young's modulus of the material, and $\mathcal{X}$ is the box domain defined by the upper and lower bounds of variables.

Some of the members that have small cross-sectional areas in the optimal solution of problem (6.11) may contribute to the reduction of structural response when uncertainty is considered. Therefore, based on the optimal shape obtained by solving problem (6.11), which is denoted by $\bar{\boldsymbol{x}}$, we choose the cross-sectional areas as the design variables and incorporate uncertainty also in nodal locations for the MOO problem (6.10), which may be written as

$$\underset{\boldsymbol{a}}{\text{Minimize}} \quad g(\boldsymbol{a}; \bar{\boldsymbol{x}}, \bar{\boldsymbol{e}}),\ g_{k:n}(\boldsymbol{a}; \boldsymbol{X}, \boldsymbol{E}) \quad (k \in \mathcal{K}) \tag{6.12a}$$

$$\text{subject to} \quad h_i(\boldsymbol{a}, \bar{\boldsymbol{x}}) \leq 0 \quad (i = 1, \cdots, n_h), \tag{6.12b}$$

$$(\boldsymbol{a}, \bar{\boldsymbol{x}}) \in \mathcal{X}, \tag{6.12c}$$

where $\boldsymbol{E}$ and $\boldsymbol{X}$ are random variable vectors of Young's modulus of material and nodal locations considering deviation from the nominal values $\bar{\boldsymbol{e}}$ and $\bar{\boldsymbol{x}}$, respectively. After obtaining the Pareto optimal solutions of problem (6.12), we remove those members with small cross-sectional areas to obtain the distinct topology of the frame.

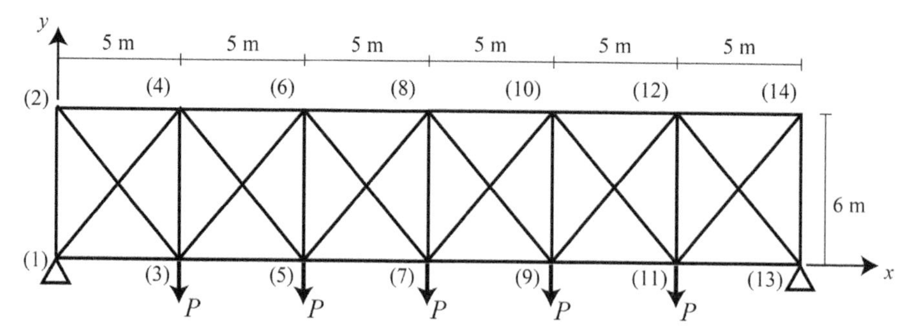

Figure 6.10 A 6 × 1 grid bridge frame model.

6.5.3 NUMERICAL EXAMPLE

An example is presented to illustrate the effectiveness of the method described in this section. Consider a 6 × 1 grid bridge frame with 14 nodes and 31 members, as shown in Fig. 6.10. The frame is pin-supported at nodes 1 and 13. Downward loads with magnitude $P = 3\times 10^6$ N are applied at nodes 3, 5, 7, 9 and 11. Each member has solid circular cross-section, and the diameter and second moment of area are expressed by its cross-sectional area. The diagonal members are not connected at their centers. Thus, considering symmetry of the frame the design variables are the cross-sectional area of 16 members and the y-coordinates of the upper nodes 2, 4, 6 and 8, i.e., the number of design variables is 20.

For verification, the total structural volume of the bridge frame $v(\boldsymbol{a}, \boldsymbol{x})$ is minimized under upper-bound constraints on downward displacements of nodes and the edge stresses of members without uncertainty. A small positive lower bound $a^{\mathrm{L}} = 1.0\times 10^{-7}\ \mathrm{m}^2$ is given for the cross-sectional area to prevent singularity of the stiffness matrix; see Shen et al. [2020] for details. Genetic algorithm (GA) implemented in Global Optimization Toolbox of MATLAB [Mathworks, 2018a] is used to solve the problem. The crossover rate is given as 0.7 and Latin hypercube sampling method is applied to enhance diversity of the initial population. The optimal solution is shown in Fig. 6.11(a), where

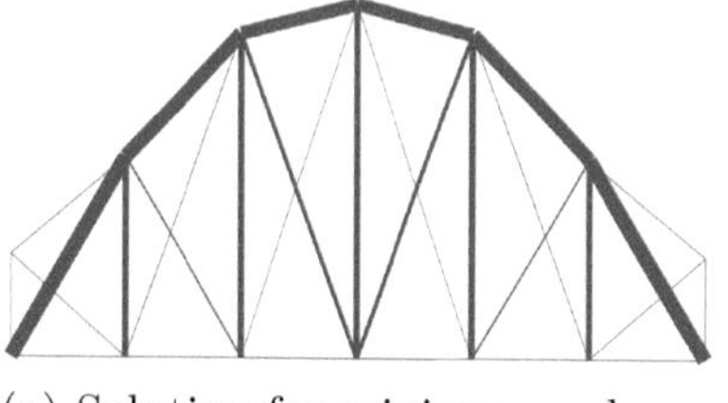

(a) Solution for minimum volume

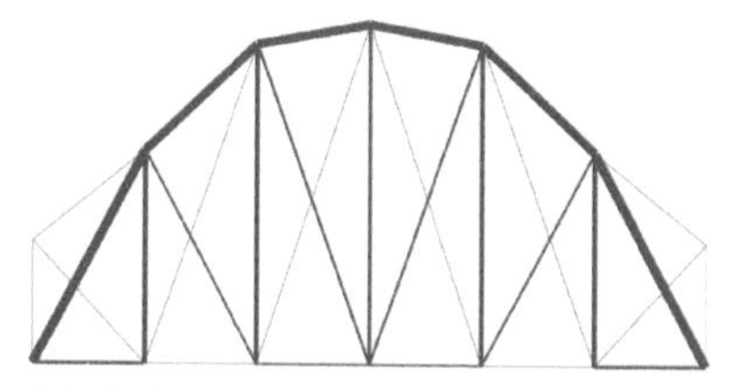

(b) Solution of problem (6.13) for minimum maximum stress

Figure 6.11 Optimal solutions of the bridge frame without uncertainty.

Table 6.3
Parameter Setting of the Bridge Frame

Parameters	Values	Parameters	Values
Nominal value $\bar{\boldsymbol{e}}$	$2.1 \times 10^{11} \cdot \mathbf{1}\,\mathrm{Pa}$	Upper bound v^{U}	$1\,\mathrm{m}^3$
Upper bound $\boldsymbol{X}^{\mathrm{U}}$	$\bar{\boldsymbol{x}} + 0.6 \cdot \mathbf{1}$ m	Lower bound $\boldsymbol{X}^{\mathrm{L}}$	$\bar{\boldsymbol{x}} - 0.6 \cdot \mathbf{1}$ m
Upper bound a^{U}	$0.1\,\mathrm{m}^2$	Lower bound a^{L}	$1.0 \times 10^{-7}\,\mathrm{m}^2$
Upper bound $\boldsymbol{E}^{\mathrm{U}}$	$1.1\,\bar{\boldsymbol{e}}$	Lower bound $\boldsymbol{E}^{\mathrm{L}}$	$0.9\,\bar{\boldsymbol{e}}$
Values of order $\mathcal{K}$	{1, 50, 100}	Confidence level α_k	0.9
Sample size n	200		

the width of each member is proportional to its radius. The optimal structural volume is $5.42\,\mathrm{m}^3$, which is about 6.4% larger than the result in Rajan [1995] for the pin-jointed truss. This is because the frame has a larger edge stress than a truss with the same cross-sectional areas due to bending deformation; hence, the frame should have a larger cross-section to satisfy the stress constraints.

Next, consider a minimization problem of the maximum stress under volume constraint. The values of parameters are listed in Table 6.3, where **1** is a vector with all entries equal to 1. The optimal solution without uncertainty is first found by solving the following problem:

$$\underset{\boldsymbol{a},\boldsymbol{x}}{\text{Minimize}} \quad \sigma^{\max}(\boldsymbol{a}, \boldsymbol{x}; \bar{\boldsymbol{e}}) = \max_{i=1,\cdots,16} \sigma_i^{\max}(\boldsymbol{a}, \boldsymbol{x}; \bar{\boldsymbol{e}}) \tag{6.13a}$$

$$\text{subject to} \quad v(\boldsymbol{a}, \boldsymbol{x}) \leq v^{\mathrm{U}}, \tag{6.13b}$$

$$a^{\mathrm{L}} \leq a_i \leq a^{\mathrm{U}} \quad (i = 1, \cdots, 16), \tag{6.13c}$$

$$y^{\mathrm{L}} \leq y_j \leq y^{\mathrm{U}} \quad (j = 2, 4, 6, 8), \tag{6.13d}$$

where $\sigma^{\max}$ is the nominal maximum stress among all the members, and $\sigma_i^{\max}$ is the maximum absolute value of stress of the ith member. The solution of problem (6.13) obtained using GA is shown in Fig. 6.11(b).

Next, the Young's modulus of each member and x- and y-coordinates of nodes 2, 4, 6, 8, 10, 12 and 14 are selected as uncertain parameters, which are assumed to be uniformly distributed in the box region bounded by their upper and lower bounds in Table 6.3. We select the maximum and median response as the desired percentile to be minimized [Jekel and Haftka, 2020], and the upper quartile, i.e., the 0.75th quantile, is also included to investigate the variation of the response. The nominal maximum value and order statistics of the stress are minimized under volume constraint. Using the optimal shape $\bar{\boldsymbol{x}}$ in Fig. 6.11(b), the cross-sectional areas are selected as design variables and thus the MOO problem is formulated as

$$\begin{aligned}
\underset{\boldsymbol{a}}{\text{Minimize}} \quad & \sigma^{\max}(\boldsymbol{a};\bar{\boldsymbol{x}},\bar{\boldsymbol{e}}),\ \sigma^{\max}_{1:200}(\boldsymbol{a};\boldsymbol{X},\boldsymbol{E}), \\
& \sigma^{\max}_{50:200}(\boldsymbol{a};\boldsymbol{X},\boldsymbol{E}),\ \sigma^{\max}_{100:200}(\boldsymbol{a};\boldsymbol{X},\boldsymbol{E}) && (6.14\text{a}) \\
\text{subject to} \quad & v(\boldsymbol{a}) \leq v^{\mathrm{U}}, && (6.14\text{b}) \\
& a^{\mathrm{L}} \leq a_i \leq a^{\mathrm{U}} \qquad (i = 1, 2, \cdots, 16). && (6.14\text{c})
\end{aligned}$$

Pareto optimal solutions of problem (6.14) are found using NSGA-II in Global Optimization Toolbox of Matlab [Mathworks, 2018a], where the solution of problem (6.13) is added in the initial population to improve convergence, and the fraction of individuals to keep on the Pareto front is 0.35 by default. A set of 70 different Pareto optimal solutions have been found through optimization by NSGA-II with 200 individuals.

The Pareto front on a plane of $\sigma^{\max}$ and $\sigma^{\max}_{1:200}$ is shown in Fig. 6.12, where the solutions on the Pareto front are indicated by PF followed by the obtained step number in the parentheses. It can be observed from Fig. 6.12(a) that the Pareto front similar to the final one is generated at step 30. Figure 6.12(b) plots the detailed view of the Pareto optimal solutions, which shows that variation of $\sigma^{\max}_{1:200}$ is greater than that of $\sigma^{\max}$. The same property has been observed for $\sigma^{\max}_{50:200}$ and $\sigma^{\max}_{100:200}$. The Pareto optimal solutions that have the smallest values of $\sigma^{\max}$, $\sigma^{\max}_{1:200}$, $\sigma^{\max}_{50:200}$ and $\sigma^{\max}_{100:200}$, respectively, are denoted by solutions A, B, C and D, and shown in Fig. 6.13, where the contour shows the maximum edge stress for the nominal values of parameters.

Note that the thin members with cross-sectional areas smaller than 0.0001 m^2 are removed from the optimal solution in Fig. 6.13, and their objective values before and after removing the thin members are listed in Table 6.4. As seen from Table 6.4, the objective values of solutions B, C and D, as well as the value of $\sigma^{\max}$ in solution A, before and after removal are very close. However, the values of $\sigma^{\max}_{1:200}$, $\sigma^{\max}_{50:200}$ and $\sigma^{\max}_{100:200}$ in solution A after removing thin members are greater than those before removal. This is mainly because solution A is unstable if the bending stiffness is neglected and the thin members are removed; hence, the stress values are sensitive to uncertainty, and the maximum stress increases rapidly leading to an asymmetric deformation. Therefore, solution A has obviously large values of other three order statistics. Solution B has the smallest value of $\sigma^{\max}_{1:200}$ and some of the members, which do not exist in solution A, have moderately large cross-sectional areas to reduce the effect of uncertainty in parameters. It has been confirmed that solutions C and D have the smallest values of $\sigma^{\max}_{50:200}$ and $\sigma^{\max}_{100:200}$, respectively, after removing members 8 and 24 in solution D. We can also see from Table 6.4 that the nominal value $\sigma^{\max}$ decreases as k is increased. However, the solutions A and D that minimize the nominal and median values, respectively, are very different, indicating that a small asymmetric property due to parameter uncertainty leads to a large increase of the maximum stress; accordingly, leads

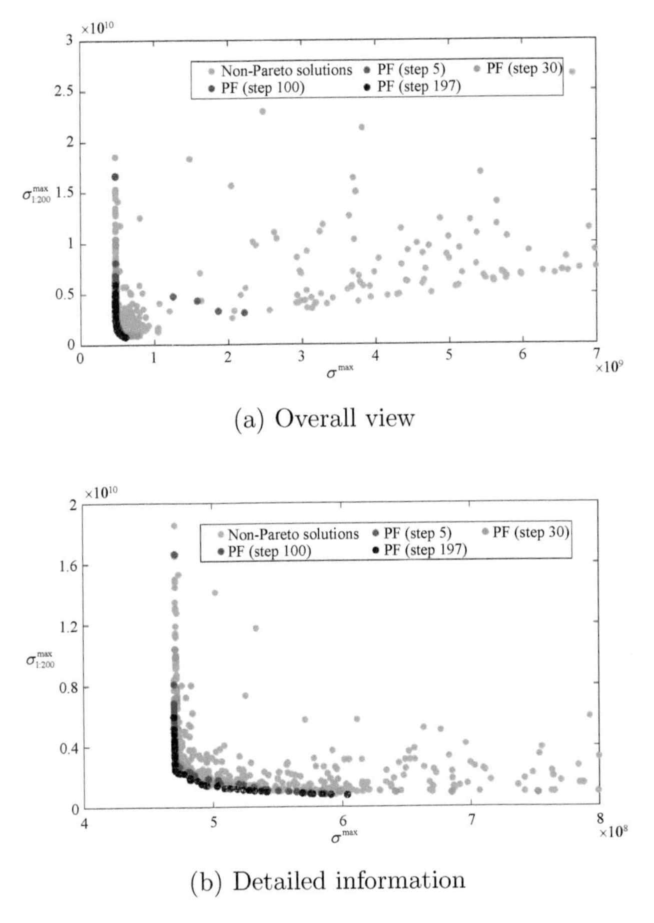

Figure 6.12 Stepwise Pareto front of objectives $\sigma^{\max}$ and $\sigma^{\max}_{1:200}$.

to a significant difference in the optimal solutions. By contrast, the extreme value $\sigma^{\max}_{1:200}$ gradually increases as k increases.

Solutions A, B, C and D have the same structural volume $1.0\,\mathrm{m}^3$, which is equal to its upper bound, and each solution minimizes one of the objective functions. However, none of them has minimum values for any of the two objective functions. Therefore, the optimal solution depends on the robustness level, and the designers can choose the most preferred solution according to the robustness level and other performance measures that have not been incorporated in the optimization problem.

To further investigate the effect of relaxing worst response to a quantile value using order statistics, the confidence level α is increased to 0.99 and the sample size is also increased to 459 to satisfy $\gamma = 0.99$ for $k = 1$; see Sec.1.4.2 for details. Note that for $n = 200$ the value of γ at $\alpha = 0.99$ corresponding

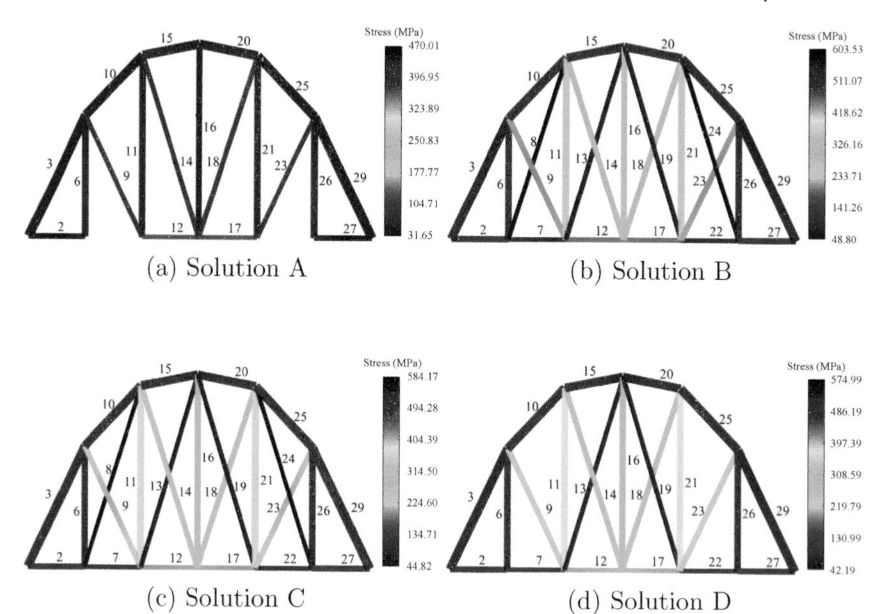

(a) Solution A (b) Solution B

(c) Solution C (d) Solution D

Figure 6.13 Pareto optimal solutions of problem (6.14).

Table 6.4

Objectives Values and Structural Volume of Solutions A, B, C and D of Problem (6.14).

(a) Before removing thin members

Solution	$\sigma^{\max}$ (Pa)	$\sigma^{\max}_{1:200}$ (Pa)	$\sigma^{\max}_{50:200}$ (Pa)	$\sigma^{\max}_{100:200}$ (Pa)	**Volume** (m^3)
A	4.7001×10^8	5.9733×10^9	2.8549×10^9	2.0861×10^9	1.0
B	6.0353×10^8	7.7531×10^8	6.9560×10^8	6.5546×10^8	1.0
C	5.8417×10^8	8.2934×10^8	6.8127×10^8	6.4418×10^8	1.0
D	5.7499×10^8	8.7466×10^8	6.9079×10^8	6.3814×10^8	1.0

(b) After removing thin members

Solution	$\sigma^{\max}$ (Pa)	$\sigma^{\max}_{1:200}$ (Pa)	$\sigma^{\max}_{50:200}$ (Pa)	$\sigma^{\max}_{100:200}$ (Pa)	**Volume** (m^3)
A	4.7080×10^8	1.7213×10^{10}	7.3785×10^9	4.1258×10^9	1.0
B	6.0929×10^8	7.7872×10^8	6.9805×10^8	6.6848×10^8	1.0
C	5.8482×10^8	8.3908×10^8	6.8469×10^8	6.4775×10^8	1.0
D	5.8070×10^8	8.8895×10^8	6.9200×10^8	6.4595×10^8	1.0

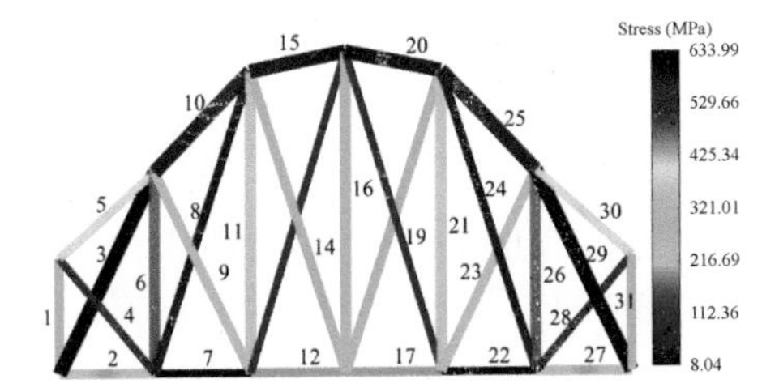

Figure 6.14 Optimal solution for minimizing $\sigma_{1:459}^{\max}(\boldsymbol{a};\boldsymbol{Y},\boldsymbol{E})$ corresponding large confidence level.

to $k = 1$ is 0.977. The configuration in Fig. 6.11(b) is used for the nominal nodal coordinates. The optimization problem for minimizing $\sigma_{1:459}^{\max}(\boldsymbol{a};\boldsymbol{Y},\boldsymbol{E})$ is solved under constraint on the total structural volume. The optimal value of $\sigma_{1:459}^{\max}$ obtained by using a GA is 7.9923×10^8 Pa, which is larger than $\sigma_{1:200}^{\max} = 7.7531 \times 10^8$ Pa of solution B due to the increase of confidence level. The optimal solution for minimizing $\sigma_{1:459}^{\max}$ is shown in Fig. 6.14, where the maximum stresses is calculated for the nominal values of parameters. It can be seen from the figure that the cross-sectional areas of members 1, 4, 5, 28, 30 and 31, which do not exist in solution B in Fig. 6.13(b), have moderately large values to ensure large robustness to the worst response, resulting in different performance on minimizing the maximum stress of the frame.

6.6 SUMMARY

Some results of robust design using order statistics have been presented for trusses and frames. It has been demonstrated that robustness level can be represented by the order of the solution for specified sample size and parameter values, and an MOO problem can be solved to obtain Pareto optimal solutions corresponding to various levels of robustness. It has been demonstrated in the examples of a small truss, a shear frame subjected to seismic loads, and a bridge-type plane frame that the most preferred solution among the Pareto optimal solutions can be selected in view of trade off between the total structural cost (volume) and the required robustness level of the structure. A design corresponding to a low robustness level has a small structural volume; however, it is very sensitive to uncertainty in the parameters and variables such as nodal locations. This way, the order statistics can be effectively used for finding Pareto optimal solutions corresponding to various levels of robustness.

ACKNOWLEDGMENT

We acknowledge the permissions from publishers for use of the licensed contents as follows:

- Figure 6.2 and Table 6.1: Reprinted from Ohsaki et al. [2019] with permission from Elsevier.
- Figures 6.7, 6.8 and 6.9: Reprinted from Ohsaki et al. [2019] with permission from Elsevier.
- Figures 6.11, 6.12, 6.13 and 6.14, and Table 6.4: Reproduced with permission from Springer Nature [Shen et al., 2020].

7 Surrogate-assisted and Reliability-based Optimization

Reliability-based design optimization (RBDO) has become an active research field to consider the cost and reliability of the structure simultaneously with developments of methodologies, the computational tools and hardwares. An RBDO problem is generally formulated to minimize the cost under deterministic and probabilistic constraints. To alleviate large computational cost of structural analysis for solving the problem, various methods have been proposed. The surrogate-assisted method is suitable for reducing the computational cost for solving the RBDO problem of complex structures. The problem may be formulated as a multi-objective optimization problem for minimizing the expected (mean) value and the variance of the objective function with a set of probabilistic constraints. The Bayesian optimization, also well-known as the efficient global optimization method, is classified into surrogate-assisted method and successfully applied to the multi-objective RBDO problems.

7.1 INTRODUCTION

As discussed in the previous chapters, uncertainty in design parameters such as nodal locations, cross-sectional sizes, material properties and static/dynamic external loads should be appropriately incorporated in the design process of engineering structures. Uncertainty can be incorporated for formulating the design problem as a robust design problem according to the specified design objectives and requirements on structural responses that are formulated as objective and constraint functions. When the sensitivity of objective function against parameter is to be reduced, robustness is quantified as variation of the objective function, and the variance minimization approach can be adopted [Doltsinis and Kang, 2004; Beyer and Sendhoff, 2007].

In the variance minimization approach, the robust design optimization (RDO) problem is formulated as a multi-objective optimization (MOO) problem for minimizing the expected (mean) value and the variance of the objective function for specified probability density functions (pdfs) or cumulative distribution functions (cdfs) [Du and Chen, 2000; Papadrakakis et al., 2005; Lee et al., 2008a; Richardson et al., 2015; Do and Ohsaki, 2021b]. The constraints are also formulated as a set of probabilistic constraints (i.e., chance constraints of the limit state functions (LSFs) [Ben-Tal et al., 2009]). There are two types of LSFs; the first type considers the joint probability over the entire system,

DOI: 10.1201/9781003153160-7

while the second type consists of a set of probabilistic constraints on the individual uncertain LSFs.

However, solving RDO problems is very difficult due to the following four issues:

(1) Computational cost is very large especially for structures with large number of degrees of freedom because each function evaluation involves structural analysis using static/dynamic frame analysis or finite element analysis.

(2) Finding feasible solutions also demands substantial computational cost because evaluation of probabilistic constraints is an NP-hard problem [Geng and Xie, 2019] and the ratio of feasible domain in the total domain may be very small even for a simple problem with linear constraints as discussed in Sec. 4.4.

(3) Evaluation of the mean and variance values of the uncertain objective function also demands large computational cost. Although there are many available methods such as Monte Carlo integration [Caflisch, 1998], polynomial chaos expansion [Rocchetta et al., 2020], Taylor series approximation [Anderson and Mattson, 2012] and Bayes-Hermite quadrature [O'Hagan, 1991], they are subject to the curse of dimensionality.

(4) It may be impossible to obtain exact optimal solutions because the feasible region defined by the probabilistic functions is generally non-convex.

Bayesian optimization (BO) [Shahriari et al., 2016; Feliot et al., 2017], also well-known as the efficient global optimization (EGO) [Jones et al., 1998], is a powerful tool for solving non-convex optimization problems with uncertain objective and/or constraint functions. BO is a sequential process consisting of Bayesian regression model called Gaussian process regression (GPR) and maximization of an acquisition function. Several types of acquisition functions are proposed for single-objective nonlinear optimization problem [Frazier, 2018]. Recently, the BO has been successfully applied to combinatorial optimization problems [Baptista and Poloczek, 2018], problems with mixed continuous/discrete variables [Zhang et al., 2020] and multi-objective structural optimization problems [Mathern et al., 2021; Shu et al., 2020].

GPR is regarded as a powerful, principled and practical probabilistic tool also in the field of machine learning, where it is called a kernel machine, e.g., [Bishop, 2006; Rasmussen and Williams, 2005]. The GPR was developed in the field of geostatistics as kriging. GPR is also mathematically equivalent under some appropriate conditions to several well known models, including Bayesian linear models, radial basis function networks and spline models. With extending application in the field of machine learning, the GPR has become a popular surrogate method for response approximation in the field of engineering design [Echard et al., 2011, 2013]. In this chapter, we present some results of GPR and BO to single and multi-objective structural optimization problems.

7.2 GAUSSIAN PROCESS MODEL AND EXPECTED IMPROVEMENT

In the process of structural design, nonlinear dynamic transient analysis is widely used, for example, in aerospace engineering and automotive engineering. In the field of seismic design of building frames, it is also called response history analysis. In particular, when we consider a minimization problem of the worst transient dynamic response, the problem is classified as a non-smooth problem. Establishing optimization techniques suitable for application to transient dynamic response is important from both theoretical and practical points of view. For this purpose, optimization process utilizing surrogate models, which is called surrogate-assisted optimization, seems to have considerable promise to reduce the computational cost. As many optimization techniques require accurate design sensitivity information, transient dynamic analysis hinders optimization for the following reasons [van Keulen et al., 2005]:

1. The response function and its derivatives are often non-smooth and/or noisy.
2. The corresponding derivatives may not be continuous although the response function is continuous.
3. The corresponding derivatives may be expensive to calculate.

Accurate and efficient optimization methods overcoming above difficulties need to be developed for the safety of structures designed considering transient-type behavior.

7.2.1 GAUSSIAN PROCESS MODEL

Hereafter, $\mathcal{N}(\boldsymbol{\mu}, \boldsymbol{\Sigma})$ or $\mathcal{N}(\boldsymbol{x}; \boldsymbol{\mu}, \boldsymbol{\Sigma})$ denotes the multivariate Gaussian (normal) distribution of a vector $\boldsymbol{x} \in \mathbb{R}^d$ with mean vector $\boldsymbol{\mu} \in \mathbb{R}^d$ and covariance matrix $\boldsymbol{\Sigma} \in \mathbb{R}^{d \times d}$. All vectors in this section are column vectors. The multivariate Gaussian distribution has a probability density given by

$$\mathcal{N}(\boldsymbol{x}; \boldsymbol{\mu}, \boldsymbol{\Sigma}) = (2\pi)^{-\frac{d}{2}} |\boldsymbol{\Sigma}|^{-\frac{1}{2}} \exp\left[-\frac{1}{2}(\boldsymbol{x} - \boldsymbol{\mu})^\top \boldsymbol{\Sigma}^{-1} (\boldsymbol{x} - \boldsymbol{\mu})\right],$$

where $|\boldsymbol{\Sigma}|$ denotes the determinant of $\boldsymbol{\Sigma}$. A Gaussian process model (GPM) is a collection of random variables, any subset of which has a joint Gaussian distribution. For a complete introduction to the Gaussian process model; see e.g., Rasmussen and Williams [2005]. The Gaussian process $G(\boldsymbol{x})$ with mean function $m(\boldsymbol{x})$ and covariance function $c(\boldsymbol{x}, \boldsymbol{x}')$ is written as

$$G(\boldsymbol{x}) \sim \mathcal{GP}\left(m(\boldsymbol{x}), c(\boldsymbol{x}, \boldsymbol{x}')\right), \tag{7.1}$$

where

$$\begin{aligned} m(\boldsymbol{x}) &= \mathbb{E}\left[G(\boldsymbol{x})\right], \\ c(\boldsymbol{x}, \boldsymbol{x}') &= \mathrm{Cov}\left[G(\boldsymbol{x}), G(\boldsymbol{x}')\right] \\ &= \mathbb{E}\left[\left(G(\boldsymbol{x}) - m(\boldsymbol{x})\right)\left(G(\boldsymbol{x}') - m(\boldsymbol{x}')\right)\right], \end{aligned}$$

and $\mathrm{Cov}[\cdot,\cdot]$ means covariance between the two variables. The mean function $m(\boldsymbol{x})$ is often but not necessarily taken to be zero. For example, $\boldsymbol{h}^\top(x)\boldsymbol{\beta}$ is also often taken as $m(\boldsymbol{x})$, where $\boldsymbol{h}(\boldsymbol{x}) = (h_1(\boldsymbol{x}), \cdots, h_q(\boldsymbol{x}))^\top$ is a q-dimensional vector of known basis functions, e.g., quadratic functions; q is the number of basis functions; and $\boldsymbol{\beta} = (\beta_1, \cdots, \beta_q)^\top$ is a vector of unknown coefficients. The covariance function is assumed to specify the covariance between the pairs of function values as

$$c(\boldsymbol{x}, \boldsymbol{x}') = \psi_0 \exp\left(-\frac{\boldsymbol{x}^\top \boldsymbol{x}'}{2\phi_1^2}\right) + \psi_2 + \psi_3 \boldsymbol{x}^\top \boldsymbol{x}', \tag{7.2}$$

where ψ_0, ψ_1, ψ_2 and ψ_3 are the hyperparameters of the GPM. This covariance function is widely used in GPMs; see e.g., Bishop [2006]. Let $\boldsymbol{y} = \left(y_1 = F(\boldsymbol{x}_1), \quad \cdots, \quad y_n = F(\boldsymbol{x}_n)\right)^\top$ denote n realizations of the process $F(\boldsymbol{x})$ at the observed points $\boldsymbol{x}_1, \cdots, \boldsymbol{x}_n$. From the definition of the Gaussian process, the joint distribution of the outputs $\boldsymbol{y}$ and the new unknown outputs Y at $\boldsymbol{x}$ according to the prior distribution is formulated as:

$$\begin{pmatrix} \boldsymbol{y} \\ Y \end{pmatrix} \sim \mathcal{N}\left(\begin{pmatrix} \boldsymbol{m}(\boldsymbol{X}) \\ m(\boldsymbol{x}) \end{pmatrix}, \begin{pmatrix} \boldsymbol{C}(\boldsymbol{X}, \boldsymbol{X}) & \mathbf{c}(\boldsymbol{X}, \mathrm{x}) \\ \mathbf{c}^\top(\boldsymbol{X}, \boldsymbol{x}) & c(\boldsymbol{x}, \boldsymbol{x}) \end{pmatrix}\right),$$

where

$$\begin{gathered} \boldsymbol{X} = (\boldsymbol{x}_1, \cdots, \boldsymbol{x}_n), \\ \boldsymbol{m}(\boldsymbol{X}) = \begin{pmatrix} m(\boldsymbol{x}_1) & \cdots & m(\boldsymbol{x}_n) \end{pmatrix}^\top \in \mathbb{R}^n, \\ \boldsymbol{C}(\boldsymbol{X}, \boldsymbol{X}) \in \mathbb{R}^{n \times n}, \quad \left[\boldsymbol{C}(\boldsymbol{X}, \boldsymbol{X})\right]_{ij} = c(\boldsymbol{x}_i, \boldsymbol{x}_j) \quad (i, j = 1, \cdots, n), \\ \mathbf{c}(\boldsymbol{X}, \boldsymbol{x}) = \begin{pmatrix} c(\boldsymbol{x}_1, \boldsymbol{x}) & \cdots & c(\boldsymbol{x}_n, \boldsymbol{x}) \end{pmatrix}^\top \in \mathbb{R}^n. \end{gathered}$$

Deriving the conditional distribution corresponding to (7.1), we arrive at the following predictive equations by GPR:

$$F(\boldsymbol{x}) = Y \mid \boldsymbol{x}, \mathcal{D} \sim \mathcal{N}\left(\hat{f}(\boldsymbol{x}), \hat{\sigma}^2(\boldsymbol{x})\right), \tag{7.3}$$

where

$$\mathcal{D} = \{\boldsymbol{X}, \boldsymbol{y}\},$$

$$\hat{f}(\boldsymbol{x}) = \mathbb{E}\left[Y \mid \boldsymbol{x}, \mathcal{D}\right] = m(\boldsymbol{x}) + \mathbf{c}(\boldsymbol{X}, \boldsymbol{x})^\top \boldsymbol{C}(\boldsymbol{X}, \boldsymbol{X})^{-1}(\boldsymbol{y} - \boldsymbol{m}(\boldsymbol{X})), \tag{7.4}$$

$$\hat{\sigma}^2(\boldsymbol{x}) = \mathrm{Var}[Y \mid \boldsymbol{x}, \mathcal{D}] = c(\boldsymbol{x}, \boldsymbol{x}) - \mathbf{c}(\boldsymbol{X}, \boldsymbol{x})^\top \boldsymbol{C}(\boldsymbol{X}, \boldsymbol{X})^{-1} \mathbf{c}(\boldsymbol{X}, \boldsymbol{x}). \tag{7.5}$$

When $m(\boldsymbol{x})$ is taken to be $\boldsymbol{h}^\top(\boldsymbol{x})\boldsymbol{\beta}$, the coefficient vector $\boldsymbol{\beta}$ is given by the least-squares solution $\hat{\boldsymbol{\beta}}$ as follows:

$$\hat{\boldsymbol{\beta}} = [\boldsymbol{H}^\top(\boldsymbol{X})\boldsymbol{C}(\boldsymbol{X},\boldsymbol{X})\boldsymbol{H}(\boldsymbol{X})]^{-1}\boldsymbol{H}^\top(\boldsymbol{X})\boldsymbol{C}^{-1}(\boldsymbol{X},\boldsymbol{X})\boldsymbol{y},$$

where $\boldsymbol{H}(\boldsymbol{X}) = (\boldsymbol{h}(\boldsymbol{x}_1), \cdots, \boldsymbol{h}(\boldsymbol{x}_n))^\top$. For GPR (7.3), we have to estimate hyperparameters $\boldsymbol{\psi} = (\psi_0, \psi_1, \psi_2, \psi_3)$ in (7.2). Various methods have been presented for this purpose. For example, maximum marginal likelihood estimator for the hyperparameters are found as

$$\begin{aligned}\hat{\boldsymbol{\psi}} &= \underset{\boldsymbol{\psi}}{\operatorname{argmax}} \log p\left(\boldsymbol{y}|\boldsymbol{X}, \boldsymbol{\psi}\right) \\ &= -\frac{1}{2}\left(\boldsymbol{y} - \boldsymbol{m}(\boldsymbol{X})\right)^\top \boldsymbol{C}\left(\boldsymbol{X},\boldsymbol{X}\right)^{-1}\left(\boldsymbol{y} - \boldsymbol{m}(\boldsymbol{X})\right) \\ &\qquad\qquad - \frac{1}{2}\log\left|\boldsymbol{C}\left(\boldsymbol{X},\boldsymbol{X}\right)\right| - \frac{n}{2}\log 2\pi.\end{aligned} \tag{7.6}$$

Then we can predict a function value $f(\boldsymbol{x})$ by using GPR (7.3) with $\hat{\boldsymbol{\psi}}$. In this chapter, we will sometimes write $Y(\boldsymbol{x};\mathcal{D})$ as $Y \mid \boldsymbol{x}, \mathcal{D}, \hat{\boldsymbol{\psi}}$. Furthermore, we will refer to $\hat{f}(\boldsymbol{x})$ as GPM of $f(\boldsymbol{x})$ in Sec. 7.3 for simplicity.

7.2.2 EXPECTED IMPROVEMENT FOR THE WORST RESPONSE FUNCTION

For a minimization problem, we can search a point which minimizes the expectation of the objective function modeled by GPR. This approach is intuitive; however, such a *greedy* algorithm does not often work. Jones et al. [1998] developed attractive sequential design strategies that add input points at each stage of the algorithms. Their algorithms are initiated by evaluating functions at preliminary experimental points. The next point is chosen to maximize the expected improvement (EI) criterion that balances for better predicted value of the function and larger uncertainty of prediction. The EI criterion is a way of escaping local minima and, given certain assumptions, will asymptotically converge to the global optimum [Locatelli, 1997].

We introduce our target problem as follows:

$$\text{Minimize} \quad f^{\mathrm{U}}\left(\boldsymbol{x}\right) = \max\left(f^{(1)}\left(\boldsymbol{x}\right), \cdots, f^{(m)}\left(\boldsymbol{x}\right)\right) \tag{7.7a}$$

$$\text{subject to} \quad \boldsymbol{x} \in \mathcal{X}, \tag{7.7b}$$

where $f^{(1)}(\boldsymbol{x}), \cdots, f^{(m)}(\boldsymbol{x})$ are the response functions, $f^{\mathrm{U}}(\boldsymbol{x})$ is the worst response function value, $\mathcal{X} \subset \mathbb{R}^d$ is a convex polytope defined as

$$\mathcal{X} = \left\{\boldsymbol{x} \in \mathbb{R}^d \mid \boldsymbol{A}\boldsymbol{x} \leq \boldsymbol{b}\right\}, \tag{7.8}$$

with a matrix $\boldsymbol{A} \in \mathbb{R}^{q\times d}$ and a vector $\boldsymbol{b} \in \mathbb{R}^q$. Directly solving problem (7.7) is expensive because the evaluation of $f^{(1)}, \cdots, f^{(m)}$ is expensive. Therefore, we use GPR as surrogates of $f^{(1)}, \cdots, f^{(m)}$. Suppose we have a dataset

$$\mathcal{D} = \{\mathcal{D}^{(k)} \mid k = 1, \cdots, m\}, \tag{7.9}$$

where

$$\mathcal{D}^{(k)} = \{\boldsymbol{X}, \boldsymbol{y}^{(k)}\}, \quad \boldsymbol{y}^{(k)} = (y_1^{(k)} = f^{(k)}(\boldsymbol{x}_1), \quad \cdots, \quad y_n^{(k)} = f^{(k)}(\boldsymbol{x}_n))^\top.$$

By using dataset (7.9), GPR denoted by

$$Y_k(\boldsymbol{x}; \mathcal{D}^{(k)}) \sim \mathcal{N}\left(\hat{y}_k, s_k^2\right) \quad (k = 1, \cdots, m) \tag{7.10}$$

is used as a surrogate of the kth response function $f^{(k)}$, where $\hat{y}_k$ and s_k^2 are defined in the same manner as $\hat{f}(\boldsymbol{x})$ and $\hat{\sigma}^2(\boldsymbol{x})$ in (7.4) and (7.5), respectively. In addition to this, we assume the random variables $Y^{(1)}, \cdots, Y^{(m)}$ are independent. Instead of the expensive problem (7.7), we consider the following problem:

$$\begin{aligned} \underset{\boldsymbol{x}}{\text{Minimize}} \quad & Y_{\max}(\boldsymbol{x}; \mathcal{D}) = \max\left(Y_1(\boldsymbol{x}; \mathcal{D}^{(1)}), \cdots, Y_m(\boldsymbol{x}; \mathcal{D}^{(m)})\right) & \text{(7.11a)} \\ \text{subject to} \quad & \boldsymbol{x} \in \mathcal{X}. & \text{(7.11b)} \end{aligned}$$

If the dataset cannot sufficiently represent features of the function, which often occurs in practice, we cannot find a meaningful solution. Sequential design is known as a promising approach to overcome this difficulty. In particular, a point selected by EGO algorithm, which balances between exploitation and exploration based on the EI criterion and sequentially improves the surrogate model, can often find a valuable solution.

We apply the EI criterion to the worst response function in problem (7.11). The best observed objective value is denoted by

$$y_{\min} = \min_{j \in \{1, \cdots, n\}} \max_{k \in \{1, \cdots, m\}} y_j^{(k)}. \tag{7.12}$$

A probabilistic improvement function from (7.12) is defined by

$$I_{\max}(\boldsymbol{x}; \mathcal{D}) = \max\{y_{\min} - Y_{\max}(\boldsymbol{x}; \mathcal{D}), 0\}. \tag{7.13}$$

Using the improvement function $I_{\max}(\boldsymbol{x}; \mathcal{D})$ in (7.13), we consider a stochastic problem:

$$\begin{aligned} \underset{\boldsymbol{x}}{\text{Maximize}} \quad & I_{\max}(\boldsymbol{x}; \mathcal{D}) & \text{(7.14a)} \\ \text{subject to} \quad & \boldsymbol{x} \in \mathcal{X}. & \text{(7.14b)} \end{aligned}$$

Problem (7.14) finds the most improvable point from the current best value. Note that $I_{\max}$ is a random variable because $Y_1, \cdots, Y_m$ are random variables. For notational simplicity, we will suppress the dependence of $I_{\max}$ and $Y_1, \cdots, Y_m$ on $\boldsymbol{x}$ and $\mathcal{D}$, i.e., $I_{\max} = I_{\max}(\boldsymbol{x}; \mathcal{D})$ and $Y_1 = Y_1(\boldsymbol{x}; \mathcal{D}), \cdots, Y_m = Y_m(\boldsymbol{x}; \mathcal{D})$. When $Y_1, \cdots, Y_m$ are assumed to be independent, the probability of improvement is given by:

$$\begin{aligned} F(i_{\max}) &= \Pr\{I_{\max} \leq i_{\max}\} \\ &= 1 - \prod_{k=1}^{m} \left[\int_{-\infty}^{y_{\min} - i_{\max}} \mathcal{N}\left(u_k; \hat{y}_k, s_k^2\right) du_k\right] \end{aligned} \tag{7.15}$$

where $i_{\max} \geq 0$ is the improvement from the current value. By differentiating (7.15) with respect to $i_{\max}$, the pdf can be found as follows:

$$p(i_{\max}) = \frac{dF(i_{\max})}{di_{\max}}$$
$$= \begin{cases} \sum_{j=1}^{m} \mathcal{N}\left(y_{\min} - i_{\max}; \hat{y}_j, s_j^2\right) \\ \qquad \prod_{\substack{k=1,\cdots,m \\ k\neq j}} \left\{\int_{-\infty}^{y_{\min}-i_{\max}} \mathcal{N}\left(u_k; \hat{y}_k, s_k^2\right) du_k\right\} & \text{if } i_{\max} > 0, \\ 1 - \prod_{k=1}^{m} \left\{\int_{-\infty}^{y_{\min}} \mathcal{N}\left(u_k; \hat{y}_k, s_k^2\right) du_k\right\} & \text{if } i_{\max} = 0. \end{cases} \tag{7.16}$$

By taking the expectation of (7.16), the EI criterion of the worst response function can be obtained as

$$\mathbb{E}[I_{\max}] = \int_0^{\infty} i_{\max}\, p(i_{\max})\, di_{\max}$$
$$= \int_{-\infty}^{y_{\min}} \sum_{j=1}^{m} (y_{\min} - y) \mathcal{N}\left(y; \hat{y}_j, s_j^2\right) \prod_{\substack{k=1,\cdots,m \\ k\neq j}} \int_{-\infty}^{y} \mathcal{N}\left(u_k; \hat{y}_k, s_k^2\right) du_k dy. \tag{7.17}$$

Thus, we finally obtain the EI criterion

$$EI_{\max}(\boldsymbol{x}; \mathcal{D}) = \mathbb{E}\left[I_{\max}(\boldsymbol{x}; \mathcal{D})\right]. \tag{7.18}$$

If $m = 1$, criterion (7.18) coincides with EI in Jones et al. [1998]; otherwise, criterion (7.18) needs numerical multiple integration because we cannot find an analytical expression of (7.17). Thus, the following problem is considered in EGO:

$$\underset{\boldsymbol{x}}{\text{Maximize}} \quad EI_{\max}(\boldsymbol{x}; \mathcal{D}) \tag{7.19a}$$
$$\text{subject to} \quad \boldsymbol{x} \in \mathcal{X}. \tag{7.19b}$$

Alternatively, an equivalent problem of (7.14) may be formulated as follows using a slack variable t:

$$\underset{\boldsymbol{x},t}{\text{Maximize}} \quad I_{\text{eq}}(\boldsymbol{x}, t; \mathcal{D}) \tag{7.20a}$$
$$\text{subject to} \quad \boldsymbol{x} \in \mathcal{X}, \quad t \geq 0, \tag{7.20b}$$

where

$$I_{\text{eq}}(\boldsymbol{x}, t; \mathcal{D}) = \begin{cases} t & \text{if } y_{\min} - \max(Y_1, \cdots, Y_m) \geq t, \\ 0 & \text{otherwise.} \end{cases}$$

Note that I_{eq} is also a random variable. When a solution of (7.20) is written as $(\hat{\boldsymbol{x}}, \hat{t})$, $\hat{\boldsymbol{x}}$ is also a solution of problem (7.14). For simplicity, we will suppress the dependence of I_{eq} on $\boldsymbol{x}$, t and $\mathcal{D}$. The probability of the equivalent improvement is written as

$$\begin{aligned} F(i_{eq}; t) &= \Pr\{I_{\text{eq}} \le i_{\text{eq}}; t\} \\ &= \begin{cases} 1 & \text{if } i_{\text{eq}} \ge t, \\ 1 - \Pr\{y_{\min} - \max(Y_1, \cdots, Y_m) \ge t\} & \text{if } t \ge i_{\text{eq}} \ge 0, \end{cases} \end{aligned} \tag{7.21}$$

where $i_{\text{eq}} \ge 0$. For $t \ge 0$, we have

$$\begin{aligned} \Pr\{y_{\min} - \max(Y_1, \cdots, Y_m) \ge t\} &= \prod_{k=1}^{m} \Pr\{Y_k \le y_{\min} - t\} \\ &= \prod_{k=1}^{m} \int_{-\infty}^{y_{\min}-t} \mathcal{N}\left(u_k; \hat{y}_k, s_k^2\right) du_k. \end{aligned}$$

By differentiating (7.21) with respect to i_{eq}, the pdf is obtained as

$$\begin{aligned} p(i_{\text{eq}} \mid t) &= \frac{dF(i_{\text{eq}}; t)}{di_{\text{eq}}} \\ &= [1 - \Pr\{y_{\min} - \max(Y_1, \cdots, Y_m) \ge t\}]\, \delta(i_{\text{eq}}) \\ &\quad + \Pr\{y_{\min} - \max(Y_1, \cdots, Y_m) \ge t\}\, \delta(i_{\text{eq}} - t), \end{aligned} \tag{7.22}$$

where $\delta(\cdot)$ denotes Dirac's delta function. By taking the expectation of (7.22), an equivalent EI criterion for the worst response function is obtained as

$$\begin{aligned} EI_{\text{eq}}(\boldsymbol{x}, t) = \mathbb{E}[I_{\text{eq}}; t] &= \int_0^{\infty} i_{eq} p(i_{eq}; t)\, di_{eq} \\ &= t \prod_{k=1}^{m} \int_{-\infty}^{y_{\min}-t} \mathcal{N}\left(u_k; \hat{y}_k, s_k^2\right) du_k. \end{aligned} \tag{7.23}$$

Evaluation of criterion (7.23) is relatively cheap compared to that of (7.17) because multiple integration does not appear in (7.23). This means that criterion (7.23) is suitable for our purpose. Thus, instead of the original problem (7.19), the following problem is to be solved:

$$\underset{\boldsymbol{x}, t}{\text{Maximize}} \quad EI_{\text{eq}}(\boldsymbol{x}, t) = t \prod_{k=1}^{m} \int_{-\infty}^{y_{\min}-t} \mathcal{N}\left(u_k; \hat{y}_k, s_k^2\right) du_k \tag{7.24a}$$

$$\text{subject to} \quad \boldsymbol{x} \in \mathcal{X}, \quad t \ge 0. \tag{7.24b}$$

Note that the solution of (7.24) may not coincide with the solution of (7.19) because expectation is taken for the different variables. Furthermore, the problem (7.24) has one dimension larger than that of the original problem (7.19).

However, the problem (7.24) is more tractable than problem (7.19) and the validity of criterion (7.23) is about the same level as criterion (7.17) in view of the statistical framework. Therefore, the strategy using the equivalent EI criterion is promising.

7.2.3 DIRECT SEARCH METHOD ENHANCED BY EGO

Due to the factors listed above, the gradient-based optimization method is not often applicable to optimization with transient dynamic analysis. In recent years, derivative-free methods that do not require accurate and expensive design sensitivities have made rapid improvement and matured. Under appropriate conditions, some of these methods guarantee convergence to a point satisfying necessary optimality conditions based on the Clarke calculus [Clarke, 1990] from arbitrary starting points. However, in many cases, the transient response function has many stationary points because of noisy and strongly non-convex properties. This means that a local solution found using a derivative-free method may be very inferior to others. Recent studies have proposed a scheme to globally explore a possibly non-convex function [Audet et al., 2008; Vaz and Vicente, 2007]. The method requires many, often computationally expensive, function evaluations, because essentially a random search moving away from a local optimum is carried out as a globalization strategy.

Rather than exploring the solution space randomly, we can utilizes the structure of the problem to improve possibility of finding the global optimal solution. Here, we introduce an approach to combine the derivative-free methods with the global strategy EGO [Jones et al., 1998]. The strategy is based on a statistical method, which will be efficient in many cases. Potentially, this combination has the following attractive properties:

(1) The method will be robust in finding a local solution, even in a situation where the problem involves noisy functions and/or design sensitivities are unavailable or too costly to evaluate, because the method inherits an efficient behavior of a direct search method (DSM).

(2) The statistical model, which uses the EI algorithm, can assist in both local and global searches, reducing the number of evaluations of the transient responses.

(3) The DSM is one of the derivative-free methods, and its global convergence is guaranteed if Condition 7.1 is satisfied [Kolda et al., 2006] as explained below.

DSM for a linearly constrained problem enhanced by EGO is applied to problem (7.7). Most of the description about the DSM is based on the algorithm in Kolda et al. [2006]; Lewis and Torczon [2000]; Kolda et al. [2003]. The reader is referred to Kolda et al. [2006] for details of the algorithm and its global convergence analysis.

A cone K is generated if there exists a set of vectors $\{\boldsymbol{v}_1, \cdots, \boldsymbol{v}_r\}$ such that

$$K = \{\lambda_1 \boldsymbol{v}_1 + \cdots + \lambda_r \boldsymbol{v}_r \mid \lambda_1 \geq 0, \cdots, \lambda_r \geq 0\}.$$

The vectors $\boldsymbol{v}_1, \cdots, \boldsymbol{v}_r$ are called generators of K. If K is a vector space, then a set of generators for K is called a positive spanning set [Davis, 1954]. We denote the feasible set by $\mathcal{X} = \{\boldsymbol{x} \mid \boldsymbol{A}\boldsymbol{x} \leq \boldsymbol{b}\}$; see (7.8). Let $\boldsymbol{a}_i^\top$ and b_i be the ith row of $\boldsymbol{A}$ and ith element of $\boldsymbol{b}$, respectively. Given $\boldsymbol{x} \in \mathcal{X}$ and $\varepsilon \geq 0$, we define

$$\mathcal{I}(\boldsymbol{x}, \varepsilon) = \left\{ i \;\middle|\; \frac{|b_i - \boldsymbol{a}_i^\top \boldsymbol{x}|}{\|\boldsymbol{a}_i\|} \leq \varepsilon \right\}.$$

The ε-normal cone $N(\boldsymbol{x}, \varepsilon)$ and its polar ε-tangent cone $T(\boldsymbol{x}, \varepsilon)$ are respectively defined to be

$$N(\boldsymbol{x}, \varepsilon) = \left\{ \sum_{i \in \mathcal{I}(\boldsymbol{x}, \varepsilon)} \xi_i \boldsymbol{a}_i \;\middle|\; \xi_i \geq 0 \right\} \cup \{\boldsymbol{0}\},$$

$$T(\boldsymbol{x}, \varepsilon) = \left\{ \boldsymbol{v} \;\middle|\; \boldsymbol{w}^\top \boldsymbol{v} \leq 0 \text{ for all } \boldsymbol{w} \in N(\boldsymbol{x}, \varepsilon) \right\}.$$

Any vector $\boldsymbol{v}$ can be written as $\boldsymbol{v} = \boldsymbol{v}_{N(\boldsymbol{x},\varepsilon)} + \boldsymbol{v}_{T(\boldsymbol{x},\varepsilon)}$, where $\boldsymbol{v}_{N(\boldsymbol{x},\varepsilon)}$ and $\boldsymbol{v}_{T(\boldsymbol{x},\varepsilon)}$ denote the projection of $\boldsymbol{v}$ onto $N(\boldsymbol{x}, \varepsilon)$ and $T(\boldsymbol{x}, \varepsilon)$, respectively.

A linearly constrained DSM enhanced by EGO is presented in Algorithm 7.1. In order to ensure convergence of Algorithm 7.1, we need the following conditions [Kolda et al., 2006]:

Condition 7.1 *There exists a constant $\kappa_{\min} > 0$, independent of k, such that for every k for which $T(\boldsymbol{x}_k, \Delta_k) \neq \{\boldsymbol{0}\}$, the set $\mathcal{G}_k$ contains generators for the cone $T(\boldsymbol{x}_k, \Delta_k)$ and satisfies*

$$\kappa(\mathcal{G}_k) = \inf_{\boldsymbol{v}_{T(\boldsymbol{x}_k, \Delta_k)} \neq \boldsymbol{0}} \max_{\boldsymbol{d} \in \mathcal{G}_k} \frac{\boldsymbol{v}^\top \boldsymbol{d}}{\left\|\boldsymbol{v}_{T(\boldsymbol{x}_k, \Delta_k)}\right\| \|\boldsymbol{d}\|} \geq \kappa_{\min}.$$

The key point of Algorithm 7.1 is that the global convergence is guaranteed using any set $\mathcal{H}_k$ as long as $\mathcal{G}_k$ satisfies Condition 7.1. This means that the search points can be chosen using EI_{eq} in (7.23). Conversely, Algorithm 7.1 guarantees the global convergence property, which means the convergence to stationary points from arbitrary starting points, if the process of generating $\mathcal{G}_k$ satisfies Condition 7.1. Thus, the method presented in Algorithm 7.1 can be interpreted as a DSM enhanced by a global strategy. There is at least one $\boldsymbol{x}$ that is a convergence point of Algorithm 7.1. If the objective function is strictly differentiable at the limit point $\boldsymbol{x}_*$, then $\boldsymbol{x}_*$ is a Clarke-Jahn stationary point [Jahn, 2007] if there exists a vector $\boldsymbol{w}$ satisfying the following equation for all $\boldsymbol{v}$:

$$\limsup_{\substack{\boldsymbol{y} \to \boldsymbol{x}_* \\ t \downarrow 0}} \frac{f(\boldsymbol{y} + t\boldsymbol{v}) - f(\boldsymbol{y})}{t} = \boldsymbol{w}^\top \boldsymbol{v}. \tag{7.25}$$

Algorithm 7.1 Linearly constrained DSM enhanced by EGO.

Require: Choose an initial guess $\boldsymbol{x}_0 \in \Omega$, and assign the stopping tolerance $\Delta_{\text{tol}} > 0$, initial step-length $\Delta_0 > \Delta_{\text{tol}}$, a forcing function for sufficient decrease $\rho(\Delta_k)$ and maximum search iteration n_{search}.

Require: Evaluate functions on initial experimental points for obtaining a dataset $\mathcal{D} = \{\mathcal{D}^{(1)}, \cdots \mathcal{D}^{(m)}\}$.

1: **Build or update surrogate**: If $k > n_{\text{search}}$, then $\mathcal{H}_k = \emptyset$ and go to Step 3. Otherwise, build GPR $Y_k(\boldsymbol{x}; \mathcal{D}^{(k)})$ $(k = 1, \cdots, m)$ in (7.10).

2: **Search a maximum equivalent EI point**: Find a probable improving point

$$(\hat{\boldsymbol{x}}_k, \hat{t}_k) = \underset{\boldsymbol{x},t}{\operatorname{argmax}} \left\{ EI_{\text{eq}}(\boldsymbol{x}, t) \,|\, \boldsymbol{x} \in \mathcal{X},\ t \geq 0 \right\},$$

and set $\mathcal{H}_k = \{\hat{\boldsymbol{x}}_k - \boldsymbol{x}_k\}$.

3: **Choose search directions**: Choose set of search directions $\mathcal{G}_k$ such that for all $\boldsymbol{d} \in \mathcal{G}_k$, $\boldsymbol{d}$ satisfies $\|\boldsymbol{d}\| = 1$ and Condition 7.1. Let $\Gamma_k = \mathcal{G}_k \cup \mathcal{H}_k$.

4: **Successful iteration**: If there exists $\boldsymbol{d}_k \in \Gamma_k$ such that $f(\boldsymbol{x}_k + \Delta_k \boldsymbol{d}_k) < f(\boldsymbol{x}_k) - \rho(\Delta_k)$ and $\boldsymbol{x}_k + \Delta_k \boldsymbol{d}_k \in \Omega$, then set $\boldsymbol{x}_{k+1} = \boldsymbol{x}_k + \Delta_k \boldsymbol{d}_k$ and $\Delta_{k+1} = \alpha_k \Delta_k$ $(\alpha_k \geq 1)$.

5: **Unsuccessful iteration**: Otherwise, set $\boldsymbol{x}_{k+1} = \boldsymbol{x}_k$ and $\Delta_{k+1} = \beta_k \Delta_k$ $(0 < \beta_k < 1)$. If $\Delta_{k+1} < \Delta_{\text{tol}}$, then terminate.

6: **Advance**: Add $\hat{\boldsymbol{x}}_k$ and $f^{(1)}(\hat{\boldsymbol{x}}_k), \cdots, f^{(m)}(\hat{\boldsymbol{x}}_k)$ to dataset $\mathcal{D}$. If the iteration k is successful, also add $\boldsymbol{x}_k$ and $f^{(1)}(\boldsymbol{x}_k), \cdots, f^{(m)}(\boldsymbol{x}_k)$ to dataset $\mathcal{D}$. Increase k as $k \leftarrow k+1$ and go to Step 1.

Unfortunately, our objective function $f^{\text{U}}(\boldsymbol{x})$ defined in (7.7a) is not a globally strictly differentiable function. This means $f^{\text{U}}(\boldsymbol{x})$ does not satisfy (7.25) at some points. We can check whether $f^{\text{U}}(\boldsymbol{x})$ is strictly differentiable at the convergence point $\boldsymbol{x}_*$. If two or more function values among $f^{(1)}(\boldsymbol{x}_*), \cdots, f^{(m)}(\boldsymbol{x}_*)$ coincide with the function values $f^{\text{U}}(\boldsymbol{x}_*)$ at the convergence point $\boldsymbol{x}_*$, then $f^{\text{U}}(\boldsymbol{x})$ is not globally strictly differentiable at $\boldsymbol{x}_*$ and the solution $\boldsymbol{x}_*$ may not be a stationary point in theory. However, $\boldsymbol{x}_*$ is often a good approximate solution in practice.

7.2.4 NUMERICAL EXAMPLE

Consider a seismic design problem of a 7-story shear frame model subjected to three input earthquake motions. The problem is defined as

$$\underset{\boldsymbol{x}}{\text{Minimize}} \quad f^{\text{U}}(\boldsymbol{x}) = \max\left(f^{(1)}(\boldsymbol{x}), \cdots, f^{(m)}(\boldsymbol{x})\right) \tag{7.26a}$$

$$\text{subject to} \quad \sum_{i=1}^{7} x_i \leq 2.1, \tag{7.26b}$$

$$0.05 \leq x_i \leq 0.4 \quad (i = 1, \cdots, 7), \tag{7.26c}$$

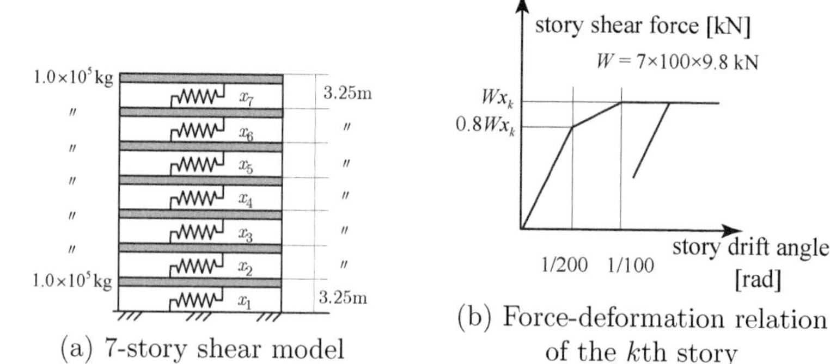

Figure 7.1 Building model.

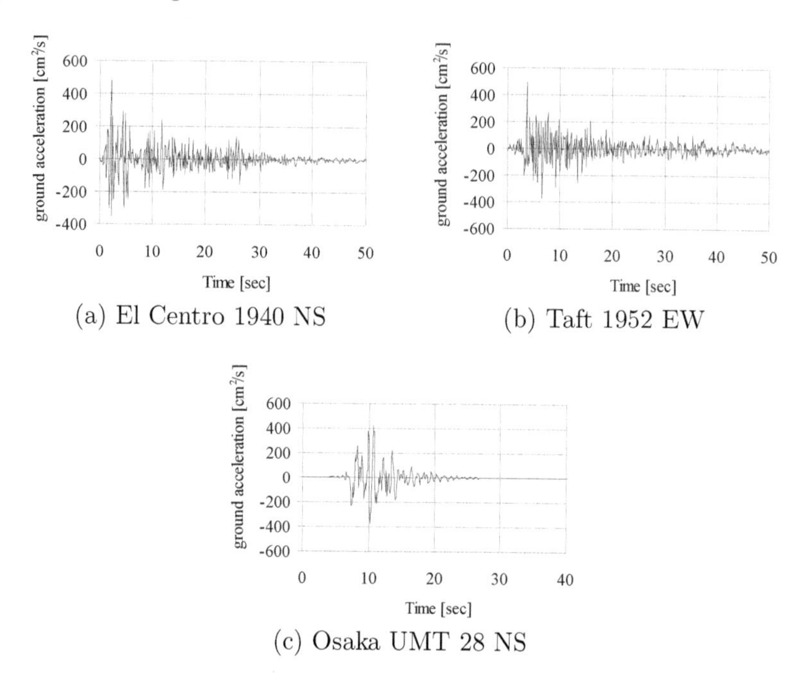

Figure 7.2 Input earthquake motions (ground acceleration).

where the design variable vector $\boldsymbol{x} = (x_1, \cdots, x_7)$ represents the story shear coefficients and $f^{(k)}$ denotes the maximum drift angle of the kth story under earthquake loading. The configuration of this 7-story shear model is shown in Fig. 7.1(a) and the relation between story drift angle and story shear force of kth story is shown in Fig. 7.1(b). The maximum story drift angle of each story is obtained by nonlinear transient dynamic analysis, where three input earthquake motions are shown in Fig. 7.2. Note that Osaka UMT28 NS is an extremely large artificial wave.

Diagonal test

We check a diagonal test as a preliminary full-dimensional problem. In this problem, the design variables are given the same value as

$$x_1 = x_2 = \cdots = x_7 = x.$$

Hence, this is a one-dimensional problem and easy to handle. We compare the following four methods in this diagonal test:

SQP: sequential quadratic programming method with finite difference approximation [Mathworks, 2009]

DSM: pure direct search method

DSME: direct search method with searching minimum expectation of the function value

DSMEI: direct search method with searching maximum EI of the function value

DSME and DSMEI use GPR, for which computer experiments are needed in advance. The initial experimental points are given by 10 equally spaced points between 0.05 and 0.4 as follows:

$$\boldsymbol{X} = \{0.0500, 0.0889, 0.1278, 0.1667, 0.2056, \\ 0.2444, 0.2833, 0.3222, 0.3611, 0.4000\}.$$

DSME uses one GPR model as a surrogate of the worst response function. By contrast, DSMEI uses seven GPR models for computing the equivalent EI criterion. In SQP and DSM, GPR models are not used; however, the best point of the experiments is given as the initial point also in SQP and DSM. For a fair comparison, we set the conditions as follows:

- The initial point of all methods is given by the best experimental point, which is 0.2833.
- We include the initial function evaluations for the computer experiments to the number of function evaluations of all the methods because the initial point is selected by the experiments also in SQP and DSM.
- Stopping criterion in all methods is only given by the number of function evaluations. Each optimization process is terminated if the number of function evaluations exceeds 100.
- We set $\rho(\Delta_k) = 0.1(\Delta_k)^2$ to a forcing function for sufficient decrease in Algorithm 7.1.
- For searching a minimum expectation point and a maximum equivalent EI point, respectively, we use a hybrid method, which is a crossover from the particle swarm method [Van Den Bergh, 2006] to SQP [Mathworks, 2009]. The hybrid approach is effective for searching a better solution in the multimodal problems.

Table 7.1
Results of the Diagonal Test

	Initial	SQP	DSM	DSME	DSMEI
Solution of x	0.2833	0.3000	0.3000	0.3000	0.1341
Objective value (1/1000 rad)	35.06	31.30	31.30	31.30	**30.98**

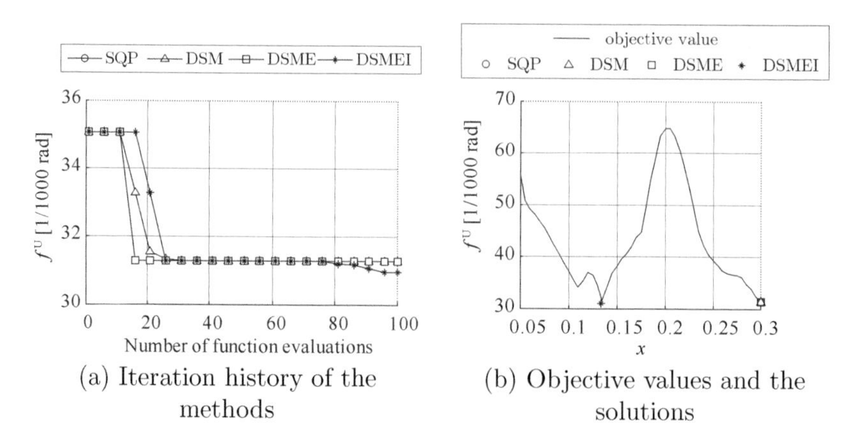

(a) Iteration history of the methods

(b) Objective values and the solutions

Figure 7.3 Results of the diagonal test.

- The outputs in the dataset are normalized with mean 0 and variance 1 in log-space and the surrogate model is built in log-space because the output value is limited to positive value only.

The results of the diagonal test are summarized in Table 7.1 and the objective function value searched by DMSEI is the best. This demonstrates the global property of the method. The plot in Fig. 7.3(a) gives the iteration history of the objective function values. DSM and DSME show the quickest convergence; however, the best convergence point is found by DSMEI. We illustrate objective function values among the entire feasible region and the solutions by the diagonal test in Fig. 7.3(b). Although the problem is multi-modal and it is difficult to find its global solution, DSMEI could find the global minimum, while the other methods could not. This implies importance of exploitation and exploration when the uncertainty is large.

Full dimensional problem

Next we consider a full dimensional problem of (7.26), i.e., a 7-dimensional problem, where the design variables are independent. We continue to compare the results obtained by SQP, DSM, DSME and DSMEI.

Table 7.2
Results of Seismic Design Problem (7.26)

	Initial	Objective value (1/1000 rad) SQP	DSM	DSME	DSMEI
Case 1	45.43	29.59	38.16	38.57	**14.18**
Case 2	45.91	**14.74**	42.44	32.47	15.08
Case 3	36.11	23.19	23.73	23.33	**14.04**
Case 4	38.73	22.19	25.03	27.45	**13.55**
Case 5	47.00	37.18	40.02	40.88	**13.49**
Average		25.38	33.88	32.54	**14.07**
Best		14.74	23.73	23.33	**13.49**
Std. dev.		7.55	7.88	6.58	**0.57**

In this problem, the conditions are assigned as follows:

- Five different sets of initial experimental points are generated by Latin-hypercube sampling. The number of experimental points is 70.
- Results are compared using above five trials with different random seeds, which are referred to as Case 1, ..., Case 5.
- For DSME and DSMEI, the initial GPR is built by using the experimental points and is updated sequentially. The maximum search iteration is given by $n_{\text{search}} = 30$; see Algorithm 7.1.
- We include the 70 computer experiments to the number of function evaluations of all the methods because the initial point in each case for all methods is given as the best point among experimental points.
- Stopping criterion is only given by the number of function evaluations. Optimization processes are terminated if the number of function evaluations exceeds 700.
- The other conditions are the same as those of the diagonal test.

The results of five different cases are shown in Table 7.2. The objective value using DMSEI is the lowest in average and in almost all cases, which means DSMEI could find the best solution. Although the result shows a slight dependency of DSMEI on initial configurations, it has attractive property for globality of search.

The plots of Fig. 7.4 give first four cases of the iteration history of the the objective function values, where DSMEI shows the best accuracy and fast convergence, while the other methods often stagnate. In this problem, SQP in Case 2 is competitive with DSMEI. If we can provide a good initial point, SQP will work. Otherwise, DSMEI might be a good choice.

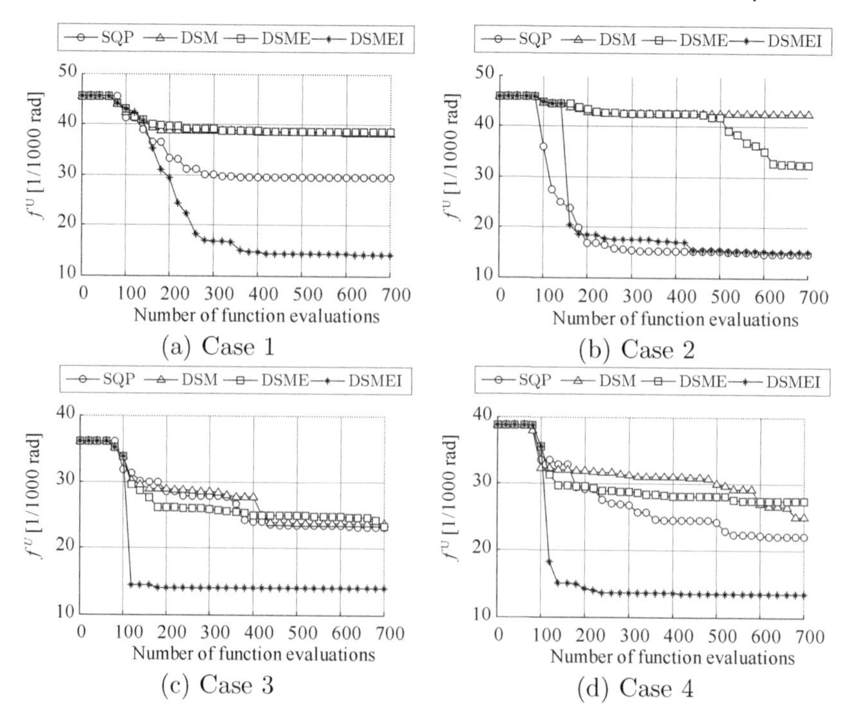

Figure 7.4 Iteration history of seismic design problem (7.26).

7.3 SEQUENTIAL MIXTURE OF GAUSSIAN PROCESS MODELS

7.3.1 FORMULATION OF DESIGN PROBLEMS

Reliability-based design optimization (RBDO) has become an active research field to consider the cost and reliability of the structure simultaneously with developments of methodologies as discussed in Sec. 5.1 as well as the computational tools and hardwares. Let $\boldsymbol{s} = (s_1, \cdots, s_{d_1}) \in \mathbb{R}^{d_1}$ and $\boldsymbol{\Theta} = (\Theta_1, \cdots, \Theta_{d_2}) \in \mathbb{R}^{d_2}$ denote the vectors of design variables and independent and identically distributed random variables (iidrvs), respectively. An RBDO problem is generally formulated to minimize the cost $c(\boldsymbol{s})$ under deterministic and probabilistic constraints as follows:

$$
\begin{aligned}
&\text{Minimize} && c(\boldsymbol{s}) && && (7.27\text{a})\\
&\text{subject to} && \Pr\{g(\boldsymbol{s};\boldsymbol{\Theta}) \leq 0\} \leq P_{\mathrm{a}}, && && (7.27\text{b})\\
& && h_j(\boldsymbol{s}) \leq 0 && (j = 1, \cdots, J), && (7.27\text{c})\\
& && s_k \in \mathcal{S}_k && (k = 1, \cdots, d_1), && (7.27\text{d})
\end{aligned}
$$

where g is the LSF with $g(\boldsymbol{s};\boldsymbol{\Theta}) \leq 0$ indicating the ith independent failure event; $\mathcal{S}_k = [s_k^{\mathrm{L}}, s_k^{\mathrm{U}}]$, where s_k^{L} and s_k^{U} are the lower and upper bounds of s_k, respectively; P_{a} is the required upper-bound probability for the ith failure event;

and $h_j(\boldsymbol{s})$ are deterministic constraint functions. To alleviate large computational cost of structural analysis for solving problem (7.27), various methods have been proposed; see Sec. 5.1 for details. The surrogate-assisted method is suitable for reducing the computational cost for solving an RBDO problem of complex structures. However, a large number of samples are needed for accurate prediction of responses for a problem with many variables [Moustapha and Sudret, 2019].

Do and Ohsaki [2021b] introduced mixture models of Gaussian processes (MGPs) for describing the LSFs in problem (7.27). A large sample set of the input variables and the corresponding structural responses is first generated, and divided into independent subsets, which have a small number of samples, using a Gaussian mixture model (GMM) clustering method. This section presents a method of sequential mixture of Gaussian process models to solve an RBDO problem by sequentially solving deterministic optimization problems.

7.3.2 MIXTURE OF GAUSSIAN PROCESSES

Let $\boldsymbol{x}_i = (\boldsymbol{s}_i, \boldsymbol{\theta}_i) \in \mathbb{R}^d$ $(d = d_1 + d_2)$ denote the d-dimensional vectors of input variables consisting of the design variables and realizations of the uncertain parameter vector $\boldsymbol{\Theta}$. The associated output values such as structural responses are denoted by $y_i \in \mathbb{R}$, and the dataset of N samples is defined as $\mathcal{D} = \{\boldsymbol{X}, \boldsymbol{y}\} = \{\boldsymbol{x}_i, y_i\}_{i=1}^N$. The GPR is used to approximate the input-output mapping $\hat{g}(\boldsymbol{x})$ for a target function $g(\boldsymbol{x})$, where $\hat{g}(\boldsymbol{x})$ is the mean function or expectation of the GPR described in (7.4). However, the GPR is applicable to relatively small number of sample points because the inversion and determinant of an $N \times N$ kernel matrix should be computed and its computational cost is $O(N^3)$; see (7.6). Therefore, for application to a larger sample set, the data in $\mathcal{D}$ are first classified into several subsets with smaller sizes than N using the GMM clustering method; see Appendix A.5 for details. We then construct a local GPR for each subset and combine all subsets to form the MGP.

Figure 7.5 illustrates the process of constructing an MGP, where $\mathcal{D}$ is classified into K small independent subsets using the GMM clustering method, i.e., $\mathcal{D} = \mathcal{D}_1 \cup \cdots \cup \mathcal{D}_K$. Note that K is equal to the number of Gaussians of the GMM; see Appendix A.5. Let $\boldsymbol{\psi}$ denotes a vector of the GMM parameters. Local GPR $\hat{g}_k$ is then generated for the kth subset. The MGP denoted by $\hat{g}$ is defined, as follows, as a weighted average of the K local GPRs:

$$\hat{g}(\boldsymbol{x}) = \sum_{k=1}^{K} w_k(\boldsymbol{x}) \hat{g}_k(\boldsymbol{x}), \tag{7.28a}$$

$$0 \leq w_k(\boldsymbol{x}) \leq 1, \quad \sum_{k=1}^{K} w_k(\boldsymbol{x}) = 1, \tag{7.28b}$$

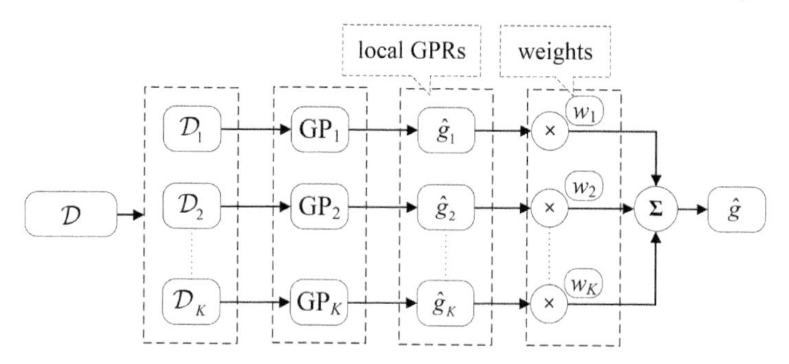

Figure 7.5 Construction of an MGP for a given sample set.

where w_k is the weight that represents the probability that the input variable $\boldsymbol{x}$ belongs to the projection of the kth subset onto the input variable space, and is computed by (A.28), i.e.,

$$w_k = \Pr\{Z_i = k \mid \boldsymbol{x}, \boldsymbol{\psi}\},$$

where Z_i indicates the index of subset in which the ith sample belongs to. Note that the original function g is a function of the design variables $\boldsymbol{s}$; however, the MGP $\hat{g}$ is regarded as a function of both of the design variables $\boldsymbol{x}$ and the parameters $\boldsymbol{\theta}$ to deal with deterministic optimization. The algorithm of GMM-based MGP is summarized in Algorithm 7.2. This way, overall accuracy of the regression model can be improved without sacrificing computational cost [Masoudnia and Ebrahimpour, 2014; Liu et al., 2020].

7.3.3 A SEQUENTIAL DETERMINISTIC OPTIMIZATION FOR RBDO

To explore the feasible region that may contain the optimal solution, and improve accuracy of the MGPs to approximate $g_i(\boldsymbol{s}; \boldsymbol{\Theta})$ in that region, optimal solution of problem (7.27) is obtained by sequentially solving deterministic optimization problems. This process is generally called sequential deterministic optimization (SDO) including the sequential optimization and reliability assessment [Du and Chen, 2004; Goswami et al., 2019; Li et al., 2019], which is used in Sec. 5.3 for robust topology optimization and has been applied to RBDO problems. This process is equivalent with that of simultaneously increasing the load effects and reducing the structural resistances in the structural design process. In particular, we shift the LFSs to a safe region from the nominal values, and sequentially update the offset values. The tth cycle

Algorithm 7.2 GMM-based MGP.

Require: $\mathcal{D} = \{\boldsymbol{x}_i, y_i\}_{i=1}^n$: sample set.
Require: $\pi_k^{(0)}$, $\boldsymbol{\mu}_k^{(0)}$ and $\boldsymbol{\Sigma}_k^{(0)}$: initial values of GMM parameters.
Require: $K_{\max}$: maximum number of Gaussians, e.g., $K_{\max} = 50$.
1: **for** $k \leftarrow 1$ to $K_{\max}$ **do**
2: Perform EM algorithm using (A.26) and (A.27); obtain π_k, $\boldsymbol{\mu}_k$ and $\boldsymbol{\Sigma}_k$.
3: Compute BIC_k using (A.29).
4: **end for**
5: Substitute $K \leftarrow \text{argmin}_k\{\text{BIC}_k, k = 1, \cdots, K_{\max}\}$.
6: Determine $\boldsymbol{\mu}_{\boldsymbol{X},k}$ and $\boldsymbol{\Sigma}_{\boldsymbol{XX},k}$ in (A.20) and (A.21), respectively.
7: Initialize $\mathcal{D}_k \leftarrow \emptyset$ $(k = 1, \cdots, K)$.
8: **for** $i \leftarrow 1$ to n **do**
9: **for** $k \leftarrow 1$ to K **do**
10: Compute $\mathbb{E}[Z_{ik}]$ for the sample $(\boldsymbol{x}_i, y_i)$.
11: **end for**
12: Substitute $k_i \leftarrow \text{argmax}_k\{\mathbb{E}[Z_{ik}], k = 1, \cdots, K\}$.
13: Update $\mathcal{D}_{k_i} \leftarrow \mathcal{D}_{k_i} \cup \{\boldsymbol{x}_i, y_i\}$
14: **end for**
15: **for** $k \leftarrow 1$ to K **do**
16: Construct local GPR $\hat{g}_k(\boldsymbol{x})$.
17: **end for**
18: **return** $\hat{g}(\boldsymbol{x})$ using (7.28) with $w_k = \Pr\{Z_i = k \mid \boldsymbol{x}, \boldsymbol{\psi}\}$ in (A.28).

of SDO problem associated with problem (7.27) is stated as follows:

$$
\begin{aligned}
&\text{Find} && \boldsymbol{s}^{(t)} = \text{argmin}_{\boldsymbol{s}}\ c(\boldsymbol{s}) && && \text{(7.29a)}\\
&\text{subject to} && \hat{g}^{(t)}(\boldsymbol{s}, \boldsymbol{\mu}_\theta) - \lambda^{(t)} \geq 0, && && \text{(7.29b)}\\
& && h_j(\boldsymbol{s}) \leq 0 && (j = 1, \cdots, J), && \text{(7.29c)}\\
& && s_k \in \mathcal{S}_k && (k = 1, \cdots, d_1), && \text{(7.29d)}
\end{aligned}
$$

where $\hat{g}^{(t)}(\boldsymbol{s}, \boldsymbol{\mu}_\theta)$ is the current value of the ith LSF approximated by the MGP, and $\boldsymbol{\mu}_\theta$ is the mean vector of the random parameters $\boldsymbol{\Theta}$; $\lambda_i^{(t)}$ is the current ith shifting value; and $\boldsymbol{s}^{(t)}$ is the solution to the current SDO problem. Note that the effect of uncertainty in the function value evaluated at the mean vector of the input variables denoted by $\boldsymbol{\mu}_x = (\boldsymbol{s}, \boldsymbol{\mu}_\theta)$ is incorporated by the offset $\lambda_i^{(t)}$ into the safe region as illustrated in Fig. 7.6. Therefore, the most probable point moves discontinuously at each SDO cycle because the solutions to the deterministic optimization problem (7.29) are generally different.

The shifting values are updated as

$$\lambda^{(t+1)} = \hat{g}^{(t)}(\boldsymbol{s}^{(t)}, \boldsymbol{\mu}_\theta) - P_{\text{a}}^{-1},$$

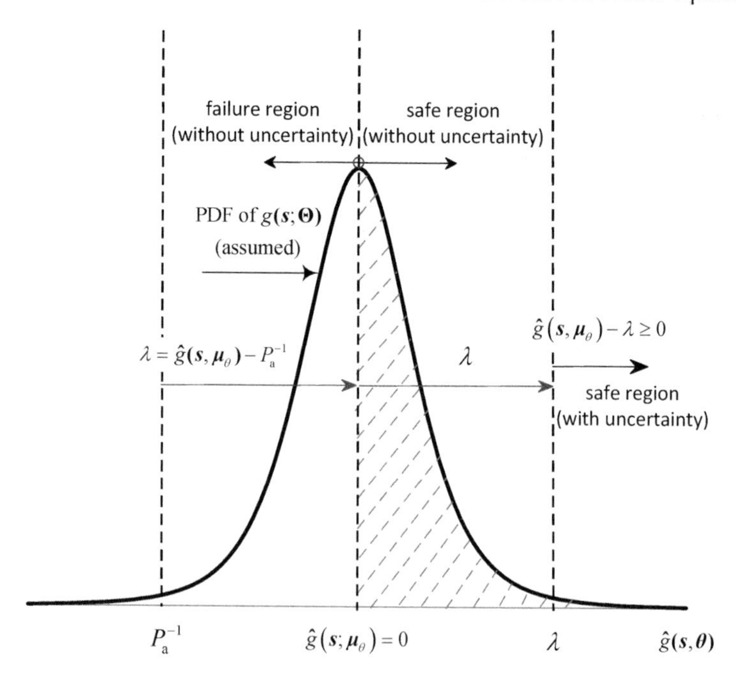

Figure 7.6 SDO problem scheme.

where P_{a}^{-1} is the inverse probability of the failure probability P_{a}. The threshold value of LSF approximated by the current MGP for specified P_{a} is estimated using the saddlepoint approximation; see Do and Ohsaki [2021b] and Appendix A.6 for details. In the following example, the central moments for the second-order Taylor series of the cumulants are estimated from 2×10^6 samples. At the initial cycle of the sequence of SDO problem, we set $\lambda^{(0)} = 0$.

To obtain a new sample for updating the MGP, structural analysis is carried out for the solution $\boldsymbol{s}^{(t)}$ of problem (7.29) to evaluated the structural response. The new sample is then added to the cluster that maximizes the expected probability value $\mathbb{E}[z_{ik}]$ in (A.24) in Appendix A.5. Thus, we carry out structural analysis only once at each cycle of SDO, and reconstruct the local GPM of only one cluster that includes the added sample for updating the MGP. The process of SDO terminates when $\left\|\lambda^{(t+1)} - \lambda^{(t)}\right\| \leq \epsilon_\lambda$ or $t > t_{\max}$ is satisfied for the specified small positive value ϵ_λ and the maximum number of SDO cycles $t_{\max}$. The algorithm of SDO is shown in Algorithm 7.3, and the whole process of RBDO utilizing SDO is summarized in Algorithm 7.3.

7.3.4 EXAMPLE: TEN-BAR TRUSS

Consider a ten-bar truss as shown in Fig. 7.7. The design variables of problem (7.27) are the cross-sectional areas of members, and their sum is the objective function to be minimized. The independent uncertain parameters $\boldsymbol{\theta}$

Algorithm 7.3 Sequential deterministic optimization for RBDO.

Require: Define the feasible region of design variables $\boldsymbol{s}$ and uncertain set of random parameters $\boldsymbol{\theta}$. Generate the samples of design variables and parameters, and carry out structural analysis to obtain the sample dataset.
1: Implement Algorithm 7.2 for obtaining the MGPs to approximate LSFs of the RBDO problem.
2: Set $t = 0$; $\hat{g}^{(0)}$ = initial MGPs; $\lambda^{(0)} = 0$.
3: **loop**
4: Solve the SDO problem to obtain $\boldsymbol{s}^{(t)}$.
5: Substitute $t \leftarrow t + 1$; compute $\lambda^{(t+1)}$.
6: **if** Stopping criterion met. **then**
7: **Break.**
8: **end if**
9: Perform structural analysis.
10: Compute $\mathbb{E}[Z_{ik}]$ values.
11: Add $\boldsymbol{s}^{(t)}$ and associated output response to the cluster that maximizes $\mathbb{E}[Z_{ik}]$; reconstruct the local GPM for that cluster.
12: **end loop**
13: Output the solution and terminate the process.

consist of the external loads P_1, P_2, P_3, Young's modulus E and the length L of the horizontal and vertical members, which have normal distributions. The mean values of P_1, P_2 and P_3 are 60, 40 and 10 kN, respectively, with coefficient of variation (CV), which is the standard deviation divided by the mean value, equal to 0.20. The mean values of E and L are 200 GPa and 1 m, respectively, and CVs are 0.10 and 0.05. The probability of the vertical displacement Δ_3 at node 3 exceeding the allowable value 4×10^{-3} m should be less than or equal to $P_{\mathrm{a}} = 6.21 \times 10^{-3}$ [Zhao and Qiu, 2013]. Therefore, the LSF is given as $g(\boldsymbol{s}, \boldsymbol{\theta}) = 4 \times 10^{-3} - \Delta_3$. The lower and upper bounds s^{L} and s^{U} for s_i are given as $1.0 \times 10^{-4}\,\mathrm{m}^2$ and $20.0 \times 10^{-4}\,\mathrm{m}^2$, respectively.

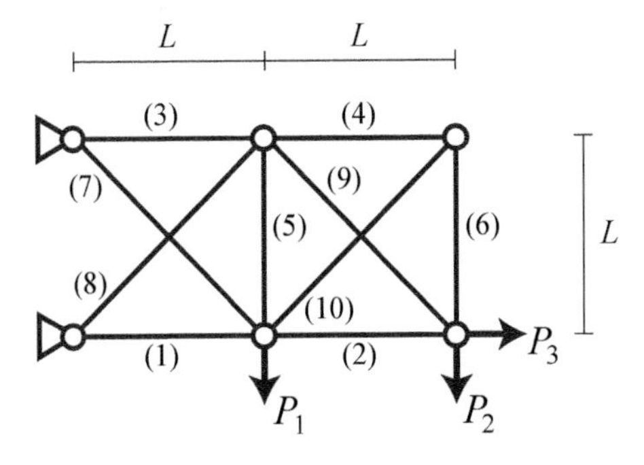

Figure 7.7 Ten-bar truss.

Table 7.3
Comparison of the Computation Time and Prediction Performance of the MGP with Those of the Global GPM.

No. samples	No. clusters	Computation time (sec.) MGP	Computation time (sec.) global GPM	Testing R^2 MGP	Testing R^2 global GPM
1×10^4	7	1386	4701	0.991	0.982
2×10^4	10	3208	32364	0.998	0.97
3×10^4	13	6383	n/a	0.995	n/a
4×10^4	18	8502	n/a	0.996	n/a
5×10^4	21	12922	n/a	0.994	n/a

Hence, the RBDO problem of the ten-bar truss is formulated as follows:

$$\text{Minimize} \quad c(\boldsymbol{s}) = \sum_{i=1}^{10} s_i \tag{7.30a}$$

$$\text{subject to} \quad \Pr\{g(\boldsymbol{s}; \boldsymbol{\Theta}) \leq 0\} \leq P_{\text{a}}, \tag{7.30b}$$

$$s^{\text{L}} \leq s_k \leq s^{\text{U}} \qquad (k = 1, \cdots, 10). \tag{7.30c}$$

The LSF g in problem (7.30) is approximated by the MGP to formulate the associated SDO problem. To find an appropriate number of samples, we first generate five sample sets with different numbers of samples, i.e., 1×10^4, 2×10^4, 3×10^4, 4×10^4 and 5×10^4.

The prediction performance of the surrogate model is assessed by the coefficient of determination R^2 for another randomly generated independent test set of 2×10^4 samples. The computation time and R^2 value are listed in Table 7.3. The results of standard global GPM are also listed for comparison purpose, and the computation time for MGP includes that for clustering the sample set. It is seen from the table that the computation time for MGP is considerably smaller than that for the global GPM. For the larger sample sets with 3×10^4, 4×10^4 and 5×10^4 samples, construction of the MGP is still possible although the global GPM could not be generated. In addition, R^2 value by MGP is larger than that by the GPM indicating a better prediction performance of the MGP. We can also see from the table that a larger number of samples does not always lead to better prediction performance. Therefore, we use the set of 2×10^4 samples with the largest R^2 value for solving the SDO problem.

To solve the SDO problem (7.30), which is a nonlinear programming problem, we use SQP algorithm available in the function *fmincon* in Optimization Toolbox of MATLAB [Mathworks, 2018b]. The gradient is computed using the finite difference approximation. Since the solution to a constrained nonlinear programming problem generally exists on the boundary of the feasible

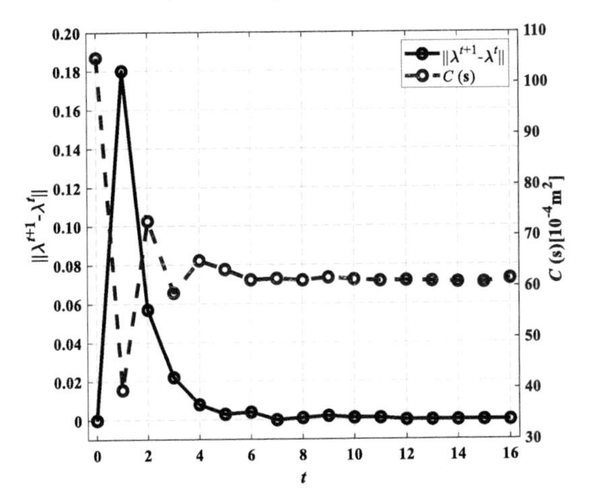

Figure 7.8 Histories of the shifting value and objective function with respect to the cycles of the SDO problem of the ten-bar truss.

region, and the probabilistic constraint is satisfied with equality, the solution may be infeasible due to a small error tolerance of the inequality constraint. Therefore, we use 70% of the threshold failure probability P_{a} in problem (7.30) to enforce overestimation of the shifting value $\lambda^{(t)}$. Figure 7.8 shows the histories of the shifting value and the objective function value with respect to the cycles of SDO problem. With $\epsilon_{\lambda} = 10^{-9}$ and $t_{\max} = 50$, the optimal solution is found at the 16th cycle after adding 15 samples. The constraint function value at the optimal solution of the deterministic optimization problem is less than the specified constraint tolerance 10^{-6}, corresponding to $\Pr\{g(\boldsymbol{s};\boldsymbol{\Theta}) \leq 0\} = 4.34 \times 10^{-3}$ (70% of the threshold value P_{a}). Table 7.4 shows the optimization results by the MGP and Zhao and Qiu [2013]. The results of MCS are also shown. We can see from the table that the objective function value by the MGP is smaller than the existing RBDO results, where FORM is used with finite element analysis (FEA) or response surface method (RSM2) for response evaluation. This result demonstrates a high efficiency of the MGP.

7.4 RELIABILITY-BASED MULTI-OBJECTIVE BAYESIAN OPTIMIZATION

7.4.1 FORMULATION OF DESIGN PROBLEMS

In view of robust design, it is preferable to minimize variance of the representative response in addition to its mean value. Therefore, in this section, an MOO approach in Do et al. [2021a] is presented for reliability-based design using the BO method.

Table 7.4
Comparison of Optimization Results of the Ten-bar Truss

Design variables $[\times 10^{-4}\,\mathrm{m}^2]$	FEA and FORM	RSM2, FORM, and MCS	MGP
s_1	10.493	10.705	10.333
s_2	5.772	5.914	5.371
s_3	14.098	14.424	13.579
s_4	1	1	1
s_5	1	1	1
s_6	1	1	1
s_7	5.46	5.531	6.418
s_8	11.586	11.853	11.273
s_9	1	1	1
s_{10}	10.958	11.223	10.508
$c(\boldsymbol{s})\ [10^{-4}\,\mathrm{m}^2]$	62.367	63.649	61.482
$\Pr\{g(\boldsymbol{s};\boldsymbol{\Theta}) \leq 0\}$	8.51×10^{-3}	6.19×10^{-3}	4.34×10^{-3}
MCS	4.22×10^{-3}	2.95×10^{-3}	5.64×10^{-3}

Consider a d_1-dimensional vector of discrete design variables $\boldsymbol{s} = (s_1, \cdots, s_{d_1}) \in \mathbb{N}^{d_1}$ representing, e.g., the indices of sections in the list $s_k \in \mathcal{S}_k$ $(k = 1, \cdots, d_1)$ of available steel sections of a building frame. Let $\boldsymbol{\Theta}$ denote a d_2-dimensional iidrvs, e.g., for the material properties, cross-sectional sizes and static/dynamic loads, and suppose the marginal pdfs of $\boldsymbol{\Theta}$ are given. Two objective functions $f_1(\boldsymbol{s})$ and $f_2(\boldsymbol{s})$ are defined as follows:

$$f_1(\boldsymbol{s}) = \mathbb{E}[f(\boldsymbol{s};\boldsymbol{\Theta})], \quad f_2(\boldsymbol{s}) = \sqrt{\mathrm{Var}[f(\boldsymbol{s};\boldsymbol{\Theta})]}$$

which are the mean and standard deviation of the uncertain objective function $f(\boldsymbol{s};\boldsymbol{\Theta})$ such as the total structural volume or mass, where $\mathbb{E}[\cdot]$ and $\mathrm{Var}[\cdot]$ denote expectation and variance with respect to the distribution of $\boldsymbol{\Theta}$, respectively, because $\boldsymbol{s}$ is not a random variable vector, and hence $f_1(\boldsymbol{s})$ and $f_2(\boldsymbol{s})$ are deterministic functions of $\boldsymbol{s}$. Let $g_i(\boldsymbol{s};\boldsymbol{\Theta})$ $(i = 1, \cdots, I)$ represent the LSFs of representative responses such as member stresses and nodal displacements. The multi-objective robust design optimization (MORDO) problem with joint probabilistic constraint is formulated as

$$\begin{aligned} &\text{Minimize} \quad f_1(\boldsymbol{s}), f_2(\boldsymbol{s}) && (7.31\text{a}) \\ &\text{subject to} \quad \Pr\{g_1(\boldsymbol{s};\boldsymbol{\Theta}) \leq 0, \cdots, g_I(\boldsymbol{s};\boldsymbol{\Theta}) \leq 0\} \geq 1 - \varepsilon, && (7.31\text{b}) \\ &\qquad\qquad s_k \in \mathcal{S}_k \quad (k = 1, \cdots, d_1), && (7.31\text{c}) \end{aligned}$$

where $\varepsilon \in (0,1)$ denotes the specified risk level. The problem with individual probabilistic constraints is also formulated as

$$\text{Minimize} \quad f_1(\boldsymbol{s}), f_2(\boldsymbol{s}) \tag{7.32a}$$

$$\text{subject to} \quad \Pr\{g_i(\boldsymbol{s};\boldsymbol{\Theta}) \leq 0\} \geq 1-\varepsilon_i \quad (i=1,\cdots,I), \tag{7.32b}$$

$$s_k \in \mathcal{S}_k \quad (k=1,\cdots,d_1), \tag{7.32c}$$

where ε_i $(0<\varepsilon_i<1)$ is the specified risk level of the ith probabilistic constraint.

7.4.2 BAYESIAN OPTIMIZATION

A sample dataset $\mathcal{D} = \{\boldsymbol{X}, \boldsymbol{y}\} = \{\boldsymbol{x}_i, y_i\}_{i=1}^N$ is generated for the input variables $\boldsymbol{x}_i = (\boldsymbol{s}_i, \boldsymbol{\theta}_i) \in \mathbb{R}^d$ $(d = d_1 + d_2)$ including parameters and the output variables $y_i \in \mathbb{R}$ representing the uncertain LSFs, where $\boldsymbol{\theta}_i$ are regarded as realizations of $\boldsymbol{\Theta}$. The appropriate number of sample sets N depends on d, for example, $N \geq 15d$ [Afzal et al., 2017]. $\mathcal{D}$ is randomly initialized using Latin-hypercube sampling (LHS), where integer values of $\boldsymbol{s}$ are determined by rounding the corresponding real values to the nearest integers. The samples are selected to be feasible in the region satisfying $g_i(\boldsymbol{s};\boldsymbol{\theta}) \leq 0$ $(i=1,\cdots,I)$.

The uncertain output variable y is approximated by the GPR $G(\boldsymbol{x})$ in (7.3), for which the posterior distribution is derived by conditioning on the sample dataset $\mathcal{D}$ and used for the process of finding an appropriate new sample point. The mean and variance of an output variable for a new set of input variables are computed as provided in (7.4) and (7.5), respectively. In the BO context, our belief about how promising each point in the input variable space can be evaluated using the acquisition function. Thus, the most promising point maximizing the acquisition function is selected as the next sampling point to refine the GPM. The common acquisition functions for a single objective optimization problem are summarized as follows:

(1) Upper confidence bound, which is a sum of mean and standard deviation at a sample point of the GPR, representing exploitation and exploration, respectively.

(2) Probability of improvement, which represents the probability that the output function increases from the current best value.

(3) EI, which represents the probability of increase of the uncertain output function value from the current best value.

Note that these values can be computed with a very small computational cost using the current GPM.

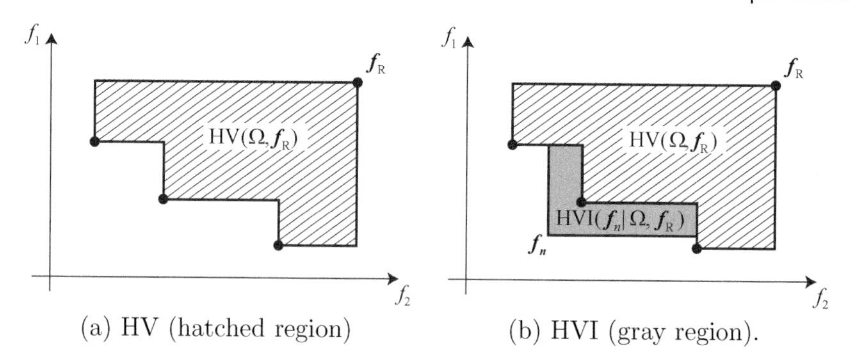

(a) HV (hatched region) (b) HVI (gray region).

Figure 7.9 Examples of the HV and HVI for a bi-objective minimization problem.

In the following, the hypervolume (HV) is used for defining the acquisition function of an MOO problem. Consider a problem with k objective functions, and suppose we have obtained M Pareto optimal solutions $\Omega = \{\boldsymbol{f}_1, \cdots, \boldsymbol{f}_M\}$ in the objective function space. A reference point $\boldsymbol{f}_{\rm R} \in \mathbb{R}^k$ dominated by all Pareto optimal solutions is assigned in the objective function space. Note that $\boldsymbol{f}_{\rm R}$ does not necessarily correspond to a particular vector of input variables. The HV is defined by Ω and $\boldsymbol{f}_{\rm R}$ as follows [Couckuyt et al., 2014]:

$$\mathrm{HV}(\Omega, \boldsymbol{f}_{\rm R}) = \Lambda\left(\{\boldsymbol{f} \in \mathbb{R}^k \mid \exists \boldsymbol{f}_m \in \Omega : \boldsymbol{f}_m \preccurlyeq \boldsymbol{f} \,\text{and}\, \boldsymbol{f} \preccurlyeq \boldsymbol{f}_{\rm R}\}\right),$$

where $\Lambda(\cdot)$ denotes the Lebesgue measure in the objective function space, $\boldsymbol{f}$ is a point in the k-dimensional objective function space, and $\boldsymbol{f}_m \preccurlyeq \boldsymbol{f}$ implies $\boldsymbol{f}_m$ dominates $\boldsymbol{f}$; see Appendix A.7 for details of MOO. Lebesgue measure corresponds to the area for $k = 2$, and the volume for $k = 3$. In the following examples, HV is computed using a sampling-based technique in MATLAB function *hypervolume* [Cao, 2008].

HV is a measure of quality of the Pareto optimal solutions, i.e., a larger HV leads to a higher quality. Figure 7.9(a) illustrates an HV for the case of three Pareto optimal solutions to a bi-objective minimization problem. A new sampling point $\boldsymbol{x}_n = (\boldsymbol{s}_n, \boldsymbol{\theta}_n)$ with the objective function $\boldsymbol{f}_n$ is generated so that the union of Ω and $\{\boldsymbol{f}_n\}$, denoted by $\{\boldsymbol{f}_n\} \cup \Omega$, leads to the HV value larger than the current value. The improvement of HV is measured by the hypervolume improvement indicator (HVI) [Emmerich et al., 2006] defined as

$$\mathrm{HVI}(\boldsymbol{f}_n \mid \Omega, \boldsymbol{f}_{\rm R}) = \mathrm{HV}(\{\boldsymbol{f}_n\} \cup \Omega, \boldsymbol{f}_{\rm R}) - \mathrm{HV}(\Omega, \boldsymbol{f}_{\rm R}).$$

The improvement is indicated by HVI > 0; hence, HVI is to be maximized as the acquisition function. Figure 7.9(b) illustrates the HVI, i.e., gray area, for an objective function vector $\boldsymbol{f}_n$ when these Pareto optimal solutions have been found for a bi-objective minimization problem.

For a constrained MOO problem, the acquisition function is to be modified to improve possibility of obtaining a feasible solution. The maximum function of the LSFs is denoted by

$$g^{\mathrm{U}}(\boldsymbol{s};\boldsymbol{\theta}) = \max\{g_i(\boldsymbol{s};\boldsymbol{\theta}), \cdots, g_I(\boldsymbol{s};\boldsymbol{\theta})\}, \tag{7.33}$$

which is approximated by

$$G^{\mathrm{U}}(\boldsymbol{s},\boldsymbol{\theta}) = \max\{G_i(\boldsymbol{s},\boldsymbol{\theta}), \cdots, G_I(\boldsymbol{s},\boldsymbol{\theta})\}, \tag{7.34}$$

where $G_i(\boldsymbol{s},\boldsymbol{\theta})$ is the GPR described in (7.3) for the constraint function $g_i(\boldsymbol{s},\boldsymbol{\theta})$. For problem (7.31), $\Delta_{1,1}(\boldsymbol{s},\boldsymbol{\theta})$ describes the probability that the point $(\boldsymbol{s},\boldsymbol{\theta})$ satisfies the joint probabilistic constraint, and is formulated as follows:

$$\Delta_{1,1}(\boldsymbol{s},\boldsymbol{\theta}) = \Pr\{G^{\mathrm{U}}(\boldsymbol{s},\boldsymbol{\theta}) \leq 0\} - 1 + \varepsilon. \tag{7.35}$$

The variable vector $\boldsymbol{x} = (\boldsymbol{s},\boldsymbol{\theta})$ to increase the chance for $\boldsymbol{s}$ to satisfy the joint probabilistic constraint of problem (7.31) is found by maximizing $\Delta_{1,1}$.

For problem (7.32), let $\Delta_{1,2}(\boldsymbol{s},\boldsymbol{\theta})$ denote the expectation that $\boldsymbol{s}$ satisfies all individual probabilistic constraints. Since $g_i(\boldsymbol{s},\boldsymbol{\theta})$ is approximated by the GPR $G_i(\boldsymbol{s},\boldsymbol{\theta})$ with mean $\hat{g}_i(\boldsymbol{s},\boldsymbol{\theta})$ and standard deviation $\hat{\sigma}_i(\boldsymbol{s},\boldsymbol{\theta})$, and all probabilistic constraints are statistically independent, $\Delta_{1,2}(\boldsymbol{s},\boldsymbol{\theta})$ is defined as follows:

$$\begin{aligned}\Delta_{1,2}(\boldsymbol{s},\boldsymbol{\theta}) &= \prod_{i=1}^{I}[\Pr\{G_i(\boldsymbol{s},\boldsymbol{\theta}) \leq 0\} - 1 + \varepsilon_i] \\ &= \prod_{i=1}^{I}\left\{\frac{1}{2}\left[1 + \operatorname{erf}\left(\frac{-\hat{g}_i(\boldsymbol{s},\boldsymbol{\theta})}{\sqrt{2}\hat{\sigma}_i(\boldsymbol{s},\boldsymbol{\theta})}\right)\right] - 1 + \varepsilon_i\right\},\end{aligned} \tag{7.36}$$

where we used the following relation:

$$\Pr\{G_i(\boldsymbol{s},\boldsymbol{\theta}) \leq 0\} = \frac{1}{2}\left[1 + \operatorname{erf}\left(\frac{-\hat{g}_i(\boldsymbol{s},\boldsymbol{\theta})}{\sqrt{2}\hat{\sigma}_i(\boldsymbol{s},\boldsymbol{\theta})}\right)\right]$$

and $\operatorname{erf}(\cdot)$ is the Gauss error function to evaluate the cdf of the Gaussian variable. If $\boldsymbol{x}$ maximizes $\Delta_{1,2}$, it also increases the chance for $\boldsymbol{s}$ to satisfy all individual probabilistic constraints of problem (7.32).

For both problems, feasibility to all deterministic constraints $h_j(\boldsymbol{s}) \leq 0$ is indicated by the following indicator function $\Delta_2(\boldsymbol{s})$:

$$\Delta_2(\boldsymbol{s}) = \begin{cases} 1 & \text{if } h_j(\boldsymbol{s}) \leq 0 \ (j = 1, \cdots, J) \\ 0 & \text{otherwise.} \end{cases}$$

Finally, to incorporate improvement criteria in the objective and constraint functions simultaneously, the following acquisition functions $\alpha_1(\boldsymbol{s},\boldsymbol{\theta})$ and $\alpha_2(\boldsymbol{s},\boldsymbol{\theta})$ are formulated for problems (7.31) and (7.32), respectively:

$$\alpha_1(\boldsymbol{s},\boldsymbol{\theta}) = \mathrm{HVI}(f \mid \Omega, \boldsymbol{f}_{\mathrm{R}})\Delta_{1,1}(\boldsymbol{s},\boldsymbol{\theta})\Delta_2(\boldsymbol{s}), \tag{7.37a}$$

$$\alpha_2(\boldsymbol{s},\boldsymbol{\theta}) = \mathrm{HVI}(f \mid \Omega, \boldsymbol{f}_{\mathrm{R}}))\Delta_{1,2}(\boldsymbol{s},\boldsymbol{\theta})\Delta_2(\boldsymbol{s}). \tag{7.37b}$$

Therefore, for MORDO problems (7.31) and (7.32), the next sampling point $\boldsymbol{x}_n = (\boldsymbol{s}_n, \boldsymbol{\theta}_n)$ is determined as the maximizer of the acquisition functions $\alpha_1(\boldsymbol{s},\boldsymbol{\theta})$ and $\alpha_2(\boldsymbol{s},\boldsymbol{\theta})$, respectively.

7.4.3 SOLUTION PROCEDURE USING BAYESIAN OPTIMIZATION

An elitist non-dominated sorting approach (NSGA-II) [Deb et al., 2002] is used to find Pareto optimal solutions of the MORDO problems. At each iteration of the BO, the function values are computed using the GPR and the Pareto optimal solutions are sorted among the set of already-generated candidate solutions $\boldsymbol{s}$, denoted by Ω_a, including the initial sample dataset and all candidate solutions generated so far.

Solution process of the MORDO problems (7.31) and (7.32) requires efficient and accurate evaluations of the mean and variance of the uncertain objective functions $f_1(\boldsymbol{s})$ and $f_2(\boldsymbol{s})$ as well as the probabilities in the probabilistic constraints for obtaining a set of feasible solutions denoted by Ω_f.

The function $f(\boldsymbol{s}, \boldsymbol{\theta})$ is approximated by GPR, and the second-order Taylor series expansion is used to evaluate $f_1(\boldsymbol{s})$ and $f_2(\boldsymbol{s})$. Furthermore, the joint and individual probabilistic constraints are estimated using a saddlepoint approximation derived utilizing GPM; see Appendix A.6 for details.

To find integer variables $\boldsymbol{s}$ and the real parameters $\boldsymbol{\theta}$ to maximize the acquisition functions $\alpha_1(\boldsymbol{s}, \boldsymbol{\theta})$ and $\alpha_2(\boldsymbol{s}, \boldsymbol{\theta})$ in (7.37a) and (7.37b), respectively, a mixed-integer non-linear programming (MINLP) problem is formulated as follows:

$$
\begin{aligned}
\text{Find} \quad & (\boldsymbol{s}_n, \boldsymbol{\theta}_n) = \operatorname{argmax}_{(\boldsymbol{s}, \boldsymbol{\theta})}\ \alpha_1(\boldsymbol{s}, \boldsymbol{\theta}) \text{ or } \alpha_2(\boldsymbol{s}, \boldsymbol{\theta}) && (7.38a)\\
\text{subject to} \quad & \mathrm{HVI}(\boldsymbol{f}_n \mid \Omega, \boldsymbol{f}_{\mathrm{R}}) \geq 0, && (7.38b)\\
& s_i \in \mathcal{S}_i \quad (i = 1, \cdots, d_1), && (7.38c)\\
& \boldsymbol{\theta} \in \mathcal{R}, && (7.38d)
\end{aligned}
$$

where the set $\mathcal{R}$ is defined by the 95% confidence intervals of the uncertain parameters $\boldsymbol{\Theta}$ in the following example.

Most of the mathematical programming based approaches for MINLP problems are not applicable to problem (7.38) because $\alpha_1(\boldsymbol{s}, \boldsymbol{\theta})$ and $\alpha_2(\boldsymbol{s}, \boldsymbol{\theta})$ are non-convex functions. Since $\mathrm{HVI}(\boldsymbol{f}_n \mid \Omega, \boldsymbol{f}_{\mathrm{R}})$ and $\Delta_2(\boldsymbol{s})$ are functions of deterministic integer variables $\boldsymbol{s}$, and the random parameters $\boldsymbol{\theta}$ appear only in $\Delta_{1,1}(\boldsymbol{s}, \boldsymbol{\theta})$ and $\Delta_{1,2}(\boldsymbol{s}, \boldsymbol{\theta})$, an optimization strategy coupling a random sampling method and simulated annealing [Kirkpatrick et al., 1983] is developed for solving problem (7.38). This strategy is an extension of a two-stage random search proposed by Do and Ohsaki [2021a], consisting of determination of $\boldsymbol{\theta}$ in the lower-level worst-case analysis followed by determining the discrete design variables $\boldsymbol{s}$ in the upper-level problem of a single-objective discrete RDO with deterministic constraints and unknown-but-bounded uncertainty in the design parameters. The solution process of problem (7.38) consists of the following four steps:

1. Randomly generate a set Ω_s consisting of a finite number of new candidate solutions $\boldsymbol{s}$.

2. Calculate $\mathrm{HVI}(\boldsymbol{f}_n \mid \Omega, \boldsymbol{f}_\mathrm{R})$ and $\Delta_2(\boldsymbol{s})$ for each solution in Ω_s, and retain in Ω_s the solutions that yield positive $\mathrm{HVI}(\boldsymbol{f}_n \mid \Omega, \boldsymbol{f}_R)$, $\Delta_2(\boldsymbol{s}) = 1$ and negative values of the constraint functions evaluated using GPM.

3. For each retained solution in Ω_s, maximize $\alpha_1(\boldsymbol{s}, \boldsymbol{\theta})$ or $\alpha_2(\boldsymbol{s}, \boldsymbol{\theta})$ using simulated annealing to find the uncertain parameters $\boldsymbol{\theta}$ associated with the specified $\boldsymbol{s}$.

4. Select the set of $\boldsymbol{s}$ in Ω_s and $\boldsymbol{\theta}$ that maximizes $\alpha_1(\boldsymbol{s}, \boldsymbol{\theta})$ or $\alpha_2(\boldsymbol{s}, \boldsymbol{\theta})$, and select it as the next sampling point $\boldsymbol{x}_n = (\boldsymbol{s}_n, \boldsymbol{\theta}_n)$.

In the first step above, two groups of new candidate solutions $\boldsymbol{x}$ are generated based on the following two different search strategies to create Ω_s:

Neighborhood search Random perturbation of the points from the current Pareto optimal solutions in the design variable space.

Global search Uniform generation of sampling points in the design domain.

The neighborhood search expects to improve the HV assuming that the neighborhood relation in the design variable space is related to that in the objective function space. To reduce the computational cost of this process when many solutions are available, Pareto optimal solutions in the objective function space are first classified into a moderately small number of clusters using the GMM [Hastie et al., 2009]; see Appendix A.5 for details. The solutions in the same cluster are regarded as the samples generated from the same Gaussian that has the mean vector at the center of cluster. Then, the point nearest to the center is selected as the representative solution, and the random perturbation is performed from the representative solution in each cluster. After the neighborhood search and global search, the members of Ω_s that already appear in the sample dataset $\mathcal{D}$ are deleted. Finally, all the members in Ω_s are combined with the current set of already-generated solutions Ω_a, and the updated Ω_a is used in the next iteration of the BO. The optimization procedure for solving problems (7.31) and (7.32) using BO is summarized in Algorithm 7.4.

7.4.4 EXAMPLE: TWO-BAR TRUSS

Consider a two-bar truss, as shown in Fig. 7.10 [Do et al., 2021a] to demonstrate the feasibility of the BO for solving an MORDO problem with individual probabilistic constraints. See Do et al. [2021a] for examples of joint probabilistic constraints. The truss is subjected to an external load P whose horizontal and vertical components P_x and P_y, respectively, satisfy $P_y = 8P_x$. The two design variables are the cross-sectional area s_1, which is the same for the two members, and the horizontal span s_2 (m), i.e., $\boldsymbol{s} = (s_1, s_2)$. The external load P, the mass density ρ and the yield stress σ_y of the truss material are considered as the uncertain parameters, i.e., $\boldsymbol{\Theta} = (P, \rho, \sigma_\mathrm{y})$. The probability distribution, mean and CV for parameters of $(P, \rho, \sigma_\mathrm{y})$ are (Lognormal,

Algorithm 7.4 Bayesian optimization (BO) procedure.

Require: Generate samples of the design variables $\boldsymbol{s}$ and the uncertain parameters $\boldsymbol{\theta}$ using LHS. Also generate the sample dataset $\mathcal{D}$ by performing the structural analysis for these samples.

1: Based on the generated sample dataset, construct GP models to approximate LSFs of the uncertain objective and probabilistic constraint functions.
2: Sort the Pareto optimal solutions from the set of already-generated candidate solutions Ω_a.
3: Terminate the BO and output the Pareto optimal solutions if one of the following criteria is satisfied: (1) the number of BO iterations reaches an upper limit, which is specified by the user to manage the trade-off between the solution quality and the computational cost for carrying out the BO; (2) the difference of the current HV and that of the previous iteration is less than a small positive value, e.g., 10^{-9}; and (3) the set Ω_s used for maximizing the acquisition function has no feasible solution. Otherwise, go to Step 4.
4: Maximize the acquisition function for each RDO problem using the optimization strategy to obtain the new sampling point $\boldsymbol{x}_n$.
5: Add $\boldsymbol{x}_n$ obtained in Step 4 to the sample dataset, determine the associated structural responses, update the GPMs for LSFs of the uncertain objective and probabilistic constraint functions, and reiterate from Step 2.

Lognormal, Normal), (104kg/m^3, 800kN, 1050MPa) and (0.20, 0.25, 0.24), respectively.

The objective functions to be minimized are the mean and standard deviation of the total mass $f(\boldsymbol{s};\boldsymbol{\theta})$ expressed as

$$f(\boldsymbol{s};\boldsymbol{\Theta}) = 10^{-4}\rho s_1\sqrt{1+s_2^2}. \tag{7.39}$$

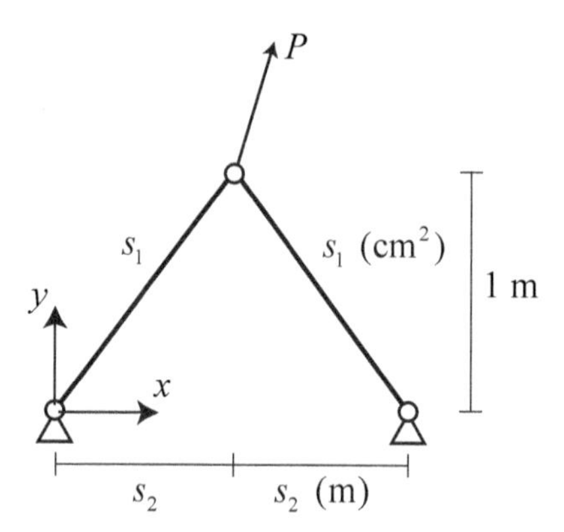

Figure 7.10 Two-bar truss.

Accordingly, $f_1(\boldsymbol{s}) = \mathbb{E}[f(\boldsymbol{s};\boldsymbol{\Theta})]$ and $f_2(\boldsymbol{s}) = \sqrt{\mathrm{Var}[f(\boldsymbol{s};\boldsymbol{\Theta})]}$ in problem (7.31). LSFs of individual probabilistic constraints $g_1(\boldsymbol{s};\boldsymbol{\Theta})$ and $g_2(\boldsymbol{s};\boldsymbol{\Theta})$ represent the axial stresses in two members expressed as

$$g_1(\boldsymbol{s};\boldsymbol{\Theta}) = \frac{5P}{\sqrt{65}s_1\sigma_{\mathrm{y}}}\sqrt{1+s_2^2}\left(8+\frac{1}{s_2}\right) - 1, \tag{7.40a}$$

$$g_2(\boldsymbol{s};\boldsymbol{\Theta}) = \frac{5P}{\sqrt{65}s_1\sigma_{\mathrm{y}}}\sqrt{1+s_2^2}\left(8-\frac{1}{s_2}\right) - 1. \tag{7.40b}$$

The MORDO problem is formulated as follows:

$$\begin{aligned} \text{Minimize} \quad & f_1(\boldsymbol{s}), f_2(\boldsymbol{s}) && (7.41a)\\ \text{subject to} \quad & \Pr\{g_1(\boldsymbol{s};\boldsymbol{\Theta}) \leq 0\} \geq 1-\varepsilon_1, && (7.41b)\\ & \Pr\{g_2(\boldsymbol{s};\boldsymbol{\Theta}) \leq 0\} \geq 1-\varepsilon_2, && (7.41c)\\ & s_1 \in \mathcal{S}_1 = \{1.0, 1.5, \cdots, 20.0\}\,\mathrm{cm}^2, && (7.41d)\\ & s_2 \in \mathcal{S}_2 = \{0.1, 0.2, \cdots, 2.0\}\,\mathrm{m}, && (7.41e) \end{aligned}$$

where $\mathcal{S}_1$ and $\mathcal{S}_2$ are the list of preassigned values of s_1 and s_2, respectively. Note that problem (7.41) has a very simple property that the mean $f_1(\boldsymbol{s})$ can be derived as a linear function of the standard deviation $f_2(\boldsymbol{s})$ as $f_1(\boldsymbol{s}) = \mathbb{E}[\rho]/\sqrt{\mathrm{Var}[\rho]})f_2(\boldsymbol{s}) = 5f_2(\boldsymbol{s})$, because $f(\boldsymbol{s};\boldsymbol{\Theta})$ in (7.39) is a linear function of ρ. Therefore, the set of Pareto optimal solutions to problem (7.41) has only one absolutely optimal solution.

A total of 200 samples are generated for the initial sample dataset, in which only 140 samples are feasible. Then, the DACE toolbox [Lophaven et al., 2002] and a second-degree polynomial mean function are used for constructing three GPMs for $f(\boldsymbol{s};\boldsymbol{\theta})$, $g_1(\boldsymbol{s};\boldsymbol{\theta})$ and $g_2(\boldsymbol{s};\boldsymbol{\theta})$. The BO is then employed to solve problem (7.41) with two risk levels, namely, $\varepsilon_1 = \varepsilon_2 = 0.1$ and $\varepsilon_1 = \varepsilon_2 = 0.05$. Here, the simulated annealing available in Global Optimization Toolbox of MATLAB [Mathworks, 2018a] is used with the default settings. At each iteration of the BO, a total of 200 candidate solutions are generated for the set Ω_s, and the upper bound for BO iterations is 20. The reference point is assigned as $f_{\mathrm{R}} = (50, 10)$ in view of the relation $f_1(\boldsymbol{s}) = 5f_2(\boldsymbol{s})$ and the maximum value of total mass computed from possible combinations of s_1 and s_2 in $\mathcal{S}_1$ and $\mathcal{S}_2$, respectively. Histories of the maximum acquisition function α_2, the corresponding HVI, and the objective functions $f_1(\boldsymbol{s})$ and $f_2(\boldsymbol{s})$ of the BO are shown in Fig. 7.11 for the two risk levels $\varepsilon_1 = \varepsilon_2 = 0.1$ and $\varepsilon_1 = \varepsilon_2 = 0.05$, where the best designs found at the 20th iteration of the BO are $\boldsymbol{s} = (8.5, 0.4)$ and $\boldsymbol{s} = (10, 0.4)$, respectively.

The exact optimal solutions to problem (7.41) are found for verification purpose. Since the sets $\mathcal{S}_1$ and $\mathcal{S}_2$ have 39 and 20 elements, respectively, $39 \times 20 = 780$ designs should be enumerated. Then, 10^5 samples of the uncertain parameters are generated to calculate $f_1(\boldsymbol{s})$, $f_2(\boldsymbol{s})$, $\Pr\{g_1(\boldsymbol{s};\boldsymbol{\Theta}) \leq 0\}$ and $\Pr\{g_2(\boldsymbol{s};\boldsymbol{\Theta}) \leq 0\}$ for each design using MCS. This process is performed three

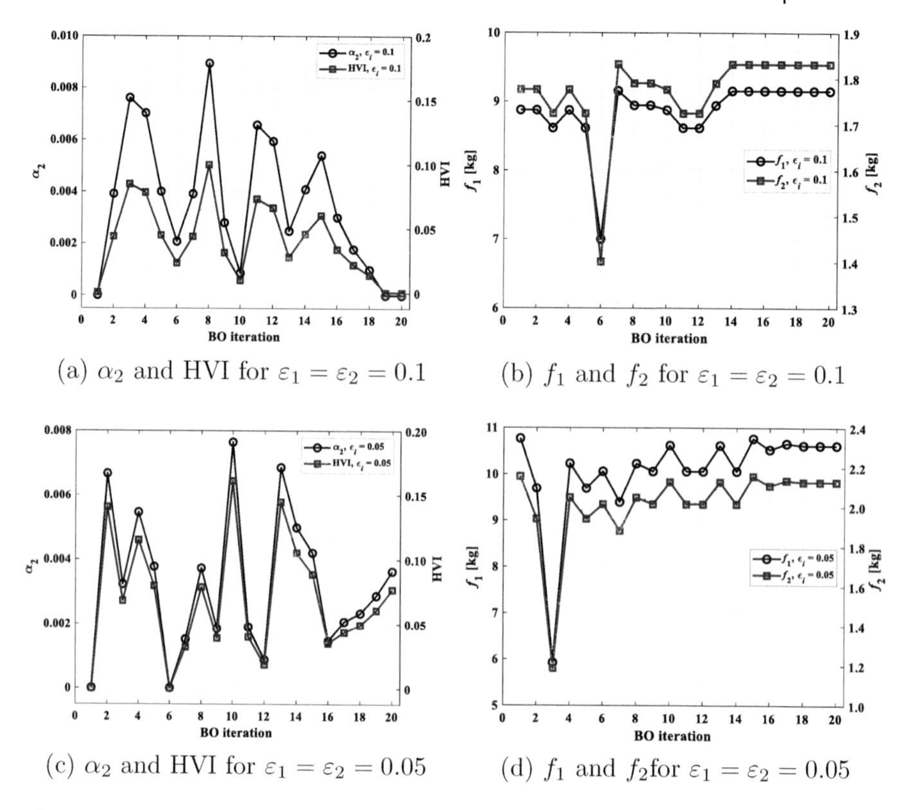

(a) α_2 and HVI for $\varepsilon_1 = \varepsilon_2 = 0.1$

(b) f_1 and f_2 for $\varepsilon_1 = \varepsilon_2 = 0.1$

(c) α_2 and HVI for $\varepsilon_1 = \varepsilon_2 = 0.05$

(d) f_1 and $f_2$for $\varepsilon_1 = \varepsilon_2 = 0.05$

Figure 7.11 Histories of the BO for the two-bar truss for two risk levels.

times for each risk level with different sets of samples. Since the solutions by the three trials are identical, they can be regarded as the exact solutions. It is confirmed from Fig. 7.12 that a good accuracy is observed between the results by the BO and the MCS, although the number of function evaluations used by the BO (1.30122×10^5) is much less than that by enumeration of designs and MCS ($3 \times 780 \times 10^5 = 2.34 \times 10^8$).

7.5 SUMMARY

Stochastic structural optimization using surrogate models have been introduced in this chapter. Since evaluation of structural responses generally demands a large computational cost, BO or EGO based on GPR is one of the most promising approaches for RBDO of real-world structures. The responses can be accurately estimated by GPR, and the samples can be sequentially added to improve the accuracy in the promising region for optimization, or to enhance exploration to find the global optimal solution. MGP can be constructed for clustered samples to improve accuracy while limiting the computational cost within an acceptable range for a large-scale problem. Hyper-

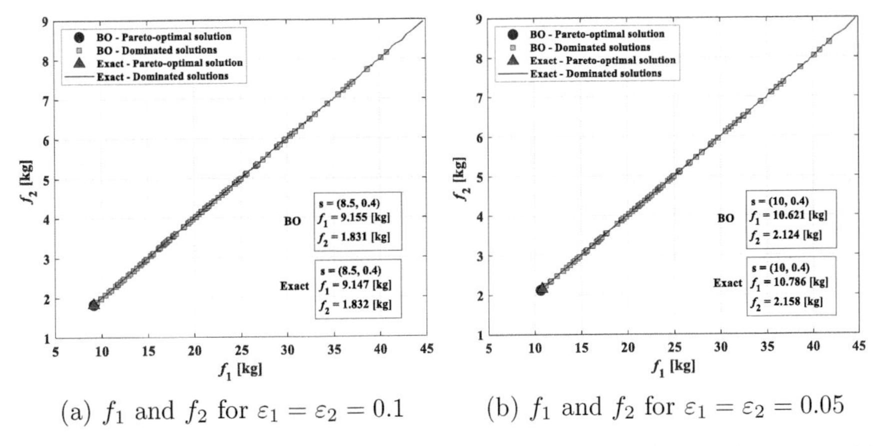

(a) f_1 and f_2 for $\varepsilon_1 = \varepsilon_2 = 0.1$ (b) f_1 and f_2 for $\varepsilon_1 = \varepsilon_2 = 0.05$

Figure 7.12 Verification of the obtained robust designs for the two-bar truss with two risk levels.

volume in the objective function space of an MOO problem is a useful measure for improvement of range and accuracy of the Pareto optimal set. BO can also be applied to multi-objective RBDO together with an approach of sequentially solving deterministic optimization problems.

ACKNOWLEDGMENT

We acknowledge the permissions from publishers for use of the licensed contents as follows:

- Figures 7.5, 7.6 and 7.8, and Tables 7.3 and 7.4: Reproduced with permission from Springer Nature [Do et al., 2021b].
- Figures 7.11 and 7.12: Reprinted from Do et al. [2021a] with permission from Elsevier.

Appendix

A.1 TEST FUNCTIONS

The test functions used in Sec. 3.3 are listed below. Note that N^{opt} is the number of local optima, and x^{L} and x^{U} are the lower and upper bounds, respectively, for the case of continuous variables in Voglis and Lagaris [2009] and Lagaris and Tsoulos [2008] with slight modification of constants:

Ackley's function ($N^{\text{opt}} = 121$):

$$f(\boldsymbol{x}) = 20 + e - 20e^{-0.2\sqrt{a}} - e^{b/2}, \quad a = \frac{1}{2}\sum_{i=1}^{2} x_i^2, \quad b = \sum_{i=1}^{2} \cos(2\pi x_i),$$

$$x^{\text{L}} = -5, \quad x^{\text{U}} = 5$$

Guillin Hill's function ($N^{\text{opt}} = 25$):

$$f(\boldsymbol{x}) = 3 + \sum_{i=1}^{2} \frac{2(x_i + 9)}{x_i + 10} \sin\left(\frac{\pi}{1 - x_i + 0.1}\right),$$

$$x^{\text{L}} = 0, \quad x^{\text{U}} = 1$$

Helder function ($N^{\text{opt}} = 85$):

$$f(\boldsymbol{x}) = -(\cos x_1 \cos x_2) \exp\left(1 - \sqrt{x_1^2 + x_2^2}/\pi\right),$$

$$x^{\text{L}} = -20, \quad x^{\text{U}} = 20$$

Piccioni's function ($N^{\text{opt}} = 28$):

$$f(\boldsymbol{x}) = -10\sin(\pi x_1)^2 - (x_1 - 1)^2(1 + 10\sin \pi x_2) - (x_2 - 1)^2,$$

$$x^{\text{L}} = -5, \quad x^{\text{U}} = 5$$

M0 function ($N^{\text{opt}} = 66$):

$$f(\boldsymbol{x}) = \left[\sin\left(2.2\pi x_1 + \frac{\pi}{2}\right) + \sin\left(\frac{\pi}{2}x_2^2 + \frac{\pi}{2}\right)\right] \frac{(2 - x_2)(3 - x_1)}{4},$$

$$x^{\text{L}} = -5, \quad x^{\text{U}} = 1$$

Test2N function ($N^{\text{opt}} = 2^n$):

$$f(\boldsymbol{x}) = \sum_{i=1}^{n} (x_i^4 - 16x_i^2 + 5x_i),$$

$$x^{\text{L}} = -4, \quad x^{\text{U}} = 4$$

A.2 DISTRIBUTION OF CONTENT RATIO

The inverse cumulative distribution function (cdf) in (4.14) is defined as follows:

$$F_Y^{-1}(u) = \inf\{y : F_Y(y) \geq u\} \quad (0 < u < 1). \tag{A.1}$$

By the right continuity of F_Y, we have $F_Y(F_Y^{-1}(u)) \geq u$ and $F_Y^{-1}(F_Y(y)) \leq y$. Thus, $u \leq F_Y(y)$ holds iff $F_Y^{-1}(u) \leq y$, because

$$u \leq F_Y(y) \Rightarrow F_Y^{-1}(u) \leq F_Y^{-1}(F_Y(y)) \leq y, \tag{A.2}$$

and

$$F_Y^{-1}(u) \leq y \Rightarrow u \leq F_Y(F_Y^{-1}(u)) \leq F_Y(y). \tag{A.3}$$

Hence, for $0 \leq F_Y(y) \leq 1$,

$$\Pr\{Y \leq y\} = F_Y(y) = \Pr\{U \leq F_Y(y)\} = \Pr\{F_Y^{-1}(U) \leq y\}, \tag{A.4}$$

where Y and U are random variables having the cumulative distribution function F_Y and standard uniform distribution, respectively, and $\stackrel{d}{=}$ stands for equality in distribution. This means that

$$Y \stackrel{d}{=} F_Y^{-1}(U). \tag{A.5}$$

Specifically, when F_Y is continuous, we have

$$F_Y(Y) \stackrel{d}{=} U. \tag{A.6}$$

From (A.6), the relation between content ratio and standard uniform distribution

$$F_Y(Y_{k,n}) \stackrel{d}{=} U_{k,n} \tag{A.7}$$

is derived as seen in (4.19), where $Y_{k,n}$ and $U_{k,n}$ are the kth order statistics for n samples on Y and U, respectively.

A.3 TABU SEARCH

Metaheuristics for solving optimization problems are categorized into population based methods such as genetic algorithm and particle swarm optimization, and those based on local search of a single solution such as simulated annealing (SA) and tabu search (TS), which is developed by Glover [1989]. The simplest local search is the deterministic approach used in Sec. 3.3. The random local neighborhood search may also be the simplest approach involving randomness. SA enhances both exploration and exploitation by allowing large variation of solution and allowing update to a non-improving solution in the early stage, and by restricting the size of neighborhood and rejecting non-improving solutions at the final stage.

Algorithm A.1 Algorithm of TS for a minimization problem.

Require: Assign initial solution $\boldsymbol{J}^{(1)}$; initialize tabu list $\mathcal{T} = \varnothing$.
1: **for** $k \leftarrow 1$ to N^{S} **do**
2: Generate N^{b} neighborhood solutions $\boldsymbol{J}_i^{\mathrm{N}}$ $(i = 1, \cdots, N^{\mathrm{b}})$ of $\boldsymbol{J}^{(k)}$.
3: Select the best solution $\boldsymbol{J}^*$ among the neighborhood solutions.
4: **if** $\boldsymbol{J}^* \in \mathcal{T}$ **then**
5: Remove $\boldsymbol{J}^*$ from the neighborhood solutions.
6: **else**
7: Add $\boldsymbol{J}^*$ to $\mathcal{T}$.
8: Remove the oldest solution in $\mathcal{T}$ if the length of the list exceeds N^{t}.
9: **end if**
10: Assign $\boldsymbol{J}^{(k+1)} \leftarrow \boldsymbol{J}^*$.
11: **if** termination conditions satisfied **then**
12: **break**
13: **end if**
14: **end for**
15: Output the best solution.

For a combinatorial problem or a mathematical programming problem with integer variables, the most strict strategy to find a local optimal solution is to enumerate all the neighborhood solutions and move to the best solution among them. However, in this case, a so-called cycling or loop can occur, where a group of neighborhood solutions are selected iteratively during several steps of local search. TS has been developed to prevent this cycling process introducing the *tabu list* containing the prohibited solutions that have already been visited [Glover and Laguna, 1997]. There are several variants of tabu list incorporating different definitions of similarity of solutions.

For a constrained integer programming problem using penalty function approach, the algorithm of TS is shown in Algorithm A.1, where N^{b} is the number of neighborhood solutions that is defined by the types of neighborhood as explained in Sec. 3.3, N^{s} is the upper bound for number of steps, N^{t} is the maximum size of tabu list, and the superscript (k) denotes a value at the kth step.

A.4 FORCE DENSITY METHOD

Force density method (FDM) has been proposed for finding the self-equilibrium shape of tension structures such as cable nets and tensegrity structures [Schek, 1974; Zhang and Ohsaki, 2015], and is equivalently called geometric stiffness method. Using the *force density*, which is the ratio of axial force to length of each member, the nodal coordinates at equilibrium to static nodal loads are obtained by solving a set of linear equations formulated utilizing the so called *force density matrix*. Basic equations of FDM are explained below for a plane truss or a cable net for complementing the methods presented in Chap. 5 for topology optimization of plane frames. In this section, the vectors are assumed to be column vectors.

Let L_i and N_i denote the length and the axial force, respectively, of the ith member of a truss or a cable net consisting of n nodes and m members. The global coordinates on the plane are denoted by (x, y). Let q_i denote the force density of member i defined as

$$q_i = \frac{N_i}{L_i} \quad (1 = 1, \cdots, m). \tag{A.8}$$

The vector $\boldsymbol{q} = (q_1, \cdots, q_m)^\top$ is called force density vector. If member i is connected to nodes j and k, then the components of connectivity matrix, or incidence matrix, $\boldsymbol{C} \in \mathbb{R}^{m \times n}$ is defined as follows:

$$C_{ji} = -1, \quad C_{ki} = 1 \quad (i = 1, \cdots, m). \tag{A.9}$$

By using $\boldsymbol{C}$ and $\boldsymbol{q}$, the force density matrix $\boldsymbol{Q} \in \mathbb{R}^{n \times n}$ is defined as [Zhang and Ohsaki, 2015]

$$\boldsymbol{Q} = \boldsymbol{C}\,\mathrm{diag}(\boldsymbol{q})\boldsymbol{C}^\top \tag{A.10}$$

Note that $\boldsymbol{Q}$ does not change when i and k in (A.9) are exchanged. Accordingly, the diagonal element Q_{jj} of $\boldsymbol{Q}$ is equal to the total amount of force densities of members connected to node j, and off-diagonal element Q_{jk} $(j \neq k)$ is equal to $-q_i$, if node k is connected to node j by member i; otherwise, $Q_{jk} = 0$.

Let $\boldsymbol{x} \in \mathbb{R}^n$ and $\boldsymbol{y} \in \mathbb{R}^n$ denote the vectors of x- and y-coordinates, respectively, of all nodes including the supports. The nodal load vectors including the reactions in x- and y-directions are denoted by $\boldsymbol{p}^x \in \mathbb{R}^n$ and $\boldsymbol{p}^y \in \mathbb{R}^n$, respectively. Using the force density matrix, the equilibrium equations in x- and y-directions have the following same form:

$$\boldsymbol{Q}\boldsymbol{x} = \boldsymbol{p}^x, \quad \boldsymbol{Q}\boldsymbol{y} = \boldsymbol{p}^y \tag{A.11}$$

See, e.g., Schek [1974], Tibert and Pellegrino [2003] and Zhang and Ohsaki [2006] for details.

The nodes are re-ordered so that the components of free nodes precede those of fixed nodes as

$$\boldsymbol{x} = \begin{pmatrix} \boldsymbol{x}_{\text{free}} \\ \boldsymbol{x}_{\text{fix}} \end{pmatrix}, \quad \boldsymbol{y} = \begin{pmatrix} \boldsymbol{y}_{\text{free}} \\ \boldsymbol{y}_{\text{fix}} \end{pmatrix}. \tag{A.12}$$

The force density matrices in two directions are combined to formulate $\tilde{\boldsymbol{Q}} \in \mathbb{R}^{2n \times 2n}$ as

$$\tilde{\boldsymbol{Q}} = \left(\begin{array}{cc|cc} \boldsymbol{Q}^x_{\text{free}} & \boldsymbol{0} & \boldsymbol{Q}^x_{\text{link}} & \boldsymbol{0} \\ \boldsymbol{0} & \boldsymbol{Q}^y_{\text{free}} & \boldsymbol{0} & \boldsymbol{Q}^y_{\text{link}} \\ \hline \boldsymbol{Q}^{x\top}_{\text{link}} & \boldsymbol{0} & \boldsymbol{Q}^x_{\text{fix}} & \boldsymbol{0} \\ \boldsymbol{0} & \boldsymbol{Q}^{y\top}_{\text{link}} & \boldsymbol{0} & \boldsymbol{Q}^y_{\text{fix}} \end{array}\right) \begin{array}{l} \}\, n_{\text{free}} \\ \\ \}\, n_{\text{fix}} \end{array}, \tag{A.13}$$

where the first two column blocks span n_{free} columns and the last two span n_{fix} columns; n_{free} and n_{fix} are the numbers of free and fixed degrees of freedom, respectively, satisfying $n_{\text{free}} + n_{\text{fix}} = 2n$.

The matrices $(\boldsymbol{Q}^x_{\text{free}}, \boldsymbol{Q}^y_{\text{free}})$, $(\boldsymbol{Q}^x_{\text{fix}}, \boldsymbol{Q}^y_{\text{fix}})$ and $(\boldsymbol{Q}^x_{\text{link}}, \boldsymbol{Q}^y_{\text{link}})$ are rearranged to the matrices $\tilde{\boldsymbol{Q}}_{\text{free}} \in \mathbb{R}^{n_{\text{free}} \times n_{\text{free}}}$, $\tilde{\boldsymbol{Q}}_{\text{fix}} \in \mathbb{R}^{n_{\text{fix}} \times n_{\text{fix}}}$ and $\tilde{\boldsymbol{Q}}_{\text{link}} \in \mathbb{R}^{n_{\text{free}} \times n_{\text{fix}}}$, respectively, and $\tilde{\boldsymbol{Q}}$ in (A.13) is rewritten as

$$\tilde{\boldsymbol{Q}} = \begin{pmatrix} \tilde{\boldsymbol{Q}}_{\text{free}} & \tilde{\boldsymbol{Q}}_{\text{link}} \\ \tilde{\boldsymbol{Q}}^\top_{\text{link}} & \tilde{\boldsymbol{Q}}_{\text{fix}} \end{pmatrix}. \tag{A.14}$$

The equilibrium equations in x- and y-directions in (A.11) are combined as

$$\begin{pmatrix} \tilde{\boldsymbol{Q}}_{\text{free}} & \tilde{\boldsymbol{Q}}_{\text{link}} \\ \tilde{\boldsymbol{Q}}^\top_{\text{link}} & \tilde{\boldsymbol{Q}}_{\text{fix}} \end{pmatrix} \begin{pmatrix} \boldsymbol{X}_{\text{free}} \\ \boldsymbol{X}_{\text{fix}} \end{pmatrix} = \begin{pmatrix} \boldsymbol{P}_{\text{free}} \\ \boldsymbol{P}_{\text{fix}} \end{pmatrix}, \tag{A.15}$$

where

$$\begin{aligned} \boldsymbol{X}_{\text{free}} &= (\boldsymbol{x}^\top_{\text{free}}, \boldsymbol{y}^\top_{\text{free}})^\top, \\ \boldsymbol{X}_{\text{fix}} &= (\boldsymbol{x}^\top_{\text{fix}}, \boldsymbol{y}^\top_{\text{fix}})^\top, \end{aligned} \tag{A.16}$$

and $\boldsymbol{P}_{\text{free}} \in \mathbb{R}^{n_{\text{free}}}$ and $\boldsymbol{P}_{\text{fix}} \in \mathbb{R}^{n_{\text{fix}}}$ are the nodal load vector and reaction force vector corresponding to $\boldsymbol{X}_{\text{free}}$ and $\boldsymbol{X}_{\text{fix}}$, respectively.

If the force densities of all members and the locations of fixed nodes (supports) are assigned, then the locations of free nodes are obtained from the following set of linear equations, which is derived from (A.15):

$$\tilde{\boldsymbol{Q}}_{\text{free}} \boldsymbol{X}_{\text{free}} = \boldsymbol{P}_{\text{free}} - \tilde{\boldsymbol{Q}}_{\text{link}} \boldsymbol{X}_{\text{fix}}. \tag{A.17}$$

Therefore, $\boldsymbol{X}_{\text{free}}$ is considered as function of $\boldsymbol{q}$.

A.5 CLUSTERING DATASET USING GAUSSIAN MIXTURE MODEL

Accuracy and computational efficiency of surrogate models may be improved by clustering the sample dataset into subsets sharing similar properties. The procedure of clustering consists of the measuring of similarity of the samples and selecting a clustering algorithm [Xu and Wunsch, 2005]. Here, we assume that two training samples are similar if they emerge from the same probability distribution function (pdf).

Let $\boldsymbol{x} \in \mathbb{R}^d$ and y denote the input variable vector and the corresponding output value, respectively, that are combined into a sample data $(\boldsymbol{x}, y)$. The vector is assumed to be column vector also in this section. The joint pdf of the sample $p(\boldsymbol{x}, y)$ is split into different Gaussian components using the Gaussian mixture model (GMM) [Hastie et al., 2009]. Each pair of Gaussian components is assigned to a subset, and the samples are distributed into subsets. Let $\phi(\boldsymbol{x}, y \mid \boldsymbol{\mu}_k, \boldsymbol{\Sigma}_k) \sim \mathcal{N}_{d+1}(\boldsymbol{\mu}_k, \boldsymbol{\Sigma}_k)$ denote the kth $(d+1)$-variate Gaussian. Its mean vector and covariance matrix are denoted by $\boldsymbol{\mu}_k$ and $\boldsymbol{\Sigma}_k$, respectively. The GMM describes the joint pdf $p(\boldsymbol{x}, y)$ by a convex combination of K Gaussians as follows:

$$p(\boldsymbol{x}, y|\boldsymbol{\psi}) = \sum_{k=1}^{K} \pi_k \phi(\boldsymbol{x}, y \mid \boldsymbol{\mu}_k, \boldsymbol{\Sigma}_k), \tag{A.18}$$

$$\sum_{k=1}^{K} \pi_k = 1, \quad 0 \le \pi_k \le 1, \tag{A.19}$$

$$\boldsymbol{\mu}_k = (\boldsymbol{\mu}_{\boldsymbol{X},k}^\top, \mu_{y,k})^\top, \tag{A.20}$$

$$\boldsymbol{\Sigma}_k = \begin{pmatrix} \boldsymbol{\Sigma}_{\boldsymbol{XX},k} & \boldsymbol{\Sigma}_{\boldsymbol{X}y,k} \\ \boldsymbol{\Sigma}_{y\boldsymbol{X},k} & \boldsymbol{\Sigma}_{yy,k} \end{pmatrix}, \tag{A.21}$$

where π_k is the mixing proportion of Gaussians, and the unknown parameters $\pi_k, \boldsymbol{\mu}_k$ and $\boldsymbol{\Sigma}_k$ are combined to a set $\boldsymbol{\psi} = \{\pi_k, \boldsymbol{\mu}_k, \boldsymbol{\Sigma}_k\}_{k=1}^K$.

Let $\boldsymbol{Z} = (Z_1, \ldots, Z_N)^\top$ denote a latent random vector consisting of the index of subset in which each of N samples belongs to; i.e., $Z_i \in \{1, \ldots, K\}$. The probability of the ith sample $(\boldsymbol{x}_i, y_i)$ belonging to the kth Gaussian (or the kth subset) is denoted by Z_{ik} as

$$Z_{ik} = \begin{cases} 1 & \text{if } Z_i = k \\ 0 & \text{otherwise.} \end{cases} \tag{A.22}$$

Although Z_i is unknown in advance, the expectation of Z_{ik}, denoted by $\mathbb{E}[Z_{ik}]$, can be determined by using Bayes' rule as follows:

$$\begin{aligned} \mathbb{E}[Z_{ik}] &= \Pr\{Z_i = k \mid \boldsymbol{x}_i, y_i, \boldsymbol{\psi}\} \\ &= \frac{p(Z_i = k|\boldsymbol{\psi})\, p(\boldsymbol{x}_i, y_i \mid Z_i = k, \boldsymbol{\psi})}{\sum_{h=1}^{K} p(Z_i = h \mid \boldsymbol{\psi})\, p(\boldsymbol{x}_i, y_i \mid Z_i = h, \boldsymbol{\psi})}, \end{aligned} \tag{A.23}$$

where $p(Z_i = k \mid \boldsymbol{\psi}) = \pi_k$ is the prior and $p(\boldsymbol{x}_i, y_i \mid Z_i = k, \boldsymbol{\psi}) = \phi(\boldsymbol{x}_i, y_i \mid \boldsymbol{\mu}_k, \boldsymbol{\Sigma}_k)$ is the likelihood. Thus, (A.23) can be rewritten as

$$\begin{aligned} \mathbb{E}[Z_{ik}] &= \Pr\{Z_i = k \mid \boldsymbol{x}_i, y_i, \boldsymbol{\psi}\} \\ &= \frac{\pi_k \phi(\boldsymbol{x}_i, y_i \mid \boldsymbol{\mu}_k, \boldsymbol{\Sigma}_k)}{\sum_{h=1}^{K} \pi_h \phi(\boldsymbol{x}_i, y_i \mid \boldsymbol{\mu}_h, \boldsymbol{\Sigma}_h)}. \end{aligned} \tag{A.24}$$

The parameter set $\boldsymbol{\psi}$ is determined by maximizing the log-likelihood $\mathcal{L}_c$ of the sample dataset, defined as follows, using the iterative expectation-maximization (EM) algorithm [Hastie et al., 2009]:

$$\mathcal{L}_c = \sum_{i=1}^{N} \sum_{k=1}^{K} \mathbb{E}[Z_{ik}] \log \left(\pi_k \phi(\boldsymbol{x}_i, y_i \mid \boldsymbol{\mu}_k, \boldsymbol{\Sigma}_k) \right). \tag{A.25}$$

Let the superscript (t) denote the value of the tth step of the EM algorithm, which consists of the following steps:

1. Assign the number of Gaussians (subsets) K, and initialize the mixing proportions $\pi_k^{(0)}$ uniformly. Initialize the mean vector $\boldsymbol{\mu}_k^{(0)}$ by a randomly selected vector from the training set. Initialize the variance matrix $\boldsymbol{\Sigma}_k^{(0)}$ as a diagonal matrix whose ith diagonal element is the variance of the ith input variable among the initial dataset. Assemble $\pi_k^{(0)}$, $\boldsymbol{\mu}_k^{(0)}$ and $\boldsymbol{\Sigma}_k^{(0)}$ into the initial parameter set as $\boldsymbol{\psi}^{(0)} = \{\pi_k^{(0)}, \boldsymbol{\mu}_k^{(0)}, \boldsymbol{\Sigma}_k^{(0)}\}_{k=1}^{K}$.

2. **E Step:** Compute $\mathbb{E}[Z_{ik}^{(t)}]$ defined as

$$\mathbb{E}[Z_{ik}^{(t)}] = \frac{\pi_k^{(t)} \phi(\boldsymbol{x}_i, y_i \mid \boldsymbol{\mu}_k^{(t)}, \boldsymbol{\Sigma}_k^{(t)})}{\sum_{h=1}^{K} \pi_h^{(t)} \phi(\boldsymbol{x}_i, y_i \mid \boldsymbol{\mu}_h^{(t)}, \boldsymbol{\Sigma}_h^{(t)})}. \tag{A.26}$$

3. **M Step:** Maximize $\mathcal{L}_c$ in (A.25) with respect to $\boldsymbol{\psi}$ to obtaining a new parameter set $\boldsymbol{\psi}^{(t)}$. Accordingly, update the components of $\boldsymbol{\psi}$ as

$$\pi_k^{(t+1)} = \frac{1}{N} \sum_{i=1}^{N} \mathbb{E}[Z_{ik}^{(t)}], \tag{A.27a}$$

$$\boldsymbol{\mu}_k^{(t+1)} = \frac{\sum_{i=1}^{N} \mathbb{E}[Z_{ik}^{(t)}] \boldsymbol{d}_i}{\sum_{i=1}^{N} \mathbb{E}[Z_{ik}^{(t)}]}, \tag{A.27b}$$

$$\boldsymbol{\Sigma}_k^{(t+1)} = \frac{\sum_{i=1}^{N} \mathbb{E}[Z_{ik}^{(t)}] \left(\boldsymbol{d}_i - \boldsymbol{\mu}_k^{(t+1)} \right) \left(\boldsymbol{d}_i - \boldsymbol{\mu}_k^{(t+1)} \right)^\top}{\sum_{i=1}^{N} \mathbb{E}[Z_{ik}^{(t)}]}, \tag{A.27c}$$

where $\boldsymbol{d}_i = (\boldsymbol{x}_i^\top, y_i)^\top$.

4. Move to the next iteration with the new parameter set $\boldsymbol{\psi}^{(t)}$.

The algorithm is assured to converge at a finite number of iterations because it always reduces $\mathcal{L}_c$ at each iteration. See Hastie et al. [2009] for details. After $\boldsymbol{\psi}$ is updated, the probability of the ith sample $(\boldsymbol{x}_i, y_i)$ belonging to the kth Gaussian, i.e., $\mathbb{E}[Z_{ik}]$ $(k = 1, \cdots, K)$, is computed using (A.24). The index k corresponding to the maximum value among these K values of Z_{ik} indicates the Gaussian (or subset) that the ith sample $(\boldsymbol{x}_i, y_i)$ belongs to. By projecting all subsets in $(d+1)$-dimensional space onto the d-dimensional space of input variables, the probability that an input variable $\boldsymbol{x}_i$ belongs to the kth subset can be determined (without knowing the corresponding output variable y) as follows:

$$\Pr\{Z_i = k \mid \boldsymbol{x}, \boldsymbol{\psi}\} = \frac{\pi_k \phi(\boldsymbol{x} \mid \boldsymbol{\mu}_{\boldsymbol{X},k}, \boldsymbol{\Sigma}_{\boldsymbol{XX},k})}{\sum_{h=1}^{K} \pi_h \phi(\boldsymbol{x} \mid \boldsymbol{\mu}_{\boldsymbol{X},h}, \Sigma_{\boldsymbol{XX},h})}, \tag{A.28}$$

where $\boldsymbol{\mu}_{\boldsymbol{X},k}$ and $\Sigma_{\boldsymbol{XX},k}$ are derived from (A.20) and (A.21), respectively.

Detailed discussions on criteria for selecting the best models of GMMs, including determination of the hyperparameters such as the number of Gaussians K can be found in Mclachlan and Rathnayake [2014]. In Sec. 7.3, the Bayesian information criterion (BIC) is chosen because its effectiveness has been confirmed in many studies in the field of the statistical analysis [Mclachlan and Rathnayake, 2014]. We simply increase K step by step from 1 to 50, and find the best number that minimizes BIC defined as [Hastie et al., 2009]

$$\mathrm{BIC} = -\mathcal{L} + \frac{1}{2} n_{\mathrm{p}} \log N, \tag{A.29}$$

where $\mathcal{L} = \sum_{i=1}^{N} \log \left(\sum_{k=1}^{K} \pi_k \phi(\boldsymbol{x}_i, y_i \mid \boldsymbol{\mu}_k, \boldsymbol{\Sigma}_k) \right)$ is the log-likelihood of the sample set, and n_{p} is the number of free parameters required for determining K Gaussians.

A.6 SADDLEPOINT APPROXIMATION

Saddlepoint approximation is an application of general saddlepoint method to the field of probability or statistical theory [Butler, 2007]. It approximates pdf of a function using a cumulant generating function. Let $\boldsymbol{\mu}_x = (\mu_{x_1}, \cdots, \mu_{x_d})$ denote the mean vector of the d random input variables $\boldsymbol{x} = (x_1, \cdots, x_d)$. The first three cumulants of the random variable $y = \hat{g}(\boldsymbol{x})$ are denoted by μ_y, $\mu_{2,y}$ and $\mu_{3,y}$, which are estimated using the second-order Taylor series expansion with respect to $\boldsymbol{x}$ at the mean $\boldsymbol{\mu}_x$ as follows [Anderson and Mattson, 2012]:

$$\mu_y \approx \hat{g}(\boldsymbol{\mu}_x) + \frac{1}{2} \sum_{i=1}^{d} \left. \frac{\partial^2 \hat{g}}{\partial x_i^2} \right|_{\boldsymbol{\mu}_x} \mu_{2,x_i}, \tag{A.30}$$

$$\begin{aligned} \mu_{2,y} \approx \sum_{i=1}^{d} \left(\left. \frac{\partial \hat{g}}{\partial x_i} \right|_{\boldsymbol{\mu}_x} \right)^2 \mu_{2,x_i} &+ \sum_{i=1}^{d} \left. \frac{\partial \hat{g}}{\partial x_i} \right|_{\boldsymbol{\mu}_x} \left. \frac{\partial^2 \hat{g}}{\partial x_i^2} \right|_{\boldsymbol{\mu}_x} \mu_{3,x_i} \\ &+ \frac{1}{4} \sum_{i=1}^{d} \left(\left. \frac{\partial^2 \hat{g}}{\partial x_i^2} \right|_{\boldsymbol{\mu}_x} \right)^2 (\mu_{4,x_i} - \mu_{2,x_i}^2), \end{aligned} \tag{A.31}$$

$$\mu_{3,y} \approx \sum_{i=1}^{d} \left(\left. \frac{\partial \hat{g}}{\partial x_i} \right|_{\boldsymbol{\mu}_x} \right)^3 \mu_{3,x_i} + \frac{3}{2} \sum_{i=1}^{d} \left(\left. \frac{\partial \hat{g}}{\partial x_i} \right|_{\boldsymbol{\mu}_x} \right)^2 \left. \frac{\partial^2 \hat{g}}{\partial x_i^2} \right| (\mu_{4,x_i} - \mu_{2,x_i}^2)$$
$$+ \sum_{i=1}^{d} \left. \frac{\partial \hat{g}}{\partial x_i} \right| \left(\left. \frac{\partial^2 \hat{g}}{\partial x_i^2} \right|_{\boldsymbol{\mu}_x} \right)^2 \left(\frac{3}{4} \mu_{5,x_i} - \frac{3}{2} \mu_{2,x_i} \mu_{3,x_i} \right)$$
$$+ \sum_{i=1}^{d} \left(\left. \frac{\partial^2 \hat{g}}{\partial x_i^2} \right|_{\boldsymbol{\mu}_x} \right)^3 \left(\frac{1}{4} \mu_{2,x_i}^3 - \frac{3}{8} \mu_{2,x_i} \mu_{4,x_i} + \frac{1}{8} \mu_{6,x_i} \right), \quad (A.32)$$

where μ_{k,x_i} $(k = 2, \cdots, 6)$ is the kth central moment of x_i.

Let K_y, $K_y^{(1)}$, $K_y^{(2)}$ and $K_y^{(3)}$ denote the cumulant generating function of y and its first, second and third derivatives, respectively. Do et al. [2021b] proposed the following form of K_y:

$$K_y(\xi) = (\mu_y - 2ab)\xi + \frac{1}{2}(\mu_{2,y} - 2ab^2)\xi^2 - 2a \log(1 - b\xi)^2,$$

where ξ is a real variable, and a and b are unknown parameters which are determined, as follows, from the relation between K_y and the first three cumulants of y, i.e., $K_y^{(1)}(0) = \mu_y$, $K_y^{(2)}(0) = \mu_{2,y}$ and $K_y^{(3)}(0) = \mu_{3,y}$, and the unique root condition for the saddlepoint equation:

$$a = \frac{2\mu_{2,y}^3}{\mu_{3,y}^2}, \quad b = \frac{\mu_{3,y}}{2\mu_{2,y}}.$$

If $\mu_{3,y} = 0$, then $b = 0$ is satisfied and $a = 0$ is assumed. Furthermore, the second cumulant $\mu_{2,y}$ is always positive. The pdf of the random variable $y = \hat{g}(x)$, denoted as $p_y(y)$, can be estimated as [Butler, 2007]

$$p_y(y) \approx \frac{\exp\left(K_y(\xi_s) - \xi_s y\right)}{\sqrt{2\pi K_y^{(2)}(\xi_s)}},$$

where ξ_s is called the saddlepoint that is the unique root to the following saddlepoint equation:

$$K_y^{(1)}(\xi) = y.$$

The probability for $\hat{g}(\boldsymbol{x}) \leq y$ is determined as [Butler, 2007]

$$\Pr\{\hat{g}(x) \leq y\} \approx \Phi(r) + \phi(r) \left(\frac{1}{r} - \frac{1}{v} \right),$$

where Φ and ϕ are the cdf and pdf, respectively, of the standard normal distribution, and r and v are given as

$$r = \text{sign}(\xi_s)\sqrt{2\left(\xi_s y - K_y(\xi_s)\right)},$$

$$v = \xi_s\sqrt{K_y^{(2)}(\xi_s)}.$$

Note that $\text{sign}(\xi_s) = 1, -1$ and 0 correspond to $\xi_s > 0$, $\xi_s < 0$ and $\xi_s = 0$, respectively. When either r or v does not exist due to negative value in the square root, $\Pr\{\hat{g}(x) \leq y\} = 0$ if $y \leq \mu_y$ and $\Pr\{\hat{g}(x) \leq y\} = 1$ if $y \geq \mu_y$. More details are shown in Do et al. [2021b].

A.7 MULTI-OBJECTIVE OPTIMIZATION PROBLEM

In the process of structural design, designers should generally consider several performance measures and cost functions which have trade-off relations. Therefore, a design problem usually turns out to be a multi-objective optimization (MOO) problem. The formulation of MOO problem also arises when the so called soft constraints, which are preferable to be satisfied, are considered and the trade-off between the objective and constraint function values is investigated. In Chap. 6, an MOO problem is solved for obtaining robust designs corresponding to several robustness levels. The solution method of MOO problem is called multi-objective programming which has been extensively studied since 1970s also in the field of structural optimization [Cohon, 1978; Stadler, 1979, 1988; Marler and Arora, 2004].

The MOO problem of minimizing the n^{f} objective functions $\boldsymbol{f}(\boldsymbol{x}) = (f_1(\boldsymbol{x}), \cdots, f_{n^{\mathrm{f}}}(\boldsymbol{x}))$ of the variable vector $\boldsymbol{x} \in \mathbb{R}^m$ is formulated as

$$\text{Minimize} \quad \boldsymbol{f}(\boldsymbol{x}) \tag{A.33a}$$

$$\text{subject to} \quad g_i(\boldsymbol{x}) \leq 0 \quad (i = 1, \ldots, n^{\mathrm{I}}), \tag{A.33b}$$

where only inequality constraints are considered for brevity. Since the absolutely optimal solution, which minimizes all the objective functions simultaneously, does not generally exist, the purpose of solving an MOO problem is to find a solution, in which at least one objective value increases if one of the remaining objective values is decreased. This solution is called Pareto optimal solution, nondominated solution, noninferior solution or compromise solution, which is formally defined as follows based on the dominance of solution:

Definition A.1 *For two feasible solutions $\boldsymbol{x}^{(1)}$ and $\boldsymbol{x}^{(2)}$ satisfying the constraints* (A.33b), *if $f_i(\boldsymbol{x}^{(1)}) \leq f_i(\boldsymbol{x}^{(2)})$ for all $i = 1, \ldots, n^{\mathrm{f}}$ and $f_j(\boldsymbol{x}^{(1)}) < f_j(\boldsymbol{x}^{(2)})$ for one of $j \in \{1, \ldots, n^{\mathrm{f}}\}$, then $\boldsymbol{x}^{(2)}$ is said to be dominated by $\boldsymbol{x}^{(1)}$, which is denoted as $\boldsymbol{x}^{(1)} \preccurlyeq \boldsymbol{x}^{(2)}$. If there is no solution that dominates a feasible solution $\boldsymbol{x}^*$, then $\boldsymbol{x}^*$ is a Pareto optimal solution, and the set of such solutions is called* Pareto front *in the objective function space.*

In the general decision process including structural design, a single solution should be selected at the final stage. Therefore, the best solution called most preferred solution is selected from the set of Pareto optimal solutions in view of another performance measure that has not been used in the formulation of MOO problem. The process of selecting the most preferred solution is summarized as follows [Marler and Arora, 2004; Ohsaki, 2010]:

1. *Approach with a priori information*
   The MOO problem is reformulated to a single-objective problem to find the best solution. For example, the linear weighted sum approach combines the objective functions to a single function, and the constraint approach moves the $n^{\mathrm{f}} - 1$ objective functions to the inequality constraints with appropriate upper bounds, leaving only one objective function. The goal programming minimizes the distance from the ideal point in the objective function space. In these approaches, prior information is needed for the appropriate values of weights, bounds and the ideal values of the objective functions.

2. *Approach without a priori information*

   (a) Enumeration of Pareto optimal solutions:
   When appropriate prior information is not available, it is natural to generate many Pareto optimal solutions and select the most preferred solution from the set of Pareto optimal solutions in view of additional (*a posteriori*) information. Owing to the recent development of computer hardware and algorithms for MOO problem, this approach is most popular for solving MOO problems in structural engineering.

   (b) Interactive approach:
   If a tentative information of preference is available, e.g., for a set of weight coefficients and the ideal objective function values, then those values are first specified and the MOO problem is converted to a single objective problem. After obtaining a single Pareto optimal solution by solving the single objective problem, the preference information is conceived as a parameter that is to be modified interactively to obtain a more preferred solution.

The most popular approach is to use multi-objective genetic algorithm (MOGA) such as NSGA-II [Deb et al., 2002] to generate a set of Pareto optimal solutions. It utilizes the Pareto ranking for the performance measure of each solution. The rank is defined as the number of solutions that dominates each solution, e.g., the rank of a Pareto optimal solution is 0.

Bibliography

Achtziger, W. (2007). On simultaneous optimization of truss geometry and topology. *Structural and Multidisciplinary Optimization 33*, 285–304.

Afzal, A., K. Y. Kim, and J. W. Seo (2017). Effects of Latin hypercube sampling on surrogate modeling and optimization. *International Journal of Fluid Machinery and Systems 10*(3), 240–253.

Anderson, T. V. and C. A. Mattson (2012). Propagating skewness and kurtosis through engineering models for low-cost, meaningful, nondeterministic design. *Journal of Mechanical Design, Transactions of the ASME 134*(10), 100911.

Andreassen, E., A. Clausen, M. Schevenels, B. S. Lazarov, and O. Sigmund (2011). Efficient topology optimization in MATLAB using 88 lines of code. *Structural and Multidisciplinary Optimization 43*(1), 1–16.

Aoues, Y. and A. Chateauneuf (2010). Benchmark study of numerical methods for reliability-based design optimization. *Structural and Multidisciplinary Optimization 41*(2), 277–294.

Apostol, T. M. (1974). *Mathematical Analysis, Second Edition*, Addison-Wesley series in mathematics, Volume 9. Addison Wesley.

Arnold, B. C., N. Balakrishnan, and H. N. H. N. Nagaraja (2008). *A First Course in Order Statistics.* Philadelphia: Society for Industrial and Applied Mathematics.

Audet, C., V. Béchard, and S. Le Digabel (2008). Nonsmooth optimization through mesh adaptive direct search and variable neighborhood search. *Journal of Global Optimization 41*(2), 299–318.

Augusti, G., A. Barratta, and F. Casciati (1984). *Probabilistic Methods in Structural Engineering.* London: Chapman and Hall.

Augustin, T. and R. Hable (2010). On the impact of robust statistics on imprecise probability models: A review. *Structural Safety 32*(6), 358–365.

Avérous, J., C. Genest, and S. C. Kochar (2005). On the dependence structure of order statistics. *Journal of Multivariate Analysis 94*(1), 159–171.

Baptista, R. and M. Poloczek (2018). Bayesian optimization of combinatorial structures. In *35th International Conference on Machine Learning, ICML 2018*, Volume 2, 782–796.

Beck, A. T., W. J. Gomes, R. H. Lopez, and L. F. Miguel (2015). A comparison between robust and risk-based optimization under uncertainty. *Structural and Multidisciplinary Optimization 52*(3), 479–492.

Bellman, R. (1957). *Dynamic Programming: A Rand Corporation Research Study*. Princeton Unversity Press.

Ben-Haim, Y. (1994). Convex models of uncertainty: Applications and implications. *Erkenntnis 41*(2), 139–156.

Ben-Haim, Y., G. Chen, and T. Soong (1996). Maximum structural response using convex model. *ASCE Journal of Engineering Mechanics 122*(4), 325–333.

Ben-Haim, Y. and I. Elishakoff (1990). *Convex Models of Uncertainty in Applied Mechanics*. Amsterdam: Elsevier.

Ben-Tal, A., L. E. Ghaoui, and A. Nemirovski (2009). *Robust Optimization*. Princeton University Press.

Bendsøe, M. P. (1989). Optimal shape design as a material distribution problem. *Structural Optimization 1*, 193–202.

Bendsøe, M. P. and O. Sigmund (2003). *Topology Optimization: Theory, Methods and Applications*. Berlin: Springer.

Berry, S. M. and J. B. Kadane (1997). Optimal bayesian randomization. *Journal of the Royal Statistical Society. Series B: Statistical Methodology 59*(4), 813–819.

Beyer, H. G. and B. Sendhoff (2007). Robust optimization: A comprehensive survey. *Computer Methods in Applied Mechanics and Engineering 196*(33-34), 3190–3218.

Bishop, C. M. (2006). *Pattern Recognition and Machine Learning*. New York: Springer New York.

Blum, C. and A. Roli (2003). Metaheuristics in combinatorial optimization. *ACM Computing Surveys (CSUR) 35*(3), 268–308.

Boender, C. G. and A. H. Rinnooy Kan (1987). Bayesian stopping rules for multistart global optimization methods. *Mathematical Programming 37*(1), 59–80.

Boland, P. J., M. Hollander, K. Joag-Dev, and S. Kochar (1996). Bivariate dependence properties of order statistics. *Journal of Multivariate Analysis 56*(1), 75–89.

Bourdin, B. (2001). Filters in topology optimization. *International Journal for Numerical Methods in Engineering 50*(9), 2143–2158.

Bruns, T. E. and D. A. Tortorelli (2001). Topology optimization of non-linear structures and compliant mechanisms. *Computer Methods in Applied Mechanics and Engineering 190*, 3443–3459.

Buot, M. (2006). Probability and computing: randomized algorithms and probabilistic analysis. *Journal of the American Statistical Association 101*(473), 395–396.

Butler, R. W. (2007). *Saddlepoint Approximations with Applications.* Cambridge: Cambridge University Press.

Caflisch, R. E. (1998). Monte Carlo and quasi-Monte Carlo methods. *Acta Numerica 7*, 1–49.

Cao, Y. (2008). Hypervolume Indicator. *File Exchange: MATLAB Central.*

Chen, X., T. K. Hasselman, and D. J. Neill (1997). Reliability based structural design optimization for practical applications. In *Collection of Technical Papers - AIAA/ASME/ASCE/AHS/ASC Structures, Structural Dynamics and Materials Conference*, Volume 4, 2724–2732.

Chen, Z., H. Qiu, L. Gao, and P. Li (2013). An optimal shifting vector approach for efficient probabilistic design. *Structural and Multidisciplinary Optimization 47*(6), 905–920.

Cheng, G. and X. Guo (1997). ε-relaxed approach in structural topology optimization. *Structural Optimization 13*, 258–266.

Choi, S. K., R. A. Canfield, and R. V. Grandhi (2007). *Reliability-Based Structural Design.* London: Springer London.

Clarke, F. H. (1990). *Optimization and Nonsmooth Analysis.* Philadelphia: Society for Industrial and Applied Mathematics.

Cohon, J. L. (1978). *Multiobjective Programming and Planning.* Mathematics in Science and Engineering, 140. New York: Academic Press.

Colson, B., P. Marcotte, and G. Savard (2007). An overview of bilevel optimization. *Annals of Operations Research 153*(1), 235–256.

Couckuyt, I., D. Deschrijver, and T. Dhaene (2014). Fast calculation of multiobjective probability of improvement and expected improvement criteria for Pareto optimization. *Journal of Global Optimization 60*(3), 575–594.

David, H. A. and H. N. Nagaraja (2003). *Order Statistics.* New Jersey: John Wiley.

Davis, C. (1954). Theory of positive linear dependence. *American Journal of Mathematics 76*(4), 733.

De, S., J. Hampton, K. Maute, and A. Doostan (2019). Topology optimization under uncertainty using a stochastic gradient-based approach. *Structural and Multidisciplinary Optimization 62*(5), 2255–2278.

Deb, K. (2001). *Multi-objective Optimization using Evolutionary Algorithms.* New York: John Wiley & Sons.

Deb, K., A. Pratap, S. Agarwal, and T. Meyarivan (2002). A fast and elitist multiobjective genetic algorithm: NSGA-II. *IEEE Transactions on Evolutionary Computation 6*(2), 182–197.

Deng, J. and M. D. Pandey (2008). Estimation of the maximum entropy quantile function using fractional probability weighted moments. *Structural Safety 30*(4), 307–319.

Der Kiureghian, A. and P. Liu (1986). Structural reliability under incomplete probability information. *Journal of Engineering Mechanics 112*(1), 85–104.

Descamps, B. and R. Filomeno Coelho (2014). The nominal force method for truss geometry and topology optimization incorporating stability considerations. *International Journal of Solids and Structures 51*(13), 2390–2399.

Dimarogonas, A. D. (1995). Interval analysis of vibrating systems. *Journal of Sound and Vibration 183*(4), 739–749.

Do, B. and M. Ohsaki (2021a). A random search for discrete robust design optimization of linear-elastic steel frames under interval parametric uncertainty. *Computers and Structures 249*, 106506.

Do, B. and M. Ohsaki (2021b). Gaussian mixture model for robust design optimization of planar steel frames. *Structural and Multidisciplinary Optimization 63*(1), 137–160.

Do, B., M. Ohsaki, and M. Yamakawa (2021a). Bayesian optimization for robust design of steel frames with joint and individual probabilistic constraints. *Engineering Structures 245*, 112859.

Do, B., M. Ohsaki, and M. Yamakawa (2021b). Sequential mixture of Gaussian processes and saddlepoint approximation for reliability-based design optimization of structures. *Structural and Multidisciplinary Optimization 64*(2), 625–648.

Doltsinis, I. and Z. Kang (2004). Robust design of structures using optimization methods. *Computer Methods in Applied Mechanics and Engineering 193*(23-26), 2221–2237.

Domingo, C., R. Gavaldà, and O. Watanabe (2002). Adaptive sampling methods for scaling up knowledge discovery algorithms. *Data Mining and Knowledge Discovery 6*(2), 131–152.

Dorea, C. C. Y. (1983). Expected number of steps of a random optimization method. *Journal of Optimization Theory and Applications 39*(2), 165–171.

Dorea, C. C. Y. and C. R. Gonçalves (1993). Alternative sampling strategy for a random optimization algorithm. *Journal of Optimization Theory and Applications 78*(2), 401–407.

Du, X. (2008). Saddlepoint approximation for sequential optimization and reliability analysis. *Journal of Mechanical Design, Transactions of the ASME 130*(1), 011011.

Du, X. and W. Chen (2000). Towards a better understanding of modeling feasibility robustness in engineering design. *Journal of Mechanical Design, Transactions of the ASME 122*(4), 385–394.

Du, X. and W. Chen (2004). Sequential optimization and reliability assessment method for efficient probabilistic design. *Journal of Mechanical Design, Transactions of the ASME 126*(2), 225–233.

Dyer, M., A. Frieze, and R. Kannan (1991). A random polynomial-time algorithm for approximating the volume of convex bodies. *Journal of the ACM 38*(1), 1–17.

Echard, B., N. Gayton, and M. Lemaire (2011). AK-MCS: An active learning reliability method combining Kriging and Monte Carlo simulation. *Structural Safety 33*(2), 145–154.

Echard, B., N. Gayton, M. Lemaire, and N. Relun (2013). A combined importance sampling and Kriging reliability method for small failure probabilities with time-demanding numerical models. *Reliability Engineering and System Safety 111*, 232–240.

Elishakoff, I., Y. W. Li, and J. H. Starnes, Jr. (1994). A deterministic method to predict the effect of unknown-but-bounded elastic moduli on the buckling of composite structures. *Computer Methods in Applied Mechanics and Engineering 111*(1–2), 155–167.

Elishakoff, I. and M. Ohsaki (2010). *Optimization and Anti-Optimization of Structures.* London: Imperial College Press.

Emmerich, M. T., K. C. Giannakoglou, and B. Naujoks (2006). Single- and multiobjective evolutionary optimization assisted by Gaussian random field metamodels. *IEEE Transactions on Evolutionary Computation 10*(4), 421–439.

Enevoldsen, I. and J. D. Sørensen (1994). Reliability-based optimization in structural engineering. *Structural Safety 15*(3), 169–196.

Feliot, P., J. Bect, and E. Vazquez (2017). A Bayesian approach to constrained single- and multi-objective optimization. *Journal of Global Optimization 67*(1-2), 97–133.

Frangopol, D. (1995). Reliability-based optimum structural design. In C. Sundararajan (Ed.), *Probabilistic Structural Mechanics Handbook*, Chapter 16, 352–387. Dordrecht: Chapman & Hall.

Frangopol, D. M. (1985). Structural optimization using reliability concepts. *ASCE Journal of Structural Engineering 111*(11), 2288–2301.

Frazier, P. I. (2018). A Tutorial on Bayesian Optimization. *arXiv preprint*, arXiv:1807.02811.

Gabrel, V., C. Murat, and A. Thiele (2014). Recent advances in robust optimization: An overview. *European Journal of Operational Research 235*(3), 471–483.

Geng, X. and L. Xie (2019). Data-driven decision making in power systems with probabilistic guarantees: Theory and applications of chance-constrained optimization. *Annual Reviews in Control 47*, 341–363.

Glover, F. (1989). Tabu search: Part I. *ORSA Journal on Computing 1*(3), 190–206.

Glover, F. and M. Laguna (1997). *Tabu Search.* Boston: Kluwer Academic Publishers.

Goswami, S., S. Chakraborty, R. Chowdhury, and T. Rabczuk (2019). Threshold shift method for reliability-based design optimization. *Structural and Multidisciplinary Optimization 60*(5), 2053–2072.

Greenwood, J. A., J. M. Landwehr, N. C. Matalas, and J. R. Wallis (1979). Probability weighted moments: Definition and relation to parameters of several distributions expressable in inverse form. *Water Resources Research 15*(5), 1049–1054.

Groenwold, A. A. and M. P. Hindley (2002). Competing parallel algorithms in structural optimization. *Structural and Multidisciplinary Optimization 24*(5), 343–350.

Gu, X., J. E. Renaud, S. M. Batill, R. M. Brach, and A. S. Budhiraja (2000a). Worst-case propagated uncertainty in multidisciplinary systems in robust design optimization. *Structural and Multidisciplinary Optimization 20*, 190–213.

Gu, X., J. E. Renaud, S. M. Batill, R. M. Brach, and A. S. Budhiraja (2000b). Worst case propagated uncertainty of multidisciplinary systems in robust design optimization. *Structural and Multidisciplinary Optimization 20*(3), 190–213.

Guest, J. K. (2009). Imposing maximum length scale in topology optimization. *Structural and Multidisciplinary Optimization 37*(5), 463–473.

Guo, X., G. Cheng, and K. Yamazaki (2001). A new approach for the solution of singular optima in truss topology optimization with stress and local buckling constraints. *Structural and Multidisciplinary Optimization 22*, 364–372.

Gurav, S. P., J. F. Goosen, and F. Vankeulen (2005). Bounded-but-unknown uncertainty optimization using design sensitivities and parallel computing: Application to MEMS. *Computers and Structures 83*(14), 1134–1149.

Hardin, B. O. and V. P. Drnevich (1972). Shear modulus and damping in soils: Design wquations and curves. *ASCE Journal of the Soil Mechanics and Foundations Division 98*(7), 667–692.

Hart, W. E. (1998). Sequential stopping rules for random optimization methods with applications to multistart local search. *SIAM Journal on Optimization 9*(1), 270–290.

Hashimoto, D. and Y. Kanno (2015). A semidefinite programming approach to robust truss topology optimization under uncertainty in locations of nodes. *Structural and Multidisciplinary Optimization 51*(2), 439–461.

Hastie, T., R. Tibshirani, and J. Friedman (2009). *The Elements of Statistical Learning*. Springer Series in Statistics. New York: Springer New York.

Hayashi, K. and M. Ohsaki (2019). FDMopt: Force density method for optimal geometry and topology of trusses. *Advances in Engineering Software 133*, 12–19.

He, W., H. Yang, G. Zhao, Y. Zeng, and G. Li (2021). A quantile-based SORA method using maximum entropy method with fractional moments. *Journal of Mechanical Design, Transactions of the ASME 143*(4).

Holmström, K., A. O. Göran, and M. M. Edvall (2007). *Tomlab Optimization. User 's Guide for Tomlab/CONOPT*. San Diego: Tomlab Optimization.

Hosking, J. R. (2007). Distributions with maximum entropy subject to constraints on their L-moments or expected order statistics. *Journal of Statistical Planning and Inference 137*(9), 2870–2891.

Hosking, J. R. M. (1990). L-Moments: Analysis and estimation of distributions using linear combinations of order statistics. *Journal of the Royal Statistical Society: Series B 52*(1), 105–124.

Hosking, J. R. M. and J. R. Wallis (1997). *Regional Frequency Analysis.* New York: Cambridge University Press.

Hromkovič, J. (2005). *Design and Analysis of Randomized Algorithms: Introduction to Design Paradigms.* Texts in Theoretical Computer Science An EATCS Series. Berlin, Heidelberg: Springer Berlin Heidelberg.

Huang, K. L. and S. Mehrotra (2015). An empirical evaluation of a walk-relax-round heuristic for mixed integer convex programs. *Computational Optimization and Applications 60*(3), 559–585.

Ito, M., N. H. Kim, and N. Kogiso (2018). Conservative reliability index for epistemic uncertainty in reliability-based design optimization. *Structural and Multidisciplinary Optimization 57*(5), 1919–1935.

Iyengar, R. N. and P. N. Rao (1979). Generation of spectrum compatible accelerograms. *Earthquake Engineering and Structural Dynamics 7*(3), 253–263.

Iyenger, R. N. and C. S. Manohar (1987). Nonstationary random critical seismic excitations. *ASCE Journal of Engineering Mechanics 113*(4), 529–541.

Jahn, J. (2007). *Introduction to the theory of nonlinear optimization.* Springer Berlin Heidelberg.

Jalalpour, M., J. K. Guest, and T. Igusa (2013). Reliability-based topology optimization of trusses with stochastic stiffness. *Structural Safety 43*, 41–49.

Jansen, M., G. Lombaert, M. Schevenels, and O. Sigmund (2014). Topology optimization of fail-safe structures using a simplified local damage model. *Structural and Multidisciplinary Optimization 49*(4), 657–666.

Jaynes, E. T. (1957). Information theory and statistical mechanics. *Physical Review 106*(4), 620–630.

Jekel, C. F. and R. T. Haftka (2020). Risk allocation for design optimization with unidentified statistical distributions. In *AIAA Scitech 2020 Forum*, Volume 1 Part F, Orlando. AIAA.

Jezowski, J., R. Bochenek, and G. Ziomek (2005). Random search optimization approach for highly multi-modal nonlinear problems. *Advances in Engineering Software 36*(8), 504–517.

Jones, D. R., M. Schonlau, and W. J. Welch (1998). Efficient global optimization of expensive black-box functions. *Journal of Global Optimization 13*, 455–492.

Joseph, G. D. (2017). Variations in the application of a budget of uncertainty optimization approach. *Structural and Multidisciplinary Optimization 55*(1), 77–89.

Kannan, R. and H. Narayanan (2012). Random walks on polytopes and an affine interior point method for linear programming. *Mathematics of Operations Research 37*(1), 1–20.

Kanno, Y. (2017). Redundancy optimization of finite-dimensional structures: Concept and derivative-free algorithm. *ASCE Journal of Structural Engineering 143*(1), 04016151.

Kanno, Y. (2019). A data-driven approach to non-parametric reliability-based design optimization of structures with uncertain load. *Structural and Multidisciplinary Optimization 60*(1), 83–97.

Kanno, Y., M. Ohsaki, and N. Katoh (2001). Sequential semidefinite programming for optimization of framed structures under multimodal buckling constraints. *International Journal of Structural Stability and Dynamics 1*(4), 585–602.

Kasai, H. (2018). SGDlibrary: A MATLAB library for stochastic optimization algorithms. *Journal of Machine Learning Research 18*, 1–5.

Kingma, D. P. and J. L. Ba (2015). Adam: A method for stochastic optimization. In *3rd International Conference on Learning Representations, ICLR 2015 - Conference Track Proceedings*, 1–15.

Kirkpatrick, S., C. D. Gelatt, Jr., and M. P. Vecchi (1983). Optimization by simulated annealing. *Science 220*(4598), 671–680.

Klir, G. and J. Folger (1988). *Fuzzy Sets, Uncertainty and Information.* Englewood Cliffs: Prentice-Hall.

Knoll, F. and T. Vogel (2009). *Design for Robustness.* Zürich: International Association for Bridge and Structural Engineering (IABSE).

Kohta, R., M. Yamakawa, N. Katoh, Y. Araki, and M. Ohsaki (2014). A design method for optimal truss structures with redundancy based on combinatorial rigidity theory. *Journal of Structural and Construction Engineering*, Transactions of AIJ, *79*(699), 583–592. (In Japanese).

Kolda, T. G., R. M. Lewis, and V. Torczon (2003). Optimization by direct search: New perspectives on some classical and modern methods. *SIAM Review 45*(3), 385–482.

Kolda, T. G., R. M. Lewis, and V. Torczon (2006). Stationarity results for generating set search for linearly constrained optimization. *SIAM Journal on Optimization 17*(4), 943–958.

Krishnamoorthy, K. and T. Mathew (2008). *Statistical Tolerance Regions: Theory, Applications, and Computation.* Wiley.

Kuschel, N. and R. Rackwitz (1997). Two basic problems in reliability-based structural optimization. *Mathematical Methods of Operations Research 46*(3), 309–333.

Lagaris, I. E. and I. G. Tsoulos (2008). Stopping rules for box-constrained stochastic global optimization. *Applied Mathematics and Computation 197*(2), 622–632.

Le, C., J. Norato, T. Bruns, C. Ha, and D. Tortorelli (2010). Stress-based topology optimization for continua. *Structural and Multidisciplinary Optimization 41*(4), 605–620.

Lee, I., K. K. Choi, L. Du, and D. Gorsich (2008a). Dimension reduction method for reliability-based robust design optimization. *Computers and Structures 86*(13-14), 1550–1562.

Lee, I., K. K. Choi, L. Du, and D. Gorsich (2008b). Inverse analysis method using MPP-based dimension reduction for reliability-based design optimization of nonlinear and multi-dimensional systems. *Computer Methods in Applied Mechanics and Engineering 198*(1), 14–27.

Lewis, R. M. and V. Torczon (2000). Pattern search methods for linearly constrained minimization. *SIAM Journal on Optimization 10*(3), 917–941.

Li, G., W. He, and Y. Zeng (2019). An improved maximum entropy method via fractional moments with Laplace transform for reliability analysis. *Structural and Multidisciplinary Optimization 59*(4), 1301–1320.

Li, G., H. Yang, and G. Zhao (2020). A new efficient decoupled reliability-based design optimization method with quantiles. *Structural and Multidisciplinary Optimization 61*(2), 635–647.

Li, X., C. Gong, L. Gu, Z. Jing, H. Fang, and R. Gao (2019). A reliability-based optimization method using sequential surrogate model and Monte Carlo simulation. *Structural and Multidisciplinary Optimization 59*(2), 439–460.

Liang, J., Z. P. Mourelatos, and J. Tu (2008). A single-loop method for reliability-based design optimisation. *International Journal of Product Development 5*(1-2), 76–92.

Lipton, R. J. and J. F. Naughton (1995). Query size estimation by adaptive sampling. *Journal of Computer and System Sciences 51*(1), 18–25.

Liu, H., Y. S. Ong, X. Shen, and J. Cai (2020). When Gaussian Process Meets Big Data: A Review of Scalable GPs. *IEEE Transactions on Neural Networks and Learning Systems 31*(11), 4405–4423.

Locatelli, M. (1997). Bayesian algorithms for one-dimensional global optimization. *Journal of Global Optimization 10*(1), 57–76.

Lophaven, S. N., J. Søndergaard, and H. B. Nielsen (2002). DACE: A MATLAB Kriging toolbox. *Technical report, Technical University of Denmark*, IMM–REP–2002–12.

Ma, J. and D. Yarats (2019). On the adequacy of untuned warmup for adaptive optimization. *arXiv preprint*, arXiv:1910.04209.

Makkonen, L. (2008). Problems in the extreme value analysis. *Structural Safety 30*(5), 405–419.

Marler, R. T. and J. S. Arora (2004). Survey of multi-objective optimization methods for engineering. *Structural and Multidisciplinary Optimization 26*(6), 369–395.

Masoudnia, S. and R. Ebrahimpour (2014). Mixture of experts: A literature survey. *Artificial Intelligence Review 42*(2), 275–293.

Mathern, A., O. S. Steinholtz, A. Sjöberg, M. Önnheim, K. Ek, R. Rempling, E. Gustavsson, and M. Jirstrand (2021). Multi-objective constrained Bayesian optimization for structural design. *Structural and Multidisciplinary Optimization 63*(2), 689–701.

Mathworks (2009). *Optimzation Toolbox User's Guide R2009b.*

Mathworks (2018a). *Global Optimization Toolbox User's Guide R2018a.*

Mathworks (2018b). *Optimization Toolbox User's Guide R2018a.*

Mathworks (2018c). *Statistics and Machine Learning Toolbox User's Guide R2018a.*

Maxwell, J. C. (1890). On reciprocal figures, frames, and diagrams of forces. *Scientific Papers 2*, 161–207.

McGuire, W., R. Gallagher, and R. Ziemian (2015). Matrix Structural Analysis, 2nd Edition. Wiley.

McKenna, F. (2011). OpenSees: A framework for earthquake engineering simulation. *Computing in Science and Engineering 13*(4), 58–66.

Mclachlan, G. J. and S. Rathnayake (2014). On the number of components in a Gaussian mixture model. *Wiley Interdisciplinary Reviews: Data Mining and Knowledge Discovery 4*(5), 341–355.

Melchers, R. E. and A. T. Beck (2017). *Structural Reliability Analysis and Prediction*, 3rd Edition. New Jersey: John Wiley & Sons.

Meng, Z. and B. Keshtegar (2019). Adaptive conjugate single-loop method for efficient reliability-based design and topology optimization. *Computer Methods in Applied Mechanics and Engineering 344*, 95–119.

Michell, A. G. M. (1904). The limits of economy in frame structures. *Philosophical Magazine Sect. 6, 8*(47), 589–597.

Moon, M. Y., H. Cho, K. K. Choi, N. Gaul, D. Lamb, and D. Gorsich (2018). Confidence-based reliability assessment considering limited numbers of both input and output test data. *Structural and Multidisciplinary Optimization 57*(5), 2027–2043.

Moore, R. (1966). *Interval analysis.* New Jersey: Prentice-Hall.

Moustapha, M. and B. Sudret (2019). Surrogate-assisted reliability-based design optimization: A survey and a unified modular framework. *Structural and Multidisciplinary Optimization 60*(5), 2157–2176.

Moustapha, M., B. Sudret, J. M. Bourinet, and B. Guillaume (2016). Quantile-based optimization under uncertainties using adaptive Kriging surrogate models. *Structural and Multidisciplinary Optimization 54*(6), 1403–1421.

Muselli, M. (1997). A Theoretical Approach to Restart in Global Optimization. *Journal of Global Optimization 10*(1), 1–16.

Navarro, J. and N. Balakrishnan (2010). Study of some measures of dependence between order statistics and systems. *Journal of Multivariate Analysis 101*(1), 52–67.

Nesterov, Y. (2004). *Introductory Lectures on Convex Optimization: A Basic Course.* Applied Optimization Series. Springer Science+Business Media New York.

Nesterov, Y. and A. Nemirovskii (1994). *Interior-Point Polynomial Algorithms in Convex Programming.* New Jersey: Society for Industrial and Applied Mathematics.

Neuman, C. P. and D. I. Schonbach (1974). Discrete (Legendre) orthogonal polynomials–A survey. *International Journal for Numerical Methods in Engineering 8*(4), 743–770.

Noh, Y., K. K. Choi, and L. Du (2009). Reliability-based design optimization of problems with correlated input variables using a Gaussian Copula. *Structural and Multidisciplinary Optimization 38*(1), 1–16.

Noh, Y., K. K. Choi, I. Lee, D. Gorsich, and D. Lamb (2011). Reliability-based design optimization with confidence level under input model uncertainty due to limited test data. *Structural and Multidisciplinary Optimization 43*(4), 443–458.

Nowak, A. S. and K. R. Collins (2012). *Reliability of Structures.* CRC Press.

O'Hagan, A. (1991). Bayes-Hermite quadrature. *Journal of Statistical Planning and Inference 29*(3), 245–260.

Ohsaki, M. (1998). Simultaneous optimization of topology and geometry of a regular plane truss. *Computers and Structures 66*(1), 69–77.

Ohsaki, M. (2001). Random search method based on exact reanalysis for topology optimization of trusses with discrete cross-sectional areas. *Computers and Structures 79*(6), 673–679.

Ohsaki, M. (2010). *Optimization of Finite Dimensional Structures.* Boca Raton: CRC Press.

Ohsaki, M. and K. Hayashi (2017). Force density method for simultaneous optimization of geometry and topology of trusses. *Structural and Multidisciplinary Optimization 56*(5), 1157–1168.

Ohsaki, M. and K. Ikeda (2007). *Stability and Optimization of Structures: Generalized Sensitivity Analysis.* Mechanical Engineering Series. New York: Springer.

Ohsaki, M. and Y. Kanno (2001). Optimum design of finite dimensional systems with coincident critical points. In *Proceedings of the 4th World Conggress of Structural and Multidisciplinary Optimization (WCSMO4)*, 21, Dalian.

Ohsaki, M. and M. Katsura (2012). A random sampling approach to worst-case design of structures. *Structural and Multidisciplinary Optimization 46*(1), 27–39.

Ohsaki, M. and M. Yamakawa (2018). Stopping rule of multi-start local search for structural optimization. *Structural and Multidisciplinary Optimization 57*(2), 595–603.

Ohsaki, M., M. Yamakawa, W. Fan, and Z. Li (2019). An order statistics approach to multiobjective structural optimization considering robustness and confidence of responses. *Mechanics Research Communications 97*, 33–38.

Ohsaki, Y. (1979). On the significance of phase content in earthquake ground motions. *Earthquake Engineering & Structural Dynamics 7*(5), 427–439.

Okasha, N. M. (2016). An improved weighted average simulation approach for solving reliability-based analysis and design optimization problems. *Structural Safety 60*, 47–55.

Paly, M. D., C. M. Bürger, and P. Bayer (2013). Optimization under worst case constraints: A new global multimodel search procedure. *Structural and Multidisciplinary Optimization 48*(6), 1153–1172.

Pandey, M. D. (2000). Direct estimation of quantile functions using the maximum entropy principle. *Structural Safety 22*(1), 61–79.

Pandey, M. D. (2001a). Extreme quantile estimation using order statistics with minimum cross-entropy principle. *Probabilistic Engineering Mechanics 16*(1), 31–42.

Pandey, M. D. (2001b). Minimum cross-entropy method for extreme value estimation using peaks-over-threshold data. *Structural Safety 23*(4), 345–363.

Papadrakakis, M., N. D. Lagaros, and V. Plevris (2005). Design optimization of steel structures considering uncertainties. *Engineering Structures 27*(9), 1408–1418.

Papoulis, A. and S. U. Pillai (2002). *Probability, Random Variables, and Stochastic Processes*. New York: McGraw-Hill.

Park, C., N. H. Kim, and R. T. Haftka (2015). The effect of ignoring dependence between failure modes on evaluating system reliability. *Structural and Multidisciplinary Optimization 52*(2), 251–268.

Park, G. J., T. H. Lee, K. H. Lee, and K. H. Hwang (2006). Robust design: An overview. *AIAA Journal 44*(1), 181–191.

Parkinson, A., C. Sorensen, and N. Pourhassan (1993). A general approach for robust optimal design. *Journal of Mechanical Design, Transactions of the ASME 115*(1), 74–80.

Pedersen, C. B. W. (2003). Topology optimization of 2D-frame structures with path-dependent response. *International Journal for Numerical Methods in Engineering 57*, 1471–1501.

Polyak, B. T. and E. N. Gryazina (2010). Markov chain Monte Carlo method exploiting barrier functions with applications to control and optimization. In *Proceedings of the IEEE International Symposium on Computer-Aided Control System Design*, 1553–1557.

Qian, N. (1999). On the momentum term in gradient descent learning algorithms. *Neural Networks 12*(1), 145–151.

Qiu, Z. (2005). Convex models and interval analysis method to predict the effect of uncertain-but-bounded parameters on the buckling of composite structures. *Computer Methods in Applied Mechanics and Engineering 194*(18-20), 2175–2189.

Qiu, Z. and I. Elishakoff (1998). Antioptimization of structures with large uncertain-but-non-random parameters via interval analysis. *Computer Methods in Applied Mechanics and Engineering 152*(3–4), 361–372.

Rajan, S. D. (1995). Sizing, shape, and topology design optimization of trusses using genetic algorithm. *Journal of Structural Engineering 121*(10), 1480–1487.

Rao, S. S. (2019). *Engineering Optimization: Theory and Practice.* New Jersey: John Wiley & Sons.

Rasmussen, C. E. and C. K. I. Williams (2005). *Gaussian Processes for Machine Learning.* Cambridge, Cambridge: MIT Press.

Reddi, S. J., S. Kale, and S. Kumar (2018). On the convergence of Adam and beyond. In *6th International Conference on Learning Representations, ICLR 2018 - Conference Track Proceedings*, 1–23.

Ribeiro, C. C., I. Rosseti, and R. C. Souza (2011). Effective probabilistic stopping rules for randomized metaheuristics: GRASP implementations. In *Lecture Notes in Computer Science (including subseries Lecture Notes in Artificial Intelligence and Lecture Notes in Bioinformatics)*, Volume 6683 LNCS, 146–160. Springer, Berlin, Heidelberg.

Richardson, J. N., R. Filomeno Coelho, and S. Adriaenssens (2015). Robust topology optimization of truss structures with random loading and material properties: A multiobjective perspective. *Computers and Structures 154*, 41–47.

Riche, R. L. and R. T. Haftka (2012). On global optimization articles in SMO. *Structural and Multidisciplinary Optimization 46*(5), 627–629.

Rocchetta, R., L. G. Crespo, and S. P. Kenny (2020). A scenario optimization approach to reliability-based design. *Reliability Engineering and System Safety 196*, 106755.

Rosowsky, D. V. and W. M. Bulleit (2002). Load duration effects in wood members and connections: Order statistics and critical loads. *Structural Safety 24*(2-4), 347–362.

Rozvany, G. I. N. (1996). Difficulties in truss topology optimization with stress, local buckling and system stability constraints. *Structural Optimization 11*, 213–217.

Rustem, B. and M. Howe (2002). *Algorithms for worst-case design and applications to risk management*, Princeton: Princeton University Press.

Schek, H. J. (1974). The force density method for form finding and computation of general networks. *Computer Methods in Applied Mechanics and Engineering 3*(1), 115–134.

Schmidt, R. M., F. Schneider, and P. Hennig (2020). Descending through a crowded valley: Benchmarking deep learning optimizers. *arXiv preprint*, arXiv:2007.01547.

Schnabel, P. B., J. Lysmer, and H. B. Seed (1972). SHAKE: A computer program for earthquake response analysis of horizontally layered sites.

Shahriari, B., K. Swersky, Z. Wang, R. P. Adams, and N. De Freitas (2016). Taking the human out of the loop: A review of Bayesian optimization. In *Proceedings of the IEEE 104*(1), 148–175.

Shen, W. and M. Ohsaki (2021). Geometry and topology optimization of plane frames for compliance minimization using force density method for geometry model. *Engineering with Computers 37*(3), 2029–2046.

Shen, W., M. Ohsaki, and M. Yamakawa (2020). Multiobjective robust shape and topology optimization of plane frames using order statistics. *Structural and Multidisciplinary Optimization 63*(1), 1–20.

Shen, W., M. Ohsaki, and M. Yamakawa (2021). Robust geometry and topology optimization of plane frames using order statistics and force density method with global stability constraint. *International Journal for Numerical Methods in Engineering 122*(14), 3653–3677.

Shen, W., M. Ohsaki, and M. Yamakawa (2022). Quantile-based sequential optimization and reliability assessment for shape and topology optimization of plane frames using L-moments. *Structural Safety 94*, 102153.

Shore, J. E. and R. W. Johnson (1980). Axiomatic Derivation of the Principle of Maximum Entropy and the Principle of Minimum Cross-Entropy. *IEEE Transactions on Information Theory 26*(1), 26–37.

Shu, L., P. Jiang, X. Shao, and Y. Wang (2020). A new multi-Objective bayesian optimization formulation with the acquisition function for convergence and diversity. *Journal of Mechanical Design, Transactions of the ASME 142*(9), 091703.

Sigmund, O. and K. Maute (2013). Topology optimization approaches: A comparative review. *Structural and Multidisciplinary Optimization 48*(6), 1031–1055.

Smith, R. L. (1984). Efficient Monte Carlo Procedures for Generating Points Uniformly Distributed over Bounded Regions. *Operations Research 32*(6), 1296–1308.

Spall, J. C. (2005). *Introduction to Stochastic Search and Optimization*, Hoboken: John Wiley & Sons.

Stadler, W. (1979). A survey of multicriteria optimization of the vector maximum problem: Part I:1776–1960. *Journal of Optimization Theory and Applications 29*(1), 1–52.

Stadler, W. (Ed.) (1988). *Multicriteria Optimization in Engineering and in the Sciences*. New York: Plenum Press.

Taguchi, G. (1957). *Experimental Designs*, Volume 1. Tokyo: Maruzen.

Taguchi, G. (1958). *Experimental Designs*, Volume 2. Tokyo: Maruzen.

Taguchi, G., S. Chowdhury, and S. Taguchi (2000). *Robust Engineering*. New York: McGraw Hill Professional Pub.

Taguchi, G. and M. S. Phadke (1989). Quality Engineering through Design Optimization. In K. Dehnad (Ed.), *Quality Control, Robust Design, and the Taguchi Method*, 77–96. Springer US.

Tibert, A. G. and S. Pellegrino (2003). Review of form-finding methods for tensegrity structures. *International Journal of Space Structures 18*(4), 209–223.

Tong, M. N., Y. G. Zhao, and Z. H. Lu (2021). Normal transformation for correlated random variables based on L-moments and its application in reliability engineering. *Reliability Engineering and System Safety 207*, 107334.

Torii, A. J., R. H. Lopez, and L. F. Miguel (2015). Modeling of global and local stability in optimization of truss-like structures using frame elements. *Structural and Multidisciplinary Optimization 51*(6), 1187–1198.

Torn, A. and A. Zhilinskas (1989). *Global optimization*. Berlin, Heidelberg: Springer-Verlag.

Tsuji, M. and T. Nakamura (1996). Optimum viscous dampers for stiffness design of shear buildings. *Structural Design of Tall Buildings 5*(3), 217–234.

Tu, J., K. K. Choi, and Y. H. Park (1999). A new study on reliability- based design optimization. *Journal of Mechanical Design, Transactions of the ASME 121*(4), 557–564.

Tugilimana, A., R. Filomeno Coelho, and A. P. Thrall (2018). Including global stability in truss layout optimization for the conceptual design of large-scale applications. *Structural and Multidisciplinary Optimization 57*(3), 1213–1232.

ur Rehman, S. and M. Langelaar (2015). Efficient global robust optimization of unconstrained problems affected by parametric uncertainties. *Structural and Multidisciplinary Optimization 52*(2), 319–336.

Valdebenito, M. A. and G. I. Schuëller (2010). A survey on approaches for reliability-based optimization. *Structural and Multidisciplinary Optimization 42*(5), 645–663.

Van Den Bergh, F. (2006). *An Analysis of Particle Swarm Optimizers*. PhD thesis, Faculty of Natural and Agricultural Science, University of Pretoria.

van Keulen, F., R. T. Haftka, and N. H. Kim (2005). Review of options for structural design sensitivity analysis. Part 1: Linear systems. *Computer Methods in Applied Mechanics and Engineering 194*(30-33), 3213–3243.

Vapnik, V. N. (2000). *The Nature of Statistical Learning Theory*. Springer New York.

Vaz, A. I. F. and L. N. Vicente (2007). A particle swarm pattern search method for bound constrained global optimization. *Journal of Global Optimization 39*(2), 197–219.

Virtanen, P., R. Gommers, T. E. Oliphant, M. Haberland, T. Reddy, D. Cournapeau, E. Burovski, P. Peterson, W. Weckesser, J. Bright, S. J. van der Walt, M. Brett, J. Wilson, K. J. Millman, N. Mayorov, A. R. Nelson, E. Jones, R. Kern, E. Larson, C. J. Carey, Ä. Polat, Y. Feng, E. W. Moore, J. VanderPlas, D. Laxalde, J. Perktold, R. Cimrman, I. Henriksen, E. A. Quintero, C. R. Harris, A. M. Archibald, A. H. Ribeiro, F. Pedregosa, P. van Mulbregt, A. Vijaykumar, A. P. Bardelli, A. Rothberg, A. Hilboll, A. Kloeckner, A. Scopatz, A. Lee, A. Rokem, C. N. Woods, C. Fulton, C. Masson, C. Häggström, C. Fitzgerald, D. A. Nicholson, D. R. Hagen, D. V. Pasechnik, E. Olivetti, E. Martin, E. Wieser, F. Silva, F. Lenders, F. Wilhelm, G. Young, G. A. Price, G. L. Ingold, G. E. Allen, G. R. Lee, H. Audren, I. Probst, J. P. Dietrich, J. Silterra, J. T. Webber, J. Slavič, J. Nothman, J. Buchner, J. Kulick, J. L. Schönberger, J. V. de Miranda Cardoso, J. Reimer, J. Harrington, J. L. C. Rodríguez, J. Nunez-Iglesias, J. Kuczynski, K. Tritz, M. Thoma, M. Newville, M. Kümmerer, M. Bolingbroke, M. Tartre, M. Pak, N. J. Smith, N. Nowaczyk, N. Shebanov,

O. Pavlyk, P. A. Brodtkorb, P. Lee, R. T. McGibbon, R. Feldbauer, S. Lewis, S. Tygier, S. Sievert, S. Vigna, S. Peterson, S. More, T. Pudlik, T. Oshima, T. J. Pingel, T. P. Robitaille, T. Spura, T. R. Jones, T. Cera, T. Leslie, T. Zito, T. Krauss, U. Upadhyay, Y. O. Halchenko, and Y. Vázquez-Baeza (2020). SciPy 1.0: fundamental algorithms for scientific computing in Python. *Nature Methods 2020 17*(3), 261–272.

Voglis, C. and I. E. Lagaris (2009). Towards "Ideal Multistart": A stochastic approach for locating the minima of a continuous function inside a bounded domain. *Applied Mathematics and Computation 213*(1), 216–229.

Wang, D., W. H. Zhang, and J. S. Jiang (2002). Truss shape optimization with multiple displacement constraints. *Computer Methods in Applied Mechanics and Engineering 191*(33), 3597–3612.

Wang, F., B. S. Lazarov, and O. Sigmund (2011). On projection methods, convergence and robust formulations in topology optimization. *Structural and Multidisciplinary Optimization 43*(6), 767–784.

Watanabe, O. (2000). Simple Sampling Techniques for Discovery Science. *IEICE Transactions on Information and Systems E83-D*(1), 19–26.

Weiji, L. and Y. Li (1994). An effective optimization procedure based on structural reliability. *Computers and Structures 52*(5), 1061–1067.

Wolpert, D. H. and W. G. Macready (1997). No free lunch theorems for optimization. *IEEE Transactions on Evolutionary Computation 1*(1), 67–82.

Xi, Z., C. Hu, and B. D. Youn (2012). A comparative study of probability estimation methods for reliability analysis. *Structural and Multidisciplinary Optimization 45*(1), 33–52.

Xing, C., D. Arpit, C. Tsirigotis, and Y. Bengio (2018). A walk with SGD. *arXiv preprint*, arXiv:1802.08770.

Xu, H. and S. Rahman (2004). A generalized dimension-reduction method for multidimensional integration in stochastic mechanics. *International Journal for Numerical Methods in Engineering 61*(12), 1992–2019.

Xu, R. and D. Wunsch (2005). Survey of clustering algorithms. *IEEE Transactions on Neural Networks 16*(3), 645–678.

Yamakawa, M. and M. Ohsaki (2016). Robust design optimization considering parameter variation of seismic characteristics using order statistics. *Journal of structural engineering*, Transactions of AIJ, *62B*, 381–386. (In Japanese).

Yamakawa, M. and M. Ohsaki (2020). Fail-safe topology optimization via order statistics with stochastic gradient descent. In *Proceedings of Asian Congress of Structural and Multidisciplinary Optimization 2020*, Online Conference, 254–260.

Yamakawa, M. and M. Ohsaki (2021). Order statistics approach to structural optimization considering robustness and confidence of responses. In N. Challamel, J. Kaplunov, and I. Takewaki (Eds.), *Modern Trends in Structural and Solid Mechanics 3*, Chapter 11, 225–241. New Jersey: John Wiley & Sons.

Yamakawa, M., M. Ohsaki, and K. Watanabe (2018). Robust design optimization of moment-resisting steel frame against uncertainties of surface ground properties using order statistics. In *Japan-China Workshop on Analysis and Optimization of Large-scale Structures*, 23–31. Structural Engineering of Buildings Laboratory, Department of Architecture and Architectural Engineering, Kyoto University, http://hdl.handle.net/2433/231246.

Yang, J. N., S. Sarkani, and F. X. Long (1990). A response spectrum approach for seismic analysis of nonclassically damped structures. *Engineering Structures 12*(3), 173–184.

Yoo, D. and I. Lee (2014). Sampling-based approach for design optimization in the presence of interval variables. *Structural and Multidisciplinary Optimization 49*(2), 253–266.

Zabinsky, Z. B., D. Bulger, and C. Khompatraporn (2010). Stopping and restarting strategy for stochastic sequential search in global optimization. *Journal of Global Optimization 46*(2), 273–286.

Zang, C., M. I. Friswell, and J. E. Mottershead (2005). A review of robust optimal design and its application in dynamics. *Computers and Structures 83*(4–5), 315–326.

Zhang, J. Y. and M. Ohsaki (2006). Adaptive force density method for form-finding problem of tensegrity structures. *International Journal of Solids and Structures 43*(18–19), 5658–5673.

Zhang, J. Y. and M. Ohsaki (2015). *Tensegrity Structures*, Mathematics for Industry, Volume 6, Springer Japan.

Zhang, Y., D. W. Apley, and W. Chen (2020). Bayesian optimization for materials design with mixed quantitative and qualitative variables. *Scientific Reports 10*(1), 1–13.

Zhao, W. and Z. Qiu (2013). An efficient response surface method and its application to structural reliability and reliability-based optimization. *Finite Elements in Analysis and Design 67*, 34–42.

Zhao, Y. G., M. N. Tong, Z. H. Lu, and J. Xu (2020). Monotonic expression of polynomial normal transformation based on the first four L-moments. *ASCE Journal of Engineering Mechanics 146*(7), 06020003.

Zhigljavsky, A. and E. Hamilton (2010). Stopping rules in k-adaptive global random search algorithms. *Journal of Global Optimization 48*(1), 87–97.

Zhigljavsky, A. and A. Žilinskas (2008). *Stochastic Global Optimization.* Springer US.

Zhigljavsky, A. and A. Žilinskas (2021). *Bayesian and High-Dimensional Global Optimization.* SpringerBriefs in Optimization. Cham: Springer International Publishing.

Zhou, M. (1996). Difficulties in truss topology optimization with stress and local buckling constraints. *Structural Optimization 11*, 134–136.

Zieliński, R. (1981). A statistical estimate of the structure of multi-extremal problems. *Mathematical Programming 21*(1), 348–356.

Zieliński, R. (2009). Optimal nonparametric quantile estimators. towards a general theory. a survey. *Communications in Statistics - Theory and Methods 38*(7), 980–992.

Index

For Product Safety Concerns and Information please contact our
EU representative GPSR@taylorandfrancis.com Taylor & Francis
Verlag GmbH, Kaufingerstraße 24, 80331 München, Germany